T. EAS

SILICONE SURFACTANTS

SURFACTANT SCIENCE SERIES

1. Nonionic Surfactants, *edited by Martin J. Schick* (see also Volumes 19, 23, and 60)
2. Solvent Properties of Surfactant Solutions, *edited by Kozo Shinoda* (see Volume 55)
3. Surfactant Biodegradation, *R. D. Swisher* (see Volume 18)
4. Cationic Surfactants, *edited by Eric Jungermann* (see also Volumes 34, 37, and 53)
5. Detergency: Theory and Test Methods (in three parts), *edited by W. G. Cutler and R. C. Davis* (see also Volume 20)
6. Emulsions and Emulsion Technology (in three parts), *edited by Kenneth J. Lissant*
7. Anionic Surfactants (in two parts), *edited by Warner M. Linfield* (see Volume 56)
8. Anionic Surfactants: Chemical Analysis, *edited by John Cross*
9. Stabilization of Colloidal Dispersions by Polymer Adsorption, *Tatsuo Sato and Richard Ruch*
10. Anionic Surfactants: Biochemistry, Toxicology, Dermatology, *edited by Christian Gloxhuber* (see Volume 43)
11. Anionic Surfactants: Physical Chemistry of Surfactant Action, *edited by E. H. Lucassen-Reynders*
12. Amphoteric Surfactants, *edited by B. R. Bluestein and Clifford L. Hilton* (see Volume 59)
13. Demulsification: Industrial Applications, *Kenneth J. Lissant*
14. Surfactants in Textile Processing, *Arved Datyner*
15. Electrical Phenomena at Interfaces: Fundamentals, Measurements, and Applications, *edited by Ayao Kitahara and Akira Watanabe*
16. Surfactants in Cosmetics, *edited by Martin M. Rieger* (see Volume 68)
17. Interfacial Phenomena: Equilibrium and Dynamic Effects, *Clarence A. Miller and P. Neogi*
18. Surfactant Biodegradation: Second Edition, Revised and Expanded, *R. D. Swisher*
19. Nonionic Surfactants: Chemical Analysis, *edited by John Cross*
20. Detergency: Theory and Technology, *edited by W. Gale Cutler and Erik Kissa*
21. Interfacial Phenomena in Apolar Media, *edited by Hans-Friedrich Eicke and Geoffrey D. Parfitt*
22. Surfactant Solutions: New Methods of Investigation, *edited by Raoul Zana*
23. Nonionic Surfactants: Physical Chemistry, *edited by Martin J. Schick*
24. Microemulsion Systems, *edited by Henri L. Rosano and Marc Clausse*
25. Biosurfactants and Biotechnology, *edited by Naim Kosaric, W. L. Cairns, and Neil C. C. Gray*
26. Surfactants in Emerging Technologies, *edited by Milton J. Rosen*
27. Reagents in Mineral Technology, *edited by P. Somasundaran and Brij M. Moudgil*
28. Surfactants in Chemical/Process Engineering, *edited by Darsh T. Wasan, Martin E. Ginn, and Dinesh O. Shah*
29. Thin Liquid Films, *edited by I. B. Ivanov*
30. Microemulsions and Related Systems: Formulation, Solvency, and Physical Properties, *edited by Maurice Bourrel and Robert S. Schechter*
31. Crystallization and Polymorphism of Fats and Fatty Acids, *edited by Nissim Garti and Kiyotaka Sato*

32. Interfacial Phenomena in Coal Technology, *edited by Gregory D. Botsaris and Yuli M. Glazman*
33. Surfactant-Based Separation Processes, *edited by John F. Scamehorn and Jeffrey H. Harwell*
34. Cationic Surfactants: Organic Chemistry, *edited by James M. Richmond*
35. Alkylene Oxides and Their Polymers, *F. E. Bailey, Jr., and Joseph V. Koleske*
36. Interfacial Phenomena in Petroleum Recovery, *edited by Norman R. Morrow*
37. Cationic Surfactants: Physical Chemistry, *edited by Donn N. Rubingh and Paul M. Holland*
38. Kinetics and Catalysis in Microheterogeneous Systems, *edited by M. Grätzel and K. Kalyanasundaram*
39. Interfacial Phenomena in Biological Systems, *edited by Max Bender*
40. Analysis of Surfactants, *Thomas M. Schmitt*
41. Light Scattering by Liquid Surfaces and Complementary Techniques, *edited by Dominique Langevin*
42. Polymeric Surfactants, *Irja Piirma*
43. Anionic Surfactants: Biochemistry, Toxicology, Dermatology. Second Edition, Revised and Expanded, *edited by Christian Gloxhuber and Klaus Künstler*
44. Organized Solutions: Surfactants in Science and Technology, *edited by Stig E. Friberg and Björn Lindman*
45. Defoaming: Theory and Industrial Applications, *edited by P. R. Garrett*
46. Mixed Surfactant Systems, *edited by Keizo Ogino and Masahiko Abe*
47. Coagulation and Flocculation: Theory and Applications, *edited by Bohuslav Dobiáš*
48. Biosurfactants: Production • Properties • Applications, *edited by Naim Kosaric*
49. Wettability, *edited by John C. Berg*
50. Fluorinated Surfactants: Synthesis • Properties • Applications, *Erik Kissa*
51. Surface and Colloid Chemistry in Advanced Ceramics Processing, *edited by Robert J. Pugh and Lennart Bergström*
52. Technological Applications of Dispersions, *edited by Robert B. McKay*
53. Cationic Surfactants: Analytical and Biological Evaluation, *edited by John Cross and Edward J. Singer*
54. Surfactants in Agrochemicals, *Tharwat F. Tadros*
55. Solubilization in Surfactant Aggregates, *edited by Sherril D. Christian and John F. Scamehorn*
56. Anionic Surfactants: Organic Chemistry, *edited by Helmut W. Stache*
57. Foams: Theory, Measurements, and Applications, *edited by Robert K. Prud'homme and Saad A. Khan*
58. The Preparation of Dispersions in Liquids, *H. N. Stein*
59. Amphoteric Surfactants: Second Edition, *edited by Eric G. Lomax*
60. Nonionic Surfactants: Polyoxyalkylene Block Copolymers, *edited by Vaughn M. Nace*
61. Emulsions and Emulsion Stability, *edited by Johan Sjöblom*
62. Vesicles, *edited by Morton Rosoff*
63. Applied Surface Thermodynamics, *edited by A. W. Neumann and Jan K. Spelt*
64. Surfactants in Solution, *edited by Arun K. Chattopadhyay and K. L. Mittal*
65. Detergents in the Environment, *edited by Milan Johann Schwuger*

66. Industrial Applications of Microemulsions, *edited by Conxita Solans and Hironobu Kunieda*
67. Liquid Detergents, *edited by Kuo-Yann Lai*
68. Surfactants in Cosmetics: Second Edition, Revised and Expanded, *edited by Martin M. Rieger and Linda D. Rhein*
69. Enzymes in Detergency, *edited by Jan H. van Ee, Onno Misset, and Erik J. Baas*
70. Structure–Performance Relationships in Surfactants, *edited by Kunio Esumi and Minoru Ueno*
71. Powdered Detergents, *edited by Michael S. Showell*
72. Nonionic Surfactants: Organic Chemistry, *edited by Nico M. van Os*
73. Anionic Surfactants: Analytical Chemistry, Second Edition, Revised and Expanded, *edited by John Cross*
74. Novel Surfactants: Preparation, Applications, and Biodegradability, *edited by Krister Holmberg*
75. Biopolymers at Interfaces, *edited by Martin Malmsten*
76. Electrical Phenomena at Interfaces: Fundamentals, Measurements, and Applications, Second Edition, Revised and Expanded, *edited by Hiroyuki Ohshima and Kunio Furusawa*
77. Polymer-Surfactant Systems, *edited by Jan C. T. Kwak*
78. Surfaces of Nanoparticles and Porous Materials, *edited by James A. Schwarz and Cristian I. Contescu*
79. Surface Chemistry and Electrochemistry of Membranes, *edited by Torben Smith Sørensen*
80. Interfacial Phenomena in Chromatography, *edited by Emile Pefferkorn*
81. Solid–Liquid Dispersions, *Bohuslav Dobiáš, Xueping Qiu, and Wolfgang von Rybinski*
82. Handbook of Detergents, *editor in chief: Uri Zoller*
    Part A: Properties, *edited by Guy Broze*
83. Modern Characterization Methods of Surfactant Systems, *edited by Bernard P. Binks*
84. Dispersions: Characterization, Testing, and Measurement, *Erik Kissa*
85. Interfacial Forces and Fields: Theory and Applications, *edited by Jyh-Ping Hsu*
86. Silicone Surfactants, *edited by Randal M. Hill*

ADDITIONAL VOLUMES IN PREPARATION

Surface Characterization Methods: Principles, Techniques, and Applications, *edited by Andrew J. Milling*

Interfacial Dynamics, *edited by Nikola Kallay*

SILICONE SURFACTANTS

edited by
Randal M. Hill

Dow Corning Corporation
Midland, Michigan

MARCEL DEKKER, INC. NEW YORK • BASEL

ISBN: 0-8247-0010-4

This book is printed on acid-free paper.

Headquarters
Marcel Dekker, Inc.
270 Madison Avenue, New York, NY 10016
tel: 212-696-9000; fax: 212-685-4540

Eastern Hemisphere Distribution
Marcel Dekker AG
Hutgasse 4, Postfach 812, CH-4001 Basel, Switzerland
tel: 41-61-261-8482; fax: 41-61-261-8896

World Wide Web
http://www.dekker.com

The publisher offers discounts on this book when ordered in bulk quantities. For more information, write to Special Sales/Professional Marketing at the headquarters address above.

Current printing (last digit)
10 9 8 7 6 5 4 3 2 1

PRINTED IN THE UNITED STATES OF AMERICA

Preface

Although silicone surfactants are a commercially important class of novel surfactants with numerous applications—ranging from their use in the manufacture of polyurethane foam to applications in coatings, household and personal care products, and foam control and as exceptional wetting agents—the literature on their properties and applications is limited and widely scattered among many journals, patents, and trade publications. Judging from the rapidly growing number of recent publications, the interest in this class of surfactants is increasing dramatically. The intent of the current volume is to bring together in one place a comprehensive introduction to the preparation, uses, and physical chemistry of silicone surfactants. As such, it should be of value both as an introduction to and as a reference source for this fascinating class of surfactants.

Polydimethyl siloxane and many copolymers containing dimethyl siloxane groups are surface active in a variety of aqueous and nonaqueous media. This book focuses primarily on those silicone polyoxyalkylene copolymers that are surface active in aqueous systems, but also includes chapters on two important nonaqueous systems—polyurethane foam and polymer blend compatibilizers. The book begins with an introductory chapter that overviews the preparation, physical chemistry, and applications of silicone surfactants. This broad perspective is followed by detailed discussions of each of these areas. Chapters 2 and 3 cover the synthesis and analysis of silicone surfactants including a number of novel silicone surfactants. Chapter 4 details the surface activity and aggregation behavior of silicone surfactants. Chapters 5–10 discuss specific applications including polyurethane foam manufacture, personal care, coatings, fabric finishes and polymer surface modifiers, foam control, and agricultural adjuvancy. Chapters 11 and

12 deal with two areas of significant recent activity—the unusual wetting behavior of the trisiloxane surfactants and the ternary phase behavior of mixtures of silicone surfactants with water and silicone oils. The emphasis throughout the volume is on understanding and insight rather than formulary presentations.

A variety of authors were enlisted to contribute different perspectives to the work, including representatives from each of the major manufacturers and academic specialists who have studied the surfactancy of silicone surfactants. Thus the work represents the collective effort and knowledge of an international group of scientists and technologists. I hope it will be valuable to those seeking to make use of silicone surfactants in diverse applications as well as to researchers seeking to better understand fundamental surfactancy phenomena by examining the differences and similarities between hydrocarbon and silicone surfactants.

Randal M. Hill

Contents

Contributors

Andrea Bungard Surfactant Division, Th. Goldschmidt AG, Essen, Germany

Richard Burow* Dow Corning Corporation, Midland, Michigan

H. Ted Davis Dean, Institute of Technology and Department of Chemical Engineering and Materials Science, University of Minnesota, Minneapolis, Minnesota

Kenneth C. Fey Advanced Materials Business Development Group, Dow Corning Corporation, Midland, Michigan

David T. Floyd Surfactant Division, Goldschmidt Chemical Corporation, Hopewell, Virginia

Burghard Grüning Surfactant Division, Th. Goldschmidt AG, Essen, Germany

Randal M. Hill Central Research and Development, Dow Corning Corporation, Midland, Michigan

H. Hoffmann Physical Chemistry I, University of Bayreuth, Bayreuth, Germany

Gary E. LeGrow* Designed Materials Development, Dow Corning Corporation, Midland, Michigan

---

*Retired.

X. Li* Department of Chemical Engineering and Materials Science, University of Minnesota, Minneapolis, Minnesota

Donald Penner Department of Crop and Soil Sciences, Michigan State University, East Lansing, Michigan

Lenin J. Petroff Designed Materials Development, Dow Corning Corporation, Midland, Michigan

Frank C. Roggenbuck Department of Crop and Soil Sciences, Michigan State University, East Lansing, Michigan

Gerd Schmaucks Research and Development, Schill & Seilacher GmbH, Hamburg, Germany

L. E. Scriven Department of Chemical Engineering and Materials Science, University of Minnesota, Minneapolis, Minnesota

Steven A. Snow Dow Corning Corporation, Midland, Michigan

Robert E. Stevens Performance Chemicals Technology, Air Products and Chemicals, Inc., Allentown, Pennsylvania

T. Stoebe Eastman Kodak Company, Rochester, New York

W. Ulbricht Physical Chemistry I, University of Bayreuth, Bayreuth, Germany

Michael D. Ward Department of Chemical Engineering and Materials Science, University of Minnesota, Minneapolis, Minnesota

Iskender Yilgör Department of Chemistry, Koç University, Istanbul, Turkey

**Current affiliation*: Applied Materials Inc., Santa Clara, California.

1
Siloxane Surfactants

RANDAL M. HILL Central Research and Development, Dow Corning Corporation, Midland, Michigan

I. INTRODUCTION

Siloxane surfactants consist of a permethylated siloxane group coupled to one or more polar groups. This class of surfactants finds a variety of uses in applications where other types of surfactants are relatively ineffective [1–3]. Siloxane surfactants have certain unique properties:

1. Their hydrophobic group is silicone, so that
2. They are able to lower surface tension to ≈ 20 dyn/cm compared with ≈ 30 dyn/cm for typical hydrocarbon surfactants, causing them to be
3. Surface active in both aqueous and nonaqueous media.

In addition,

4. They are prepared by different chemistries, yielding molecular structures of different types and ranges [4,5], which are often fluid to very high molecular weights [6].

Siloxane surfactants were introduced to the marketplace in the 1950s for the manufacture of polyurethane foam [7]. Soon afterward other applications were invented for them [8]. Nonaqueous surface activity is the basis for their use in polyurethane foam manufacture, as demulsifiers in oil production, and as defoamers in fuels. Their ability to lower surface tension leads to wetting and spreading applications. Different molecular structures and high molecular weights make them useful as novel emulsifiers. Silicones impart a unique dry-lubricity feel to surfaces such as textiles, hair, and skin. Since siloxane surfactants incorporate silicone in a water-soluble or water-dispersible form, they represent a convenient means for putting silicone on a surface by way of an aqueous formulation.

The molecular origin of the principal difference between hydrocarbon and siloxane surfactants is illustrated in Fig 1. The surface active character of siloxane surfactants is due to the methyl groups, the —O—Si—O—Si— backbone simply serves as a flexible framework on which to attach the methyl groups [9–12]. The surface energy of a methyl-saturated surface is about 20 dyn/cm [9], and this is

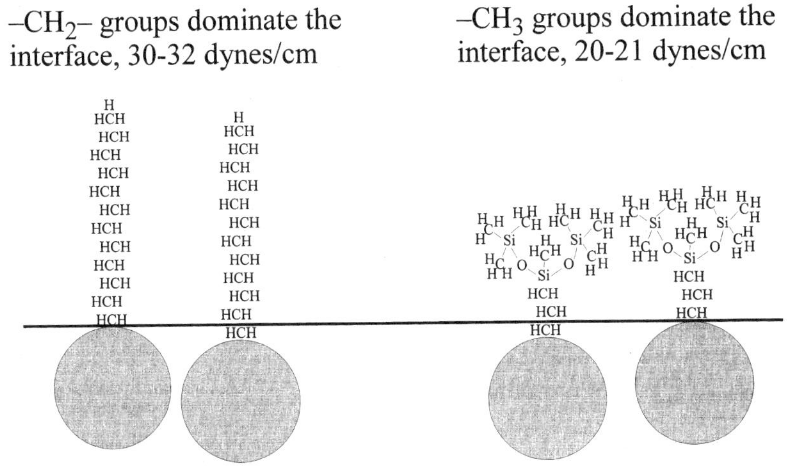

FIG. 1 Comparison of the surface character of hydrocarbon versus siloxane surfactants.

also the lowest surface tension achievable using siloxane surfactants. In contrast, most hydrocarbon surfactants consist of alkyl, or alkylaryl hydrophobes, which contain mostly —CH_2— groups, and pack loosely at the air–liquid interface. The surface energy of such a surface is dominated by the methylene groups, and for this reason hydrocarbon surfactants typically achieve surface tensions of about 30 dyn/cm or higher [9]. Thus, the lower surface tensions given by siloxane surfactants can be traced directly to molecular structure, the unusual flexibility of the siloxane backbone, and the different surface energies of —CH_3 versus —CH_2—.

Siloxane surfactants are similar to hydrocarbon surfactants in many common features of surfactancy [2,13–15]:

1. There is a break in their surface tension versus log concentration curve reflecting the onset of self-association (such as micelle formation).
2. Critical aggregation concentrations (cac) vary with molecular structure in the same way—within a homologous series, proportionately larger hydrophobic groups lead to smaller cac values.
3. They show similar patterns of self-association in aqueous solution, forming aggregates and liquid crystal phases, of the same types and following the same trends with molecular structure.
4. Siloxane surfactants incorporating polyoxyalkylene groups also show inverse temperature solubility and cloud points.

This last point requires some clarification: in the dilute concentration range, many siloxane surfactants form cloudy lamellar phase dispersions that are unrelated to the existence of a cloud point as it is usually understood [13].

Substantial advances in our understanding of this class of surfactants in recent years have covered their aqueous aggregation behavior, their ternary phase behavior with silicone oils, and their ability to promote rapid wetting of hydrophobic substrates. This chapter attempts to describe the structure, preparation, and surfactancy properties of this fascinating class of surfactants incorporating these recent advances. A brief discussion of some common applications also is given to illustrate how the unusual properties of siloxane surfactants are used. Detailed treatments of synthesis, superwetting, aqueous aggregation, and ternary phase behavior, and selected application topics are given elsewhere in this volume.

II. MOLECULAR STRUCTURES AND NOMENCLATURE

Polydimethylsiloxane (PDMS) is itself surface active [16]. Gruning and Koerner [4] suggest a broad definition of siloxane surfactants to include all surface active copolymers containing a siloxane entity. We prefer to limit the scope to molecules with well-defined and well-separated hydrophilic and hydrophobic parts. Copolymers and terpolymers based on the PDMS backbone can be used to modify interfacial properties (one such use is illustrated in Fig. 2) and could be included under the umbrella of siloxane surfactants. There is a significant literature on such materials [17–23], and they are discussed by Yilgor in this volume [24]. This introduction focuses primarily on siloxane surfactants that are useful in aqueous systems. Most siloxane surfactants are copolymers of PDMS and polyalkylene oxides of intermediate molecular weight. Nonaqueous surface chemistry is briefly discussed, as well as a selection of nonaqueous applications. The primary nonaqueous application, polyurethane foam manufacture, is reviewed elsewhere in this volume.

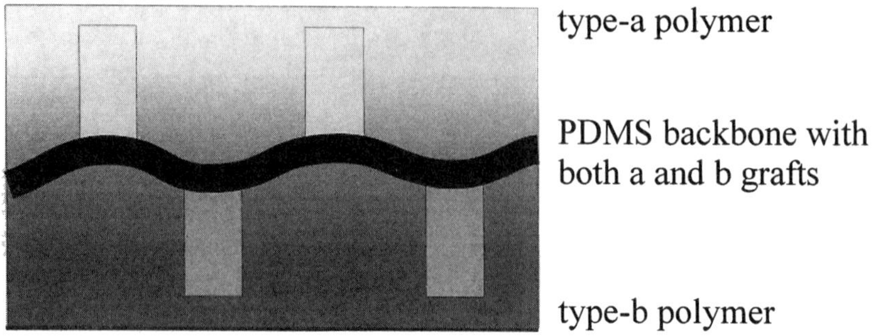

FIG. 2 Silicone terpolymer used as a polymer blend compatibilizing agent.

TABLE 1 Polar Groups in Siloxane Surfactants

Type of polar group	Examples
Nonionic	Polyoxyethylene
	Polyoxyethylene/polyoxypropylene[a]
	Saccharides (glucose, maltose)
Anionic	Sulfate
Cationic	Quaternary ammonium salts
Zwitterionic	Betaines

[a]Polyoxypropylene is hydrophobic.

The hydrophobic portion of siloxane surfactants is the permethylated siloxane group. Hydrophilic groups of many different types can be incorporated—some examples are given in Table 1. Throughout this chapter we will use the notation E_n, or EO for the polyoxyethylene group, and EO/PO for the polyoxyethylene/polyoxypropylene group. The three most common molecular structures for siloxane surfactants are rake-type copolymers (also called comb or graft copolymers), ABA copolymers (in which the "B" represents the silicone portion), and trisiloxane surfactants. These are illustrated in Figs. 3, 4, and 5, respectively. Cyclosiloxane hydrophobic groups can also be made [25], as illustrated in Fig. 6. Structures incorporating branching (T units in the siloxane backbone; see the notation below) and AB-type copolymers have also been made [14,25]. Alkyl groups (such as C_6H_{13}—), fluorocarbon groups [26], and combinations of such groups with polar groups have been attached to the siloxane backbone to make materials that are surface active toward a variety of surfaces and interfaces.

Siloxane polyoxyalkylene copolymers are known to industry by several names, including siloxane polyethers (SPEs), polyalkylene oxide silicone copolymers,

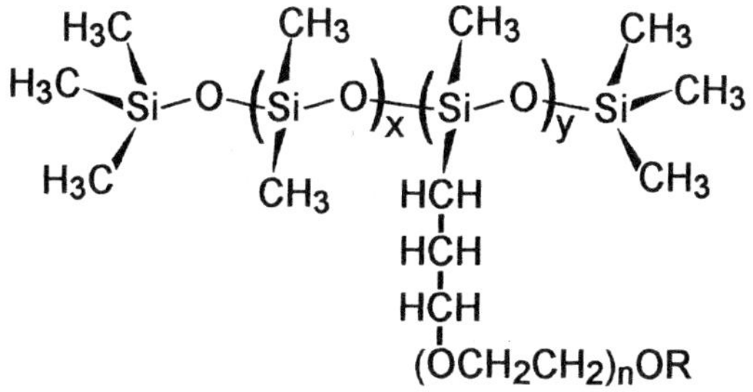

FIG. 3 Molecular structure of a rake-type siloxane surfactant. Also called a comb or graft copolymer; R stands for an end-capping group such as —H, —CH_3, or —O(O)CCH_3.

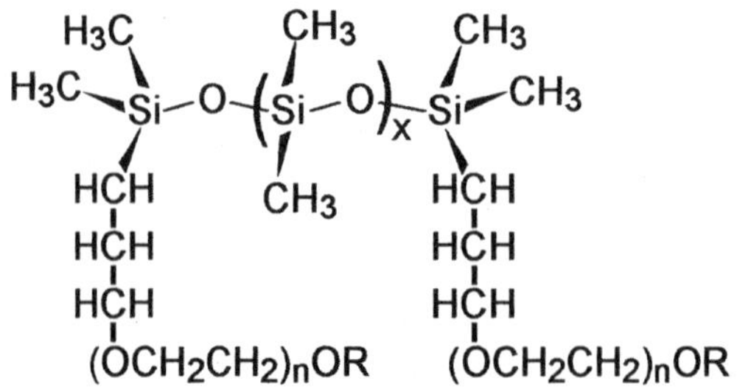

FIG. 4 Molecular structure of an ABA-type siloxane surfactant. Also called α-ω, or bolaform surfactants.

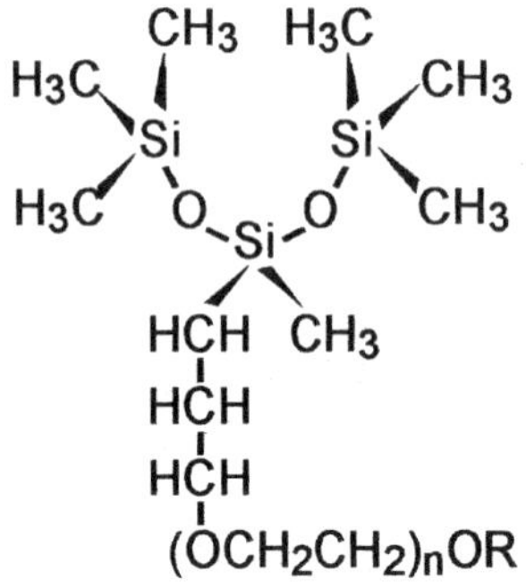

FIG. 5 Molecular structure of a trisiloxane surfactant.

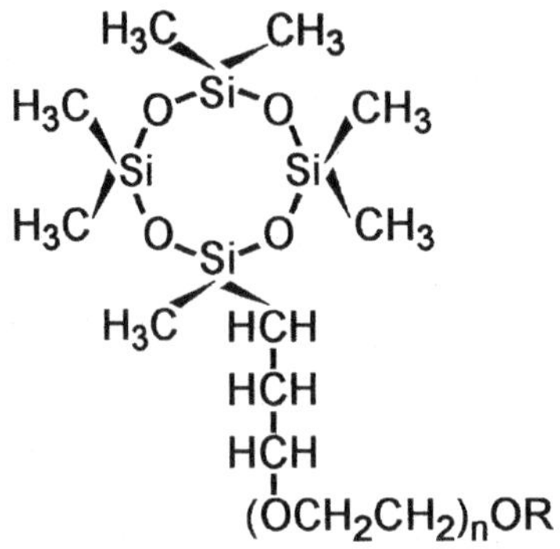

FIG. 6 Molecular structure of a cyclosiloxane surfactant.

silicone poly(oxyalkylene) copolymers, and silicone glycol copolymers (or surfactants) [27]. The International Cosmetic Ingredient Nomenclature is dimethicone copolyol [28].

Liquid crystal side chain polymers based on the polysiloxane backbone comprise a technologically important variation of the comb-type structure [29]. These polymers form thermotropic liquid crystals that are useful for liquid crystal display applications [30–33]. A few such polymers have been made that are water dispersible and for which aqueous surfactancy properties have been reported [34,35].

III. SYNTHESIS AND CHEMISTRY

Siloxane surfactants are prepared by attaching one or more polar organic groups to a permethylated siloxane backbone [5]. There are three common ways to do this:

1. Directly by the reaction of $\equiv SiOR^1$ and R^2OH (transetherification).
2. Directly by hydrosilylation, the reaction of SiH and (for example) $H_2C{=}CHCH_2R$, where R is a polar organic group.
3. Indirectly by attaching a reactive group such as an epoxy to the siloxane backbone using the reaction described in item 2 and then using that reactive group to attach a polar group.

The chemistry of these reactions is discussed in the wider context of organosilicon chemistry by Plumb and Atherton [5], Noll [36], and Clarson and Semlyn [37]. Details particularly relevant to preparing siloxane surfactants are discussed by Gruning and Koerner [4], who also catalog a large number of possible modifying groups. Nonionic siloxane surfactants are usually prepared by means of the first and second routes, and ionic siloxane surfactants by the third route.

A detailed discussion of the preparation and analysis of siloxane surfactants is given elsewhere in this volume by Legrow and Petroff [38]. This brief review, however, illustrates the primary chemistries involved, with particular attention to some of the possible by-products and side reactions. Virtually all published work on siloxane surfactants has used commercially available samples, or samples that were similar in composition. Such materials are polydisperse and can differ significantly depending on the details of preparation (e.g., at what point, if at all, the cyclics are removed—see below). This is one of the reasons for the occasional appearance of published data on nominally identical samples that are not in agreement.

A. Preparation of the Siloxane Backbone

The first step in the synthesis of a siloxane surfactant is to prepare a siloxane backbone containing reactive sites (SiOH, SiOR, or SiH) at which to attach the polar groups. There are two ways to do this—cohydrolysis of the appropriate chlorosilanes and the equilibration reaction.

For example, to prepare a rake-type siloxane surfactant using route 2 in Sec. III, an SiH functional siloxane backbone could be prepared by cohydrolysis of the appropriate chlorosilanes [36,37] according to reaction (1):

$$2(CH_3)_3SiCl + x\,(CH_3)_2SiCl_2 + y\,(CH_3)HSiCl_2 + H_2O \rightarrow MD_xD'_yM + HCl \quad (1)$$

This backbone could also be prepared by equilibration of the appropriate proportions of end-cap and monomer units according to reaction (2):

$$MM + xD + yD' \rightleftharpoons MD_xD''_yM \quad (2)$$

Reaction (2), which is called the equilibration reaction, is catalyzed by either acid or base. ABA structures, or mixed ABA and rake-type structures, can be prepared by substituting M′M′ for all or some of the MM in reaction (2). Branched siloxane backbones can be prepared by including T groups. The MDTQ notation is defined in Table 2 [39].

For example, the structure of one of the trisiloxane superwetter surfactants is written as follows: M(D′E_8OH)M. The R group on the D′ is specified as an eight-unit polyoxyethylene group, E_8, with a hydroxy end cap. The propyl spacer between the EO chain and the siloxane is not shown explicitly. There are three features of the equilibration reaction that may influence the properties of siloxane surfactants. First, during equilibration, both linear and cyclic species are formed. However, only low molecular weight cyclics are formed to a significant degree, ranging from the trimer, D_3, to the hexamer, D_6. Thus, as the molecular weight of the desired silicone polymer increases, the distribution becomes distinctly bimodal. Second, when a mixture of D and D′ units is equilibrated, the cyclics that

TABLE 2 MDTQ Notation for Siloxane Building Block Units

M	$Me_3SiO_{1/2}$—	A trimethyl end-cap unit
D	—Me_2SiO—	The basic dimethyl unit
T	—$MeSiO_{3/2}$—	A three-way branch point unit
Q	—SiO_2—	A four-way branch point unit
M′	$Me_2(R)SiO_{1/2}$—	A substituted trifunctional end-cap unit
D′	—Me(R)SiO—	A substituted difunctional unit
T′	—$(R)SiO_{3/2}$—	A substituted three-way branch point unit
Me	—CH_3	
R	H, or (after hydrosilylation) some nonmethyl organic group such as —$CH_2CH_2CH_2(OCH_2CH_2)_nOH = E_nOH$	

Source: Ref. 39.

are formed include cocyclics containing both D and D′ units. The cyclics are relatively volatile and may be stripped out before the polar groups are attached (thereby removing some D′ units), afterward (removing only the cyclics that contain no D′units), or not at all. Third, the number of D′ units per chain is a distribution that includes some chains with no D′ units.

Thus, the backbone of a rake-type siloxane copolymer may possess three distinct types of polydispersity: (1) a (possibly bimodal) distribution of chain lengths, (2) a distribution in the number of D′ units per chain, and (3) random location of the D′ units along the chain. For an ABA-type copolymer, only monomodal polydispersity of type 1 is present.

B. Transetherification

The first nonionic siloxane surfactants were prepared by coupling alkoxymethylsiloxane polymers and hydroxy-terminated polyoxyalkylenes using the transetherification reaction [41]:

$$\equiv SiOR^1 + R^2OH \rightarrow \equiv SiOR^2 + R^1OH \qquad (3)$$

R^1 is usually —CH_3 or —CH_2CH_3 and R^2 is a polyalkylene oxide. The ≡ denotes that there are three other bonds to the silicon that are not explicitly shown. This reaction yields products in which the hydrophilic groups are linked to the siloxane hydrophobe through an Si—O—C linkage. These materials are useful in polyurethane foam manufacture and other nonaqueous applications, but in water this linkage hydrolyzes (rapidly at nonneutral pH) to generate silanol and alcohol [40].

C. Hydrosilylation

More hydrolytically stable siloxane–polyoxyalkylene surfactants can be prepared by the hydrosilylation of methyl siloxanes containing Si–H groups with vinyl functional polyoxyalkylenes [5,36,37]:

$$\equiv SiH + CH_2{=}CHCH_2(OCH_2CH_2)_nOR \rightarrow \equiv Si(CH_2)_3(OCH_2CH_2)_nOR \qquad (4)$$

This reaction is usually catalyzed using chloroplatinic acid (Speier's catalyst) [5], which can also cause isomerization of the terminal double bond to the less reactive internal position. To make sure that all the reactive sites on the siloxane are used, the reaction may be carried out with an excess of allyl polyether. Under aggressive reaction conditions, —OH groups on the polyether can also react with SiH generating Si—O—C bonds, which hydrolyze in water. Commercial allyl polyethers usually also contain species with twice the nominal molecular weight, and an —OH group on both ends (sometimes called diol).

D. Two-Step Synthesis Using Reactive Intermediate

Another way to attach a polar group to the siloxane backbone is to replace the SiH group with a reactive organic group, then attach the polar group to that group [5,36,37,41]. This is useful when the reaction conditions of hydrosilylation are incompatible with the desired polar group. Such a two-step synthesis is the preferred method of preparing ionic siloxane surfactants, although nonionic surfactants have also been prepared by this route [42]. For example, quaternization of halide functional siloxanes leads to cationic siloxane surfactants:

$$\equiv Si(CH_2)_3X + NR_3 \rightarrow \equiv Si(CH_2)_3N^+R_3X^- \tag{5}$$

in which X = halide [43]. Anionic siloxane surfactants can be prepared by the sulfonation of epoxy-functional silicones as follows [44]:

$$\begin{aligned} &\equiv Si(CH_2)_3OCH_2CH(O)CH_2 + NaHSO_3 \rightarrow \\ &\equiv Si(CH_2)_3OCH(OH)CH_2SO_3^-Na^+ \end{aligned} \tag{6}$$

Zwitterionic siloxane surfactants can be prepared by the ring opening of cyclic sultones by amino-functional silicones as follows [45]:

$$\begin{aligned} &\equiv Si(CH_2)_3N(CH_3)_2 + O(CH_2)_3SO_2 \rightarrow \\ &\equiv Si(CH_2)_3N^+(CH_3)_2(CH_2CH_2CH_2SO_3^-) \end{aligned} \tag{7}$$

Examples of other reactions suitable for the preparation of ionic siloxane surfactants are given by Snow et al. [41,43,45], Maki et al. [47], and Gruning and Koerner [4]. Many examples of such ionic surfactants have been prepared and investigated, and some are available commercially [47,48].

Wagner et al. prepared a series of siloxane surfactants with nonionic carbohydrate polar groups [49,50]. Initially, they prepared siloxanyl-modified glycosides by means of a low yield, multistep procedure involving protecting the hydroxyl groups of the sugar [50]. Their more recent work is based on a four-step reaction procedure with much better yield in which an R_3SiH is reacted with allyl glycidyl ether to attach an epoxy group, which is then reacted with a diamine to yield a primary amine in the terminal position. The primary amine is subsequently reacted with a lactone to form a gluconamide [51]. Wagner et al. also prepared ionic variations of carbohydrate functional siloxanes [52]. Jonas and Stadler [53], Wersig et al. [54], and Schmaucks [55] have also used multistep syntheses to obtain novel nonionic silicon-based surfactants.

E. Organophilic Siloxanes and Prepolymers

The same basic chemistry also can be used to incorporate organophilic or fluorocarbon functionality into a siloxane surfactant [4]. Roidl [56], Wolfes [57], and Rentsch [58] describe the preparation and properties of siloxane copolymers incorporating both hydrophilic and organophilic groups. The use of siloxane polymers with long chain alkyl groups ($> C_{10}$) attached to the siloxane backbone, called silicone waxes, in personal care applications is discussed by Schaefer et al, [47]. Such waxes are somewhat soluble in organic media and can be surface active in (for example) mineral oil. They are claimed to improve spreadability, luster, and pigment dispersion [59–62].

F. Carbosilane Surfactants

The surfactants discussed up to this point have been based on permethylated siloxane species containing a Si—O—Si backbone. Other structures can be prepared containing a permethylated Si—C—Si backbone (called a polysilmethylene or carbosilane), or a permethylated Si—Si backbone (called a polysilane). Such structures are not subject to hydrolysis, making them potentially useful in harsher environments, or in applications that require long-term chemical stability of the surfactant. Maki et al. [63] prepared several carbosilane and polysilane surfactants and characterized their surface activity in comparison with analogous siloxane surfactants. Renauld and coworkers [64–66] prepared analogs to the trisiloxane surfactants in which the Si—O—Si—O—Si backbone is replaced with Si—C—Si—C—Si. Klein et al. [67–71] prepared surface active silane derivatives with the following structure by hydrosilylation of trimethylsilane:

$$(CH_3)_3SiCH_2CH_2CH_2(OCH_2CH_2)_nOH \qquad (8)$$

Wagner et al. [51] recently used their four-step synthesis procedure [50] to prepare carbohydrate functional carbosilane and polysilane surfactants.

G. Hydrolytic Stability

The hydrolytic stability of siloxane surfactants (especially the trisiloxane surfactants) has often been discussed and is viewed by some as a serious problem [72,73]. The Si—O—Si linkage is susceptible to hydrolysis in the presence of water:

$$\equiv Si{-}O{-}Si\equiv + H_2O \rightleftharpoons \equiv Si{-}OH + HO{-}Si\equiv \qquad (9)$$

This equilibrium is catalyzed by acid or base, and is slow near pH 7.0 [4,72]; it is essentially a reequilibration of the siloxane backbone [see Eq. (2) above]. Among organosilicone chemists, it is well known that residual acidity or basicity of glassware surfaces can catalyze this reaction—requiring that careful work be done either in plasticware or glassware that has been rigorously treated with a hydrophobizing agent such as octyltrichlorosilane.

There are few published analytical data regarding the rate of hydrolysis of siloxane surfactants in aqueous media. For instance, Gradzielski et al. [2] observed that certain trisiloxane surfactants "hydrolyzed completely" within a few weeks, whereas the polymeric siloxane surfactants they studied were stable for at least a few months. However, no quantitative analytical results were presented. Stürmer et al. [3] noted that the aqueous phase behavior of their $M(D'E_7OH)M$ changed somewhat after a short period of time, a response they attributed to hydrolysis. This change took substantially longer to occur in D_2O than in H_2O. Hill et al. [14] and He et al. [15] studied the same class of surfactants and did not observe such short time changes. Experiments on the trisiloxane surfactants carried out in buffered solutions in plasticware show no short-term evidence of degradation. [Note that the Phosphate buffer consisted of 0.1 M potassium dihydrogen phosphate plus 0.1 M NaOH to adjust pH to 7.0.]

The influence of hydrolysis on the surfactancy of a siloxane polyether copolymer will depend on the proportion of trimethyl, M, groups in the siloxane backbone, and on the way in which randomly cutting and reassembling the chain affects the overall hydrophile–lipophile balance of the mixture. For trisiloxane surfactants, two-thirds of the siloxane hydrophobe consists of M groups. Hydrolysis forms trimethylsilanol, which may recombine to form hexamethyldisiloxane, MM [4], as shown in the following reactions.

$$(CH_3)_3Si{-}O{-}SiR(CH_3){-}O{-}Si(CH_3)_3 + H_2O \rightleftharpoons (CH_3)_3Si{-}OH + HO{-}Si(CH_3)_3 \tag{10}$$

$$(CH_3)_3Si{-}OH + HO{-}Si(CH_3)_3 \rightleftharpoons (CH_3)_3Si{-}O{-}Si(CH_3)_3 + H_2O \tag{11}$$

Trimethylsilanol is water soluble and not surface active. MM is insoluble; either it segregates into the micelle core, or it phase-separates. Polymeric surfactants with long linear segments might be expected to form cyclosiloxanes.

The rate of hydrolysis presumably depends on the degree of contact between the aqueous solution and the siloxane portion of the surfactant molecules–water molecules, and Si—O—Si bonds must be in contact for a reaction to take place. Since surfactants self-associate to form aggregates in which water is largely excluded from the hydrophobic aggregate core, it is reasonable to assume that hydrolysis predominantly involves unassociated surfactant molecules [72]. This suggests that the rate of hydrolysis should depend on the concentration of unassociated surfactant molecules, which is always less than or equal to the critical aggregation concentration (cac, or cmc). An induction period is sometimes observed during which the concentration of surface active species decreases to the cac [72]. Only after this has occurred do surface tension and wetting properties markedly deteriorate. Thus, concentrated solutions may appear to be stable for long periods of time. Polymeric siloxane surfactants are generally more hydrolytically stable because their cac is small, and because cutting and recombining the siloxane chains only slowly generates species that are either volatile, insoluble, or much more or much less hydrophobic than the starting material. The rates of these reactions are not known, but they appear to be slow near pH 7. Knoche et al. [72] determined a half-life exceeding 40 days for dilute $M(D'E_{7.5}OMe)M$ at pH 7.

Under strongly acid or basic conditions or at sustained temperatures above 70°C, hydrolysis leads to rapid loss of surfactancy for most siloxane surfactants [72].

IV. SURFACE ACTIVITY

The surface and interfacial activity of siloxane surfactants is reviewed in this volume by Hoffmann and Ulbricht [74], and only an overview is presented here.

A. Nonaqueous Systems

The surface activity of PDMS in nonaqueous media such as the polyols used in polyurethane foam manufacture results from a combination of the low surface energy of the methyl-rich siloxane species and insolubility determined by molecular weight [10,11,40]. The solubility of a polymer in a given solvent is a combination of solvent–solute interactions and the molecular weight of the polymer [75]. Low molecular weight PDMS is soluble in organic solvents such as toluene. As the molecular weight increases, the solubility decreases. In the molecular weight range in which the polymer is just marginally soluble, it shows notable surface activity, even pro-foaming [76]. Thus, the lyophobicity (i.e., phobicity to the solvent) of the permethylated siloxane backbone in nonaqueous media is partly determined by molecular weight. This means that a higher molecular weight siloxane surfactant will have a greater tendency to segregate to the surface, even with the same proportions of solvent-loving and solvent-hating groups. In nonaqueous media this can mean the difference between significant surface activity and solubility.

Schmidt [77] presents some results illustrating the ability of siloxane polyether copolymers to lower the surface tension of a variety of organic liquids including mineral oil and several polyols from values of about 25–30 dyn/cm to values near 21 dyn/cm. Kendrick and Owen [11,12] show how siloxane surfactants lower the surface tension of polyols used in the manufacture of polyurethane foams. They conclude that the polysiloxane chain is looped away from the surface with the loops anchored to the surface by the polyether segments. To lower surface tension to the 20–21 dyn/cm value, it was found to be necessary to have dimethylsiloxane segments at least 10 monomer units long between each pendant polyether group; in other words, the D/D′ ratio of the siloxane backbone must be at least 10.

B. Aqueous Systems

Figure 7 plots surface tension versus surfactant concentration for the polymeric siloxane surfactant DC 190.* This surfactant is a moderately high molecular

---

*Surface tensions were measured using the Wilhelmy plate method. Solutions were prepared by weight using DI water and measured the same day. Especially below the cac, equilibrium times were long (>2 h).

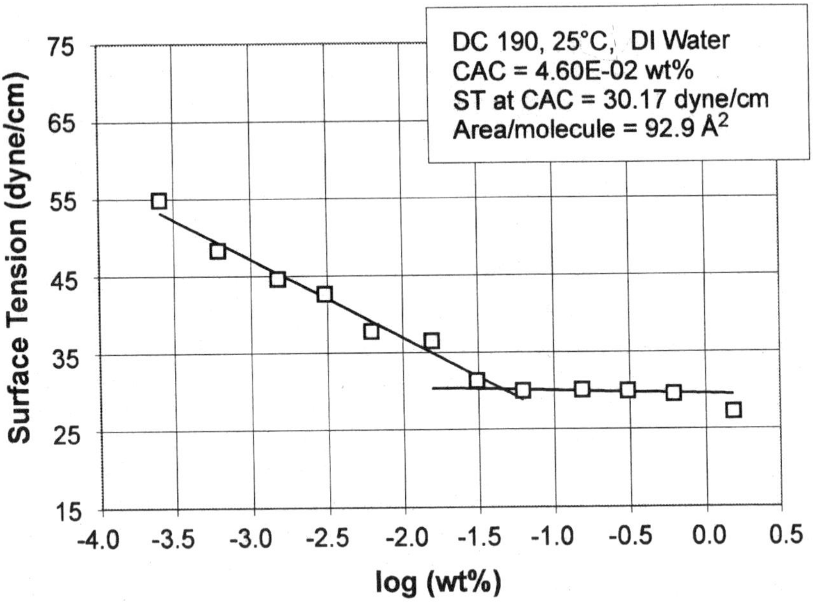

FIG. 7 Surface tension versus surfactant concentration for the polymeric siloxane surfactant DC 190, approximate structure = $MD_{103}(D'E_{18}P_{18})_{10}M$.

weight silicone–polyoxyalkylene graft copolymer in which the polyoxyalkylene groups consist of a random copolymer of EO and PO. The surface tension falls linearly until the break identified as the cac. Above this concentration, the surface tension remains constant at about 30.2 dyn/cm. The area per molecule, derived from the slope of the pre-cac region, is relatively small for such a large molecule. The small area probably indicates that the surfactant is forming a multilayer at the interface. Gibbs plots for two trisiloxane surfactants are shown in Fig. 8. Both trisiloxane surfactants lower surface tension to about 21 dyn/cm. The cac and the surface tension at the cac increase with increasing EO chain length, as they do for conventional hydrocarbon surfactants. The area per molecule also increases, indicating that its value is determined by the packing of the EO groups rather than the trisiloxane hydrophobe. Vick [27] presented surface tension results for three homologous series of rake-type polymeric nonionic siloxane surfactants. He found that the surface tension (of 1% aqueous solutions) increased with increasing size of the siloxane hydrophobe and with increasing EO chain length, but not with weight fraction of EO. This is somewhat different from the trends found in nonaqueous media described above. Siloxane surfactants containing mixed EO/PO groups generally gave higher surface tensions than those with only EO groups. The lowest surface tension values were about 21 dyn/cm, which is also the

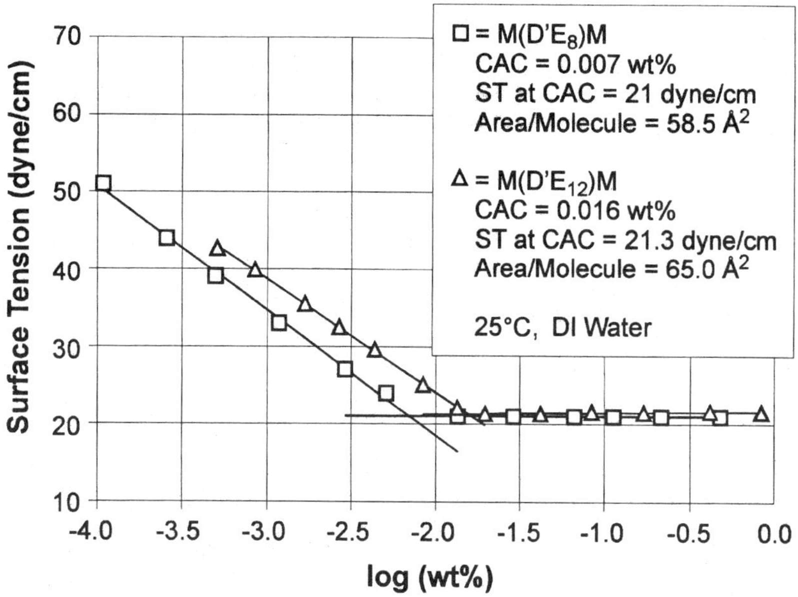

FIG. 8 Surface tension versus surfactant concentration for two trisiloxane surfactants.

surface tension of PDMS. This is considerably below values achievable using organic surfactants, but higher than fluorocarbon surfactants. Values for solutions of polymeric siloxane surfactants are typically between 20 and 35 dyn/cm.

Kanner et al. [40] prepared a number of low molecular weight siloxane polyoxyethylene surfactants and characterized their surfactancy and wetting properties. The most surface active siloxane surfactants were those with the smallest siloxane groups having two to five silicon atoms. Branching of the siloxane hydrophobe and variation of the EO chain length had only minor effects on the surface activity. Replacing methyl groups with longer alkyl groups resulted in higher surface tensions.

Gentle and Snow [78] investigated the adsorption of a homologous series of trisiloxane surfactants M(D′E_nOH)M with $n = 4$–20. They found that the cac, surface tension at the cac, and area per molecule varied with molecular structure in a way that was consistent with their "umbrella" model for the shape of the trisiloxane hydrophobe at the air–water interface. The log(cac) and the surface tension at the cac both increased linearly with EO chain length.

Pandya et al. [79] investigated the surfactancy of a group of nonionic siloxane surfactants. All the surfactants they investigated gave spherical micelles that grew slowly with increasing temperature. The ability of the micelles to solubilize oils was characterized in terms of the size of the hydrophobic and hydrophilic groups. The surfactants they investigated exhibited cloud points, which varied with the presence of electrolytes (KI, KCNS, KBr, KCl, and KF) in the same manner observed for common polyoxyalkylene surfactants.

Adsorption isotherms for a rake-type EO/PO siloxane surfactant onto several silicas were determined by Hasan and Huang [80]. They found that the surfactant adsorbed the most strongly on silicas with the least negative surface charge density. This is consistent with the observation that many surfactants adsorb weakly on polar surfaces such as silica [81]. Hydrogen bonding interactions between the polyoxyalkylene groups and SiOH cause the surfactant to adsorb with its hydrophilic group closest to the surface. Since this leaves the hydrophobe in contact with the water, a second layer of surfactant must adsorb to form a bilayer, and this tends to require relatively high surfactant concentrations. Schmaucks et al. [82] prepared several isomerically pure, branched and straight chain cationic siloxane surfactants with chain lengths of 2, 3, and 4. They found aqueous surface tensions (at the cac) from 25 dyn/cm for the disiloxanes to 19 dyn/cm for the branched tetrasiloxanes. The area per molecule at the interface was larger for the branched surfactants, and all were comparable to values derived from x-ray crystal structure analysis. Neither value supports the often-mentioned "umbrella" shape of the trisiloxane surfactants.

Maki et al. [46] reported the synthesis, surface tension, foaming, and bacteriostatic properties of several cationic organosilicon surfactants with methyl siloxane (Si—O—Si), methyl silmethylene, or carbosilane (Si—C—Si), and methyl silane (Si—Si) hydrophobic groups. Surface tensions were lowered to 22–23 dyn/cm, several of the synthesized materials were excellent foamers, and some exhibited bacteriostatic activity.

Maki et al. [42,63,83–85] reported the synthesis and surface activity of several nonionic organosilicon surfactants with methyl siloxane, methyl silmethylene (or carbosilane), and methyl silane hydrophobic groups. Nonionic surfactants containing one, two, and three silicon atoms in the hydrophobic group were synthesized to study the relation between the structures and the surface active properties. Surface tensions of the aqueous solutions of the siloxanes were 20–21 dyn/cm, while the silmethylenes were 2–4 dyn/cm higher. The silmethylenes were somewhat more effective than the siloxanes at lowering interfacial tensions against silicone oil. Interfacial areas per molecule were calculated and the results discussed in terms of chain configurations. The siloxanes gave the best wetting on low energy hydrophobic surfaces such as a polyethylene film. Similar cationic surfactants were also prepared and characterized [86,87].

Wagner et al. [88] characterized the surface activity and wetting ability of a series of siloxane carbohydrate derivatives they had previously synthesized [49,50]. Most of the surfactants gave aqueous surface tensions of 20–22 dyn/cm with little sensitivity to either the nature of the polar group or the spacer group. Wetting ability was not as good as that of the trisiloxane polyoxyethylene surfactants (the superwetters, see below) and became worse with increasing hydrophilicity. Wagner et al. [89] also investigated the effect of the spacer group between the polyether and the permethylated backbone on surface activity and found that relatively large groups had surprisingly weak effects.

C. Interfacial Tension Lowering

Agents that lower aqueous surface tension will also generally lower the interfacial tension between an aqueous solution and oil. The degree of interfacial tension lowering depends on compatibility between the hydrophobic groups of the surfactant and the oil. Schwarz and Reid [90] found interfacial tensions of 2–3 dyn/cm for aqueous solutions of nonionic and ionic siloxane surfactants (structures not given) against mineral oil, and smaller values, 0–2 dyn/cm, against a silicone oil. Stürmer et al. [3] give values of about 1 dyn/cm for the interfacial tensions of aqueous solutions of trisiloxane surfactants against *n*-decane. Gradzielski [2] found somewhat higher values, up to 10 dyn/cm, for the polymeric siloxane surfactants they investigated. Kanellopoulos and Owen [91] investigated the adsorption of a group of siloxane surfactants at the silicone oil–water interface and found interfacial tensions down to about 4 dyn/cm. Maki et al. [42] found interfacial tensions between silicone oils and aqueous siloxane surfactant solutions in the range 3–6 dyn/cm. Bailey et al. [92] found interfacial tensions "which approach zero" for solutions of AB copolymer siloxane surfactants against silicone oils. Svitova et al. [93] measured interfacial tensions between a series of normal alkanes and 0.5 wt% solutions of $M(D'E_8OH)M$. Their results, which are replotted in Fig. 9, showed that the interfacial tension varied linearly between about 0.025 dyn/cm for hexane to 0.4 dyn/cm for hexadecane. These small values indicate that trisiloxane surfactants are very effective at lowering interfacial tensions against low energy hydrocarbon substrates, which is consistent with their ability to cause spreading over such substrates.

The dependence of surface and interfacial tensions on molecular weight indicates that the conformation of polymeric siloxane surfactants at interfaces contributes to the surface energy. A polymeric surfactant adsorbed at an interface will adopt a conformation consisting of loops and tails and contact points. The unusual flexibility of the siloxane backbone presumably influences the conformation. The presence of loops and tails, and the degree to which loops and tails extend into the adjacent phase, contribute to the interfacial energy. The relationship between molecular structure, interfacial conformation, and ability to lower surface tension should be investigated for siloxane surfactants.

D. Wetting and Spreading

Wetting of a solid surface by a liquid is a basic component in many natural processes and commercial technologies. Some examples include the spreading of liquid droplets such as coatings or inks on substrates ranging from clean metals to

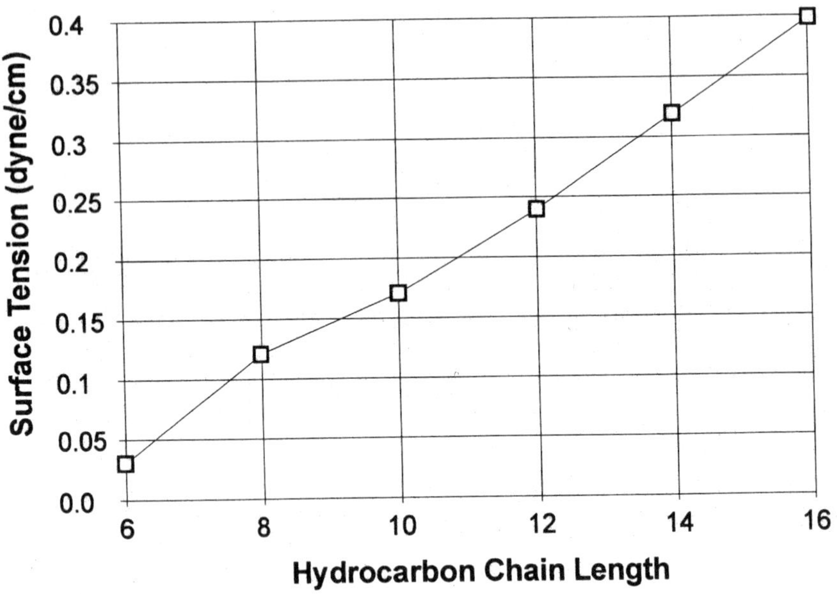

FIG. 9 Interfacial tensions at 25°C, 0.5 wt% M(D′E_8OH)M in water against normal alkanes. Data replotted from Ref. 93.

plastics, the penetration of inks into porous substrates such as paper, and the spreading of pesticide formulations on waxy weed leaf surfaces. Recent work on the unusual spreading behavior of the trisiloxane surfactants produced new insights into the spreading of surfactant solutions, but also demonstrated the inadequacy of existing models.

The ability of siloxane surfactants to promote spreading plays an important role in their use in paints and coatings [94,95], personal care products [27], textiles [96], and the oil industry [97], and as adjuvants for pesticides [73,98]. Gradzielski et al. [2] attributed the good wetting properties of siloxane surfactants to low adhesive forces between individual molecules in interfacial films. Vick [27] found that the time to wet by the Draves wetting test depended on the size of the siloxane hydrophobe and the length of the EO group—the most rapid wetting was observed for the surfactants with the shortest siloxane groups and the smaller EO groups.

Figure 10 shows a liquid droplet on a solid surface. The surface and interfacial tensions are drawn as force vectors acting on the three-phase contact line. At equilibrium, the forces must balance as described by the Young equation,

$$\gamma_{sv} = \gamma_{ls} + \gamma_{lv} \cos \theta \tag{12}$$

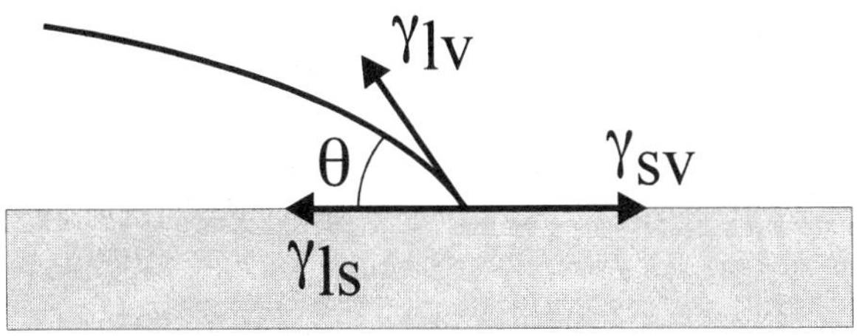

FIG. 10 A liquid droplet on a solid surface: γ_{lv} and γ_{sv} are the surface tensions of the liquid and solid, γ_{ls} is the interfacial tension between the liquid and solid, and θ is the contact angle of the liquid on the solid.

in which γ_{sv} is the surface tension of the solid, γ_{ls} is the interfacial tension between the solid and liquid, γ_{lv} is the surface tension of the liquid, and θ is the contact angle. If the contact angle is zero, the surface and interfacial tensions may sum to a positive value, indicating that there is an unbalanced Young's force at the three-phase contact line which should lead to spreading. This situation is often described in terms of the spreading coefficient, S,

$$S = \gamma_{sv} - (\gamma_{lv} + \gamma_{ls}) \tag{13}$$

A positive spreading coefficient means that spreading is energetically favored. Surfactant lowers both γ_{ls} and γ_{lv}, shifting the spreading coefficient in the positive direction. Neither γ_{sv}, nor γ_{ls} can be directly measured experimentally for solid surfaces, but a Zisman procedure [99] can be used to determine the critical surface tension of wetting. A surfactant solution with a surface tension higher than this cannot wet the surface. It has frequently been observed that the surface tension of surfactant solutions does not correlate very well with either wetting time or extent of spreading [100]. Equation (12) shows that spreading requires both a low surface tension for the surfactant solution and a low interfacial tension between between the liquid and the substrate. Fluorocarbon surfactant solutions with surface tensions of about 15 dyn/cm do not spread on low energy hydrocarbon surfaces because the interfacial tensions between fluorocarbon surfactant solutions and hydrocarbon substrates are large.

1. Superwetting

The unique ability of certain low molecular weight siloxane surfactants to promote spreading of dilute aqueous solutions on hydrophobic surfaces such as Parafilm® or polyethylene was discovered in the 1960s [25,39,90,91] and has been the subject of numerous patents and papers since [101–106]. This wet-out on low energy surfaces is called "superwetting" or "superspreading." The ability of the trisiloxane surfactants to promote spreading on waxy weed leaf surfaces such as velvetleaf and lambsquarters is the basis of their use as herbicide adjuvants or wetting agents [73,98,107]. This application is reviewed by Penner et al. [108] in this volume. Davis and coworkers have recently published several studies elaborating the unusual wetting behavior of the trisiloxanes, and relating this behavior to the larger problem of surfactant enhanced spreading. Stoebe et al. review these results elsewhere in this volume [109]. The following discussion attempts to summarize the background and recent results, and to offer direction for the technologist seeking to apply the insights.

Kanner et al. [40] observed that "certain methylsiloxane–polyether copolymers of low surface tension in aqueous solution rapidly spread to a thin film (wet out) on low energy hydrophobic surfaces such as polyethylene and polystyrene," that "the best wetting agents are based on siloxane hydrophobes containing 2 to 5 silicon atoms and have surface tensions of 20–21 dynes per cm," and further, that "these wetting agents are also characterized by a rather specific solubility balance . . . for optimum wetting, a surfactant must have a limited but finite solubility in water." The molecular structures described by Kanner et al. [40] included several disiloxane (MM′) based surfactants, and a cyclosiloxane (D_3D') based surfactant, but not the trisiloxane (MD′M) based materials that were eventually commercialized. Murphy et al. [110] found that the trisiloxane surfactant Silwet® L-77 gave lower equilibrium and dynamic surface tensions and spread better on paraffin wax film than corresponding hydrocarbon surfactants. The structure of Silwet L-77 is $M(D'E_{7.5}OMe)M$ [111]. Ananthapadmanabhan et al. [112,113] investigated the spreading behavior of several siloxane surfactants on Parafilm and found a relationship between turbidity and spreading, but not between equilibrium surface tension and spreading. These authors stated that "the structure itself of the SS1 molecule plays a governing role in determining its superior properties." Much of the literature dealing with the spreading of the trisiloxane surfactants describes their spreading properties in terms of spread area or a spreading index (which is a ratio to some benchmark condition). The spread area is usually defined as the area of a spread droplet after some fixed length of time (not always specified). At short times, spread area represents an estimate of spreading *rate* (dA/dt), while at longer times it is a measure of the *extent* of spreading.

Zhu et al. [104,114] used a video method to measure actual spreading rates of $M(D'E_{7.5}OMe)M$ on Parafilm. They found that the spread area increases linearly with time to a plateau value that is proportional to the surfactant concentration—the droplets spread until the inventory of surfactant to cover the air–water and water–substrate interfaces is exhausted. Thus, the *extent* of spreading (in the limit of long time) simply reflects the number of moles of surfactant present. This is why it is preferable to keep the *rate* and *extent* of spreading clearly separated. Zhu also found a maximum in spreading rate as a function of surfactant concentration and showed that the spreading rate was linearly related to dispersion turbidity and sensitive to humidity. The dependence on humidity was taken to indicate the need for a preexisting water film on the surface, and the rapid spreading was attributed to Marangoni effects. The critical surface tension of wetting of Parafilm is between 22 and 33 dyn/cm.* Both $M(D'E_{7.5}OMe)M$ and $M(D'E_{12}OH)M$, a more hydrophilic trisiloxane surfactant, have aqueous surface tensions above their cac of about 20.5 dyn/cm, but $M(D'E_{12}OH)M$ does not wet out on Parafilm. Zhu did not find any hydrocarbon surfactants that wet out on Parafilm.

Lin et al. [105,115], working with $M(D'E_8OH)M$, confirmed the concentration maximum using a quartz crystal microbalance (QCM) to measure spreading rates and showed that there is also a maximum in spreading rate as a function of substrate surface energy. These workers found spreading rates on surfaces of intermediate surface energy to be about 90 mm^2/s compared with the most rapid spreading rates on Parafilm of about 7 mm^2/s. They also discovered that $M(D'E_{12}OH)M$ spreads with similar characteristics except that it is somewhat slower at its maxima and does not spread on Parafilm.

2. Surfactant-Enhanced Spreading

Stoebe et al. [106,116,117] used an image analysis method to measure spreading rates for solutions of several trisiloxane and hydrocarbon, nonionic and ionic surfactants as a function of surfactant concentration, substrate surface energy, relative humidity, and temperature. They confirmed that spread area increases linearly with time, and that there are maxima versus both concentration and substrate surface energy. They found that the sensitivity to humidity depended on the type of substrate—for Parafilm it is strong, but for smoother surfaces it is much weaker. Within a homologous series, surfactants that form turbid vesicle-containing dispersions spread faster than those that form clear micellar solutions. Most importantly, Stoebe demonstrated that these characteristic features are shared by siloxane and some hydrocarbon, nonionic, and ionic surfactants. These results are reviewed in detail elsewhere in this volume [109].

Since interfacial tensions between aqueous solutions and solid substrates are not experimentally accessible, Svitova et al. [93] measured dynamic surface and interfacial tensions for trisiloxane surfactant solutions against normal alkanes and compared the calculated spreading coefficients with observed spreading. They found that a positive value of the spreading coefficient calculated from dynamic interfacial tensions at about a one-second time scale correlated well with spreading. They also found that vesicle forming $M(D'E_8OH)M$ gave faster surface tension fall rates than micelle forming $M(D'E_{12}OH)M$, consistent with their relative spreading rates. Ananthapadmanabhan [112] also observed that turbid SS1 has a faster surface tension fall rate than the more soluble SS2 and attributed this effect

*The critical surface tension of wetting of a methyl-rich wax surface such as hexatriacontane is about 22 dyn/cm. Branched polyethylene gives 33 dyn/cm. Parafilm, which is a high molecular weight alkane wax, should lie in between these extremes.

to a "higher mobility" on the part of SS1. Stoebe et al. [118] have recently observed spreading by trisiloxane and hydrocarbon surfactants on mineral oil. The behavior is similar to that on solid surfaces except that spreading rates are faster by a factor of 10–20 and there does not appear to be a concentration maximum.

Ruckenstein [119] argues that spreading of trisiloxane surfactants is facilitated "by adsorption of bilayers on the hydrophobic surface." Rosen and Song [120] studied the effect of adding certain hydrocarbon surfactants to trisiloxane surfactants and found that in some cases the extent of spreading for the mixture was larger than for the trisiloxane alone. Both these papers assume that it is the rate or extent of spreading that requires explanation.

The observed spreading behavior challenges existing models in a number of respects. Previous work with spreading of pure liquids, and with spreading of surfactant solutions on water-wettable substrates such as clean glass, demonstrated the important role of a precursor film in the spreading process. It is not clear that such a film can form for an aqueous surfactant solution on a hydrophobic substrate. Discussions of the wetting behavior of the trisiloxane surfactants have consistently found some relationship between spreading and the presence of turbidity. Stoebe et al. [106] have shown that turbidity is not essential for surfactant-enhanced spreading, but there are at least two examples of homologous series for which the turbid vesicle dispersions spread faster [109,115] and give faster surface tension fall rates [93]. These cases are difficult to reconcile with both intuition and existing models for surfactant transport. Larger particles should move more slowly. The existence of the maxima versus concentration and substrate surface energy point to competitive processes other than viscous dissipation. The nature of these competitive processes remains uncertain. Cazabat and coworkers [121,122] have studied the spreading of drops of pure liquid trisiloxane surfactants on hydrophilic and hydrophobic surfaces. This work could illuminate the competitive processes because it demonstrates that interactions between the surfactant and the substrate are quite different on high and low energy surfaces. Shanahan et al. [123] showed how such differences can influence spreading rates.

In spite of these issues, the technologist seeking to apply the insights from this work to practical problems can learn a great deal from it. Spreading rates continue to increase far beyond the critical aggregation (or micellization) concentration (cac or cmc) of the surfactant—consistent with dynamic surface tension measurements, which also show that surface tension fall rates continue to increase well beyond the cac. Applications that demand rapid spreading usually benefit from higher surfactant concentrations. However, the results indicate that at some point this trend reverses. The technologist must also determine whether extent of spreading, or rate of spreading, is critical in a given application. Several recent patents [124,125] and articles [120] claim that blends of surfactants can increase the extent of spreading (recall that extent of spreading is simply proportional to the number of moles of surfactant present [114]).

Spreading of surfactant solutions ceases as the substrate becomes more hydrophobic—the trisiloxanes are able to wet out on more hydrophobic surfaces than most hydrocarbon surfactants because the spreading coefficient remains positive to smaller values of γ_{SV}. A number of hydrocarbon surfactants are available that give surface and interfacial tensions substantially lower than can be obtained using the surfactants that are most effective for emulsification and dispersion [126]. Some of these, such as Aerosol OT [117], are known to be very effective wetting agents. Many coating and spreading problems involve mechanically forcing the liquid over the solid. In such problems, dynamic contact angles, or receding contact angles, are more useful than spreading rates. However, for problems in which the coating must rapidly wet out on the surface without mechanical assistance, it is necessary to use surfactants such as the trisiloxanes, which give sufficiently small surface and interfacial tensions with the substrate to achieve rapid spreading.

A serious problem arises when surfactants that give good spreading must be incorporated into products that have been formulated for optimal dispersion stability—most coatings, for example. It is well known that surfactant mixtures, including blends of siloxane and hydrocarbon surfactants, can behave synergistically [126,127]. However, synergistic mixing, as determined by equilibrium properties such as cmc or surface tension, is a poor guide to a mixture that will also spread rapidly. For example, most hydrocarbon surfactants are antagonistic to the spreading of the trisiloxanes [100]. This probably reflects the fundamentally different properties of surfactants that are good dispersants and those that are good wetting agents.

E. Mixtures of Siloxane and Hydrocarbon Surfactants

The surface activity of mixtures of surfactants is well documented [127], but mixtures of siloxane surfactants and hydrocarbon surfactants have received little attention. Bailey et al. [128] claimed generally synergistic behavior for a wide range of combinations of siloxane surfactants with hydrocarbon surfactants. Hill [129] reported mixed cmc results and detergency measurements for several siloxane surfactants with anionic, cationic, and nonionic hydrocarbon surfactants. The behavior of the mixtures varied from strongly synergistic to antagonistic depending on the ionic character of the hydrocarbon surfactant. Ohno et al. [130] found that a siloxane surfactant strongly lowered the surface tension of sodium dodecyl sulfate solutions but had little effect on the cmc.

Petroff [131] disclosed the use of zwitterionic siloxane surfactants with a range of hydrocarbon surfactants. Mixtures of anionic and cationic hydrocarbon surfactants with siloxane polyethers that have terminal carboxylic groups on the polyether have also been patented [132]. Policello [133] acknowledges that the nonyl phenol ethoxylate and alkyl ethoxylate nonionic surfactants, which Bailey [128] claimed were synergistic, in fact "interfere with the spreading characteristics of the organosilicone compounds" and went on to identify certain EO/PO block copolymer surfactants as non interfering. Murphy et al. [100] and Policello et al. [134] acknowledge that "many of the surfactants commonly used in agrichemical formulations antagonize the surface activity of trisiloxanes." In contrast, these two papers also claim that certain hydrocarbon surfactants "have been shown to synergize the spreading and dynamic surface tension" of the trisiloxane surfactant, Silwet L-77. The claimed synergy appears to relate to extent of spreading, not rate of spreading—the claim is that the mixture spreads to a larger area. Mixtures with short chain hydrocarbon surfactants have also been claimed [135].

Rosen and Song [120] measured spreading factors (extent of spreading, not rate) for mixtures of two trisiloxane surfactants with several hydrocarbon surfactants. Some of the hydrocarbon surfactants increased the extent of spreading and some markedly decreased it. They attributed the effects to "molecular attractive interaction between trisiloxane surfactants and additives" and dynamic surface tension effects.

F. Stabilization of Colloidal Dispersions

1. Emulsification

The use of siloxane surfactants as emulsifiers is reviewed in this volume by Gruning and Bungard [136]. Bailey reported emulsification of silicone [39] and hydrocarbon oils [37] by means of low molecular weight siloxane surfactants such as the trisiloxane superwetters. Certain siloxane surfactants are highly effective emulsifiers for water-in-oil (especially water-in-silicone oil) emulsions. They form exceptionally physically stable water-in-oil emulsions incorporating 40–90 wt% aqueous phase [138] that are used in personal care applications such as antiperspirants, skin care products, and color cosmetics [139–142]. Copolymers containing silicone and polyoxyalkylene groups have been claimed as emulsifiers for silicone oils [143], while terpolymers also incorporating alkyl groups are preferred for use with hydrocarbon oils [144,145]. Their effectiveness has been attributed to the combination of strong adsorption and steric stabilization due to their polymeric nature and to the viscoelastic properties of the surfactant film formed by these surfactants at the oil–water interface. The siloxane surfactants used for this purpose are very hydrophobic, containing only a few mole percent hydrophilic functionality. The terpolymers are visualized as situated at the (organic) oil–water interface with the alkyl groups solubilized in the oil phase and the hydrophilic groups in the water. The siloxane backbone serves to link multiple alkyl–hydrophile pairs together and prevent them from desorbing from the interface.

Dahms and Zombeck [140] discuss the stabilization of water-in-oil systems using siloxane copolymers and terpolymers. Adsorption of the siloxane surfactants is greater at lower concentrations than comparable conventional surfactants, explaining the efficiency of their materials. The effects of electrolytes and antiperspirant salts on adsorption and stability were investigated, as well as the relationship between internal phase volume and rheology. These authors also discuss the preparation of multiple emulsions using siloxane surfactants.

Gasperlin et al. discuss the microstructure [138] and rheology [146] of semisolid water-in-oil (white petrolatum) emulsions using a commercial silicone–alkyl–polyoxyalkylene terpolymer surfactant. They found the interfacial tension between the aqueous and oil phases reduced to 1–2 dyn/cm. Polarized light microscopy and freeze-fracture transmission electron microscopy (TEM) showed no evidence of ordered or liquid crystal phases. A water droplets-in-oil phase with a broad size distribution was found over a wide range of water content. Two types of water were detected: bulk water and water incorporated in surfactant aggregates. These emulsions gave typical shear thinning behavior.

Smid-Korbar et al. investigated oil-in-water emulsions of several organic and silicone oils using three polymeric siloxane surfactants [147]. All three of the siloxane surfactants were immiscible with the organic and silicone oils, from which they concluded that this class of surfactants should be considered both hydrophobic and lipophobic. The ionic surfactant formed the most stable emulsions followed by the EO surfactant followed by the EO/PO surfactant. The order of stability correlated with efficacy of surface tension lowering and cmc trends.

There is a extensive patent literature dealing with emulsification [148–154]. The focus of much of this art is the preparation of transparent gels consisting of water-in-volatile silicone oil emulsions. These formulations are often referred to in the art as microemulsions because of their transparency, which is usually achieved by index of refraction matching. Compositions giving transparent thermodynamically stable liquid crystal gels and microemulsions have been claimed by Hill [155–158].

Schmidt [6.77] and Vick [27] observe that calculating or measuring hydrophile–lipophile balance (HLB) values for siloxane surfactants yields quantities that do not correctly predict emulsifier performance, at least when siloxane surfactants are used to emulsify hydrocarbon oils. The emulsification of silicone finishes for textile treatment purposes by means of siloxane surfactants, including some nonionic examples and one anionic example, has been claimed [77].

2. Foaming and Defoaming

The ability of siloxane surfactants to stabilize polyurethane foam is the basis for their largest commercial use, and their function as foam stabilizers in this nonaqueous application has been extensively investigated and is reviewed elsewhere in this volume [7]. The aqueous foaming behavior of nonionic siloxane surfactants is similar to that of conventional hydrocarbon nonionic surfactants [159]—they are low to moderate foamers [27,96]. Vick found that the foam height generated by a series of siloxane surfactants in the Ross Miles test decreased rapidly with increasing size of the siloxane hydrophobe, and increased with increasing EO content. In all cases, the EO/PO-containing siloxane surfactants gave a light, coarse foam, compared with the more dense, fine-celled foams generated by the all-EO surfactants. Certain sulfobetaine siloxane surfactants have been claimed as foam boosters [160,161].

Siloxane surfactants are used as defoamers in hydrocarbon fuels [162–165] and as demulsifiers in gas and oil production [97,166]. Their performance as defoamers and demulsifiers is believed to be related to their ability to enter and spread at the air–oil or oil–water interfaces. Siloxane surfactants that contain polyoxyalkylene groups may have a cloud point, especially those with mixed EO and PO groups. Above the cloud temperature they function as defoamers, just as other EO/PO copolymers do [27,28,167].

3. Colloid Stabilization

Only a few studies have been reported of the use of siloxane surfactants to stabilize other types of colloidal dispersion. Gritskova et al. [168] investigated the use of surface active organosilicon compounds to stabilize dispersions of polystyrene particles. They synthesized several copolymers containing dimethylsiloxane groups together with polyoxyalkylene-, urethane-, and hydroxyl-containing groups. The more water soluble of these gave the lowest interfacial tensions, but the most hydrophobic examples gave the most stable dispersions with narrow particle size distributions.

V. AQUEOUS AGGREGATION BEHAVIOR

Surfactants are useful because they adsorb at interfaces to lower surface and interfacial tension, impart viscoelastic surface properties, provide a repulsive barrier that stabilizes emulsion droplets and foam films, and modify adhesion and friction. Surfactants also self-associate, especially in aqueous solution, to form a variety of aggregates ranging from globular, wormlike, and disk-shaped micelles to bilayer structures such as vesicles. Attractive interactions between the aggregates lead to condensation into liquid crystal phases [169,170]. Figure 11 illustrates self-association of surfactants in aqueous solution to form small globular micelles. Both the type of aggregate or liquid crystal phase formed by a particular surfactant and the progression of liquid crystal phases formed with increasing surfactant concentration, temperature, and salt level can be rationalized by means of a simple model based on molecular packing [171,172]. For example, $M(D'E_{12}OH)M$, which has a relatively large polar group, tends to form highly curved aggregates such as globular micelles. These self-association structures are important because they control the rheology and freeze–thaw stability of formulations [173], and they strongly influence interfacial tension and the ability to form and stabilize emulsions and microemulsions.

Most early papers on siloxane surfactants deal with surface properties (e.g., surface tension, adhesion, lubrication, skin feel). Only recently has the self-association of this class of surfactants in aqueous solution received much attention. It was argued at one time that because of the fluidity of the siloxane species, this class of surfactants should not form "crystalline monolayers" or liquid crystal phases [4,40]. We review here several recent studies of the aqueous aggregation properties of siloxane surfactants. These studies show that siloxane surfactants, ranging from the trisiloxanes to large polymeric structures, form a wide range of self-association structures including globular and wormlike micelles, vesicles, and normal and inverted liquid crystal (LC) phases.

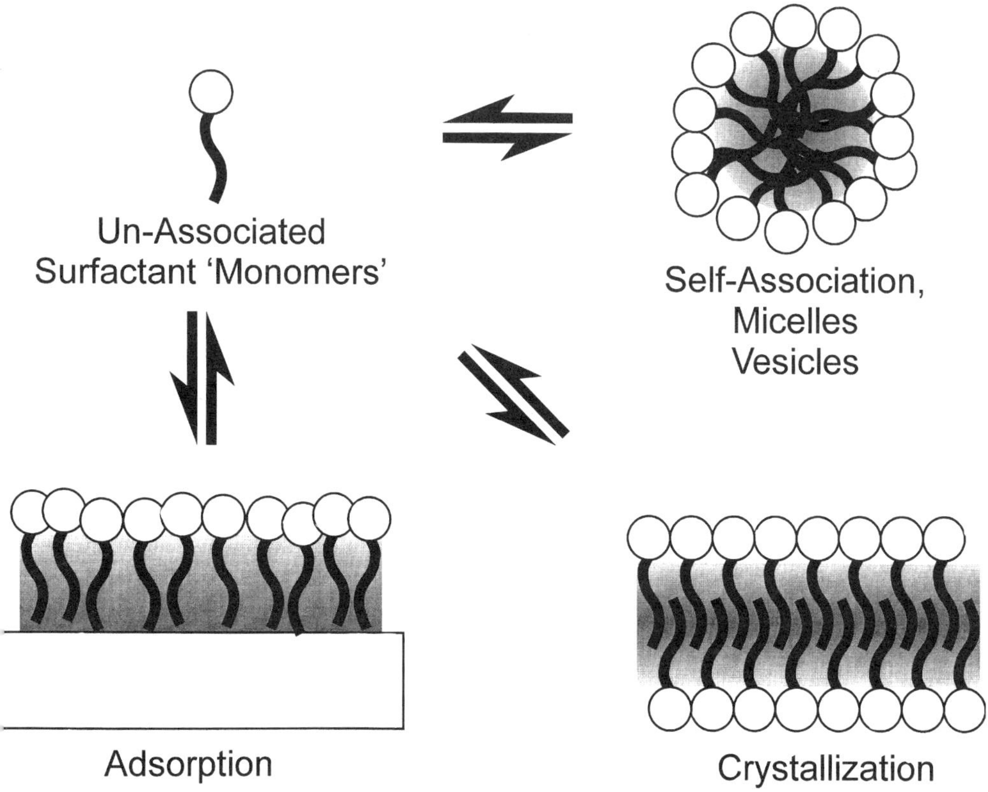

FIG. 11 Surfactants in solution adsorb at interfaces, crystallize, and self-associate to form micelles and vesicles.

Two other important aspects of surfactant phase behavior are the cloud point and the physical form of the neat surfactant at ambient temperature. Siloxane polyoxyalkylene nonionic surfactants become less soluble in water with increasing temperature just as conventional hydrocarbon surfactants do [15,27]. The temperature at which they become insoluble, and the solution becomes cloudy, is called the cloud temperature, or cloud point [174,175]. The general dependence of the cloud point on the weight fraction EO [27] and the end cap of the EO group [14] appears to be similar to that of hydrocarbon surfactants. The effects of electrolytes also follow the usual relationships [176]. Unlike hydrocarbon surfactants, these relationships cannot be simply related to molecular structure using the HLB system [27], although HLB values are sometimes quoted [28]. No simple empirical relationship analogous to the HLB system has been developed yet to predict the emulsifying or cloud point behavior of siloxane surfactants [27].

Nonionic siloxane surfactants, like other silicones, remain fluid to high siloxane molecular weights [6]. One implication of this property is that this group of surfactants do not generally show either a Krafft point or a "gel" point for aqueous lamellar phase dispersions [177]. Siloxane surfactants that incorporate EO groups longer than about 16 EO units tend to form soft waxes that melt at relatively low temperatures because of crystallization of the polyoxyethylene groups. Ionic siloxane surfactants usually form waxy crystals.

A. Nonionic Polyoxyethylene Siloxane Surfactants

1. The Trisiloxane Surfactants

The trisiloxane surfactants are an attractive series to study because they are low molecular weight materials, and they are polydisperse only in the polyoxyethylene portion. The aqueous phase behavior of the homologous series of surfactants, $M(D'E_nOR)M$, has now been reported for n = 5, 6, 7.5/8, 10, 12, 16, and 18 [2,14,15,78,79]. The polyoxyethylene portion of the n = 5 species was monodisperse [15]. For n = 7.5/8, three different end-capping groups were also studied (—H, —CH_3, and —$C(O)CH_3$) [14]. The effect of polydispersity in the EO chain length is apparently to shift the phase behavior toward more hydrophilicity [15,180].

The hydrophobicity of the trisiloxane MD′M group is comparable to that of a linear $C_{12}H_{25}$ group [2]. However, its shape is quite different—it is shorter and wider—its length is about 9.7 Å compared with 15 Å for $C_{12}H_{25}$, while its volume is larger: 530 Å^3 compared with 350 Å^3 for $C_{12}H_{25}$. The general patterns of phase behavior of the trisiloxane surfactants are similar to that of the C_iE_j series [169,172],* but because of the different size and shape of the MD′M hydrophobe,

*The C_iE_j surfactants are the linear alkyl ethoxylates such as $C_{12}E_5 = C_{12}H_{25}(OCH_2CH_2)_5OH$.

the phase behavior is shifted to longer EO groups. The phase behavior of this series of surfactants progresses from bilayer structures such as vesicles, and the L_3 and L_α liquid crystal phases for $n = 6$–8, to micelles, and the L_α and H_1 liquid crystal phases for 10–12, to micelles and H_1 liquid crystal phase for $n = 16$–18. The liquid crystal phases extend well above ambient temperature for $n = 6$–8 and $n = 16$–18 but are found only at low temperatures for $n = 10$–12.

Figure 12 shows the phase diagram for MDM′E_8OH [14]. This trisiloxane surfactant forms a large region of lamellar phase liquid crystal that extends toward lower concentrations above 45°C. This phase diagram is similar to that of $C_{12}E_5$ except that there is no cloud phase boundary cutting across the lamellar region. At low concentrations (< 1%), in the range where the trisiloxane surfactants function as superwetters, this surfactant forms a dispersion of bilayer vesicles. Hill et al. [14] showed that the aqueous phase behavior of this "linear" trisiloxane surfactant

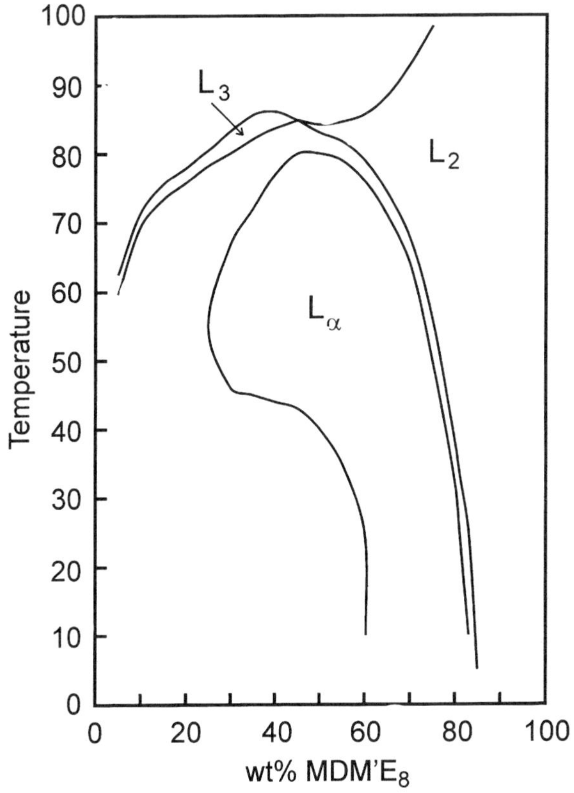

FIG. 12 Phase diagram of MDM′E_8OH in water. Redrawn from Ref. 14.

was essentially identical to that of the "branched" version, M(D′E_8OH)M. This was unexpected, since the shape of these two molecules should be quite different. This illustrates the important role of the flexibility of the Si—O—Si chain in the behavior of siloxane surfactants—the siloxane group tends to adopt a configuration that allows it to accommodate the preferred cross-sectional area of the polar group—in this case the coil of the hydrated EO chain. Stürmer et al. [3] found evidence of a highly swollen lamellar phase at about 3–5% surfactant for the closely related M(D′E_8OMe)M–water system.

The phase diagram for M(D′E_{12}OH)M is shown in Fig. 13 [15]. This surfactant forms clear isotropic solutions in water at all compositions near room temperature [178]. Below 10°C a hexagonal (H_1) and a lamellar (L_α) liquid crystal phase are found. The minimum of the lower critical solution temperature (LCST) curve is about 45°C. He et al. [178] used a combination of electron microscopy, x-ray and

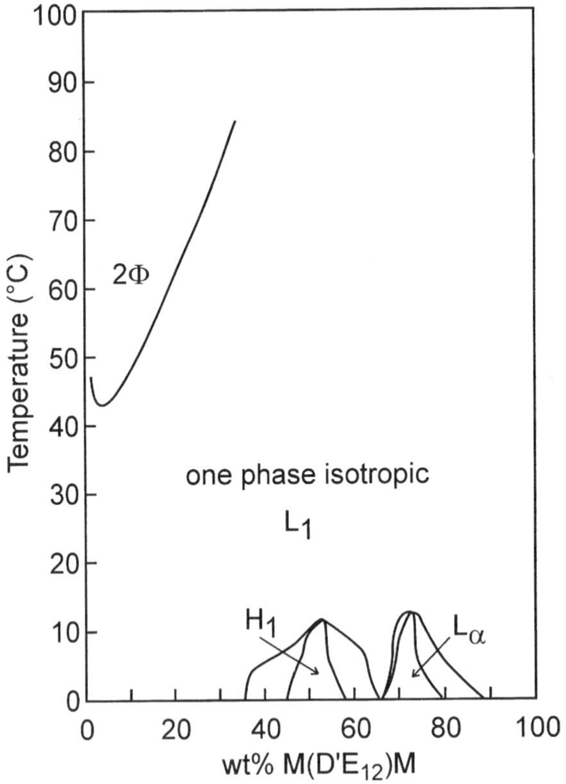

FIG. 13 Phase diagram of M(D′E_{12}OH)M in water. Redrawn from Ref. 15.

neutron scattering, rheology, and NMR self-diffusion measurements to investigate the microstructure of the isotropic solutions formed by this system near room temperature. They conclude that this system forms small globular micelles at low concentrations that evolve to entangled wormlike micelles and then to a random bilayer structure that persists to 100% surfactant. Doumaux [180] found a small region of L_3 (the "sponge" phase) above the cloud temperature curve. Li et al. [179] found a similar region for $M(D'E_{10}OH)M$, which they attributed to the polydispersity of the surfactant.

2. Polymeric Siloxane Surfactants

The polymeric siloxane surfactants are not only polydisperse (see discussion above), but the details of the preparation can yield materials with substantially different behavior. A few of the recent papers discussing their aqueous phase behavior have dealt with homologs with varying EO chain length, but even these have explored only a small range of the possible molecular structure variations. Most studies have presented results for an unrelated set of rake-type copolymers with little attempt to systematize the results into trends. To make sense of such data, studies should be carried out with systematic variation of the following:

1. Siloxane DP (or molecular weight)
2. D/D′ ratio (or average number of polar groups per molecule
3. Polyoxyalkylene chain length
4. EO/PO ratio

The presence of cyclic siloxanes should also be characterized. Studies of EO and EO/PO containing siloxane surfactants should be separate because these two types of siloxane surfactant show quite different behaviors. Differences between random and block copolymers of EO/PO may also need to be considered.

Lühmann and Finkelmann [34,35] investigated the lyotropic* LC phase behavior of several amphiphilic liquid crystal side chain polymers in which an amphiphilic molecule was attached to each monomer unit of a siloxane backbone. They found that these polymers formed the usual progression of cubic–hexagonal–lamellar phases. Yang and Wegner [181,182] investigated several ABA-type siloxane surfactants and found a solubility boundary at high temperature, a broad mesophase region below this boundary, and the usual progression of cubic–hexagonal–lamellar phases. Thus, in spite of the disorder associated with the random coiled configuration of the polymer chains, siloxane surfactants self-associate in water to form the usual micelles, vesicles, and highly ordered liquid crystal phases.

---

*"Lyotropic" distinguishes liquid crystal phases that form in the presence of a solvent from thermotropic liquid crystals, which do not require solvent.

Kanner et al. [40] reported solubility data for several polymeric and nonpolymeric siloxane surfactants. More detailed aqueous phase behavior studies have recently been reported for a rather diverse group of rake-type and ABA polymeric siloxane surfactants [2,3,13,183]. Dilute solutions of these surfactants contain wormlike micelles and vesicles. Both globular and tubular vesicles have been observed. Vesicle surfaces often appear rough, suggesting a high degree of bilayer flexibility.

Figure 14 shows the phase diagram of a rake-type siloxane surfactant, $MD_{22}(D'E_8)_2M$ [13]. This surfactant forms the inverse hexagonal liquid crystal phase, H_2, between 45 and 85% over a wide temperature range. Figure 15 shows the phase diagram of the EO_{12} homolog, $MD_{22}(D'E_{12})_2M$ [13]. This more

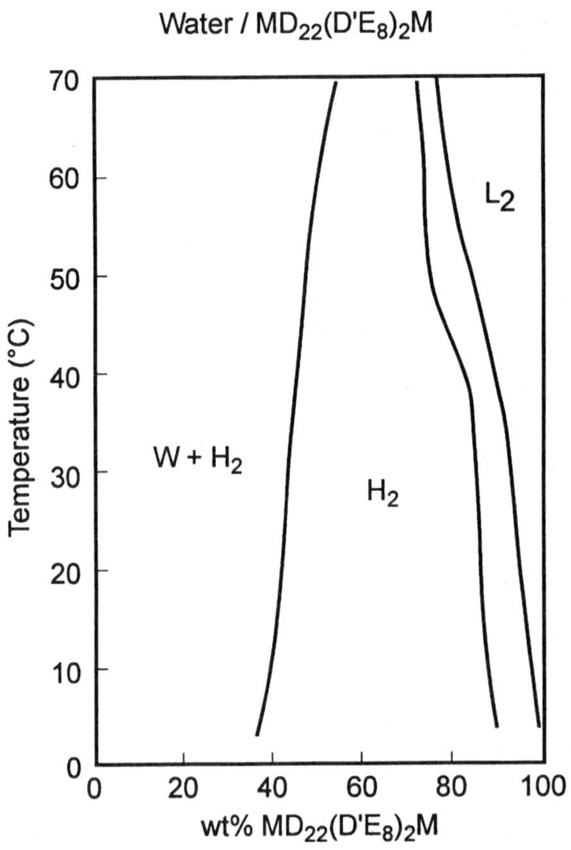

FIG. 14 Phase diagram of $MD22(D'E_8)_2M$ in water. Redrawn from Ref. 13, where the molecular structure is given as E_7.

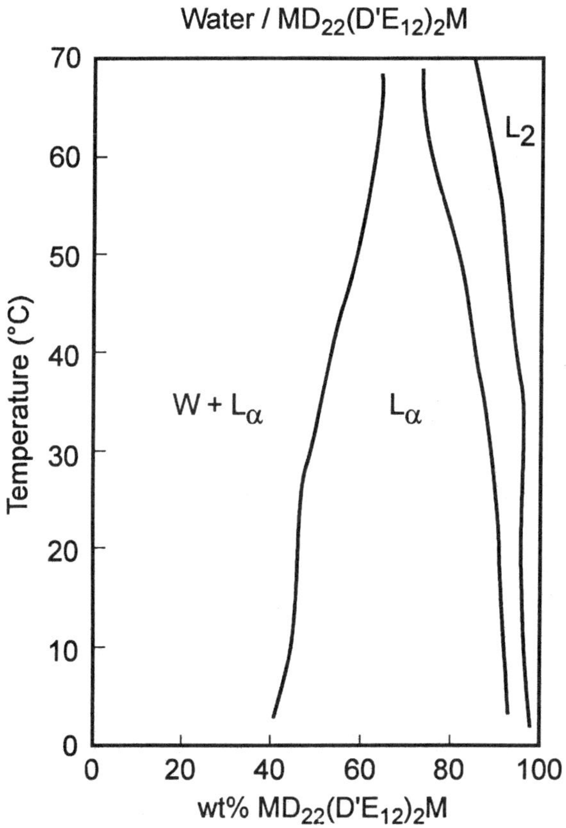

FIG. 15 Phase diagram of $MD_{22}(D'E_{12})_2M$ in water. Redrawn from Ref. 13.

hydrophilic surfactant forms lamellar phase liquid crystal, which also shows little temperature sensitivity. Low concentrations of this surfactant form a dispersion of globular and tubelike vesicles. Figure 16 shows a cryo-TEM micrograph [184] of an aqueous dispersion of a somewhat higher molecular weight siloxane surfactant. This surfactant forms an unusual diversity of globular and tubelike, unilamellar and multilamellar vesicles as well as wormlike micelles [13,183].

The ability of such branched and polydisperse molecular structures to form ordered structures such as bilayers and liquid crystal phases is attributed to the unusual flexibility of the siloxane backbone [13]. Hill et al. [13] applied the surfactant shape parameter calculation [171,172] to their polymeric siloxane surfactants and demonstrated that the predictions of the calculation were consistent with the observed trends in phase behavior.

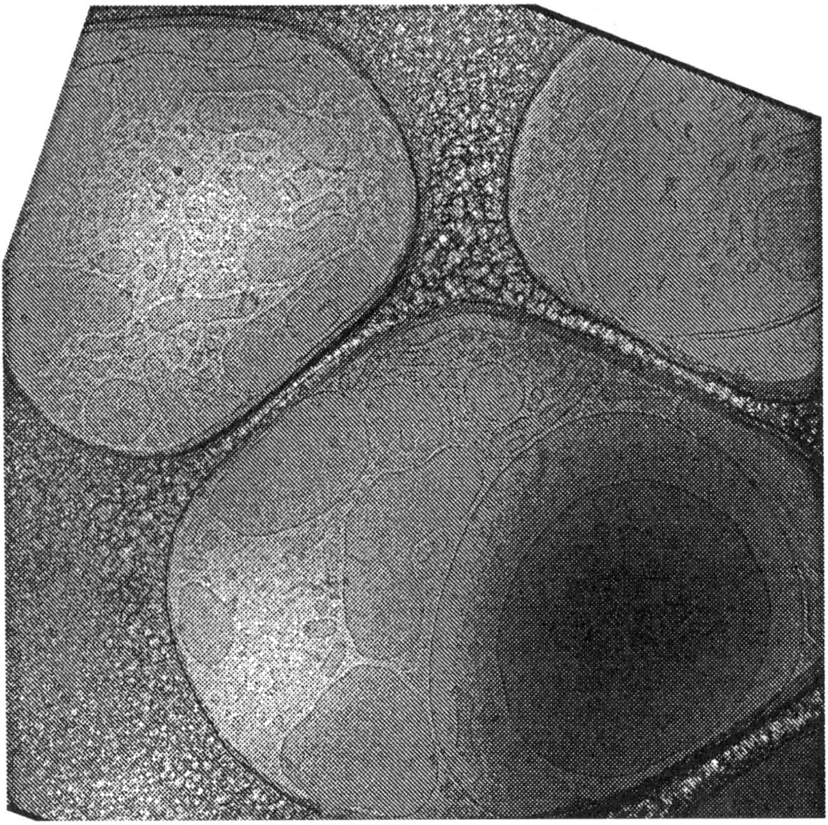

FIG. 16 Cryo-TEM micrograph of a 5 wt% solution of the rake-type siloxane polyoxyethylene copolymer surfactant $MD_{103}(D'E_{12})_{10}M$.

Stürmer et al. [3] reported the aqueous aggregation behavior of four relatively low molecular weight rake-type polymeric siloxane surfactants, including two nonionics and a zwitterionic, and a cationic example. Three of these surfactants formed bilayer structures. The phase boundaries of the cationic and the zwitterionic were insensitive to temperature, while the lamellar phase formed by the nonionic example melted at about 50°C. The more hydrophilic nonionics form globular micelles and a hexagonal phase. Stürmer et al. [3] point out that an unassociated polysiloxane surfactant molecule in water should adopt a shell conformation in which the siloxane backbone folds up in the center to minimize hydrophobic contact with water and the polar groups form an outer shell. This

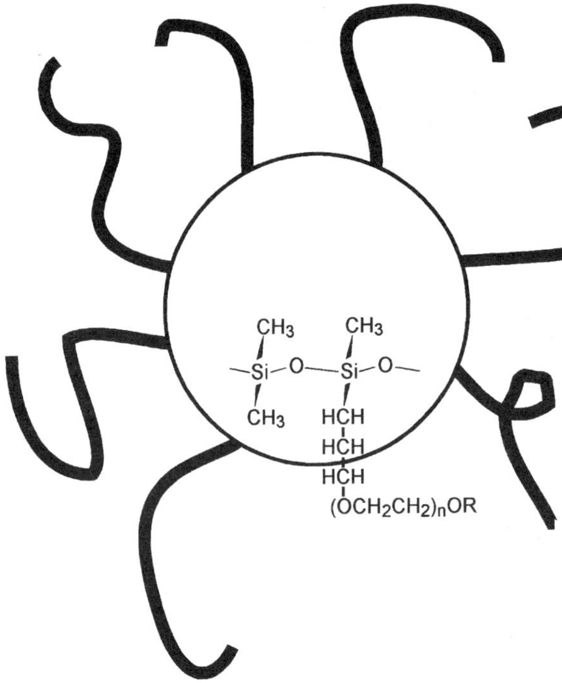

FIG. 17 Schematic representation of a polysiloxane surfactant molecule in water.

shell model of the conformation of a polymeric siloxane surfactant in water is illustrated in Fig. 17. From micellar radii results Gradzielski et al. [2] concluded that the siloxane chain must be coiled in the aggregates. The flexibility of the siloxane backbone would be expected to facilitate this folding.

B. EO/PO-Based Siloxane Surfactants

Siloxane surfactants incorporating copolymers (often random copolymers) of EO and PO are widely used in polyurethane foam manufacture and in personal care applications, but to date no studies have been reported of their aggregation behavior. The aqueous phase behavior of this class of siloxane surfactants tends to be quite different from that of the polyoxyethylene-based materials discussed above. The lower HLB examples are simply water insoluble, while the more hydrophilic examples are miscible with water in all proportions below their cloud temperature. We have not observed formation of ordered microstructures such as liquid crystal phases for this class of siloxane surfactants. For example, phase maps of Dow Corning 190 in DI water and 5% NaCl are shown in Figs. 18 and 19. Below the

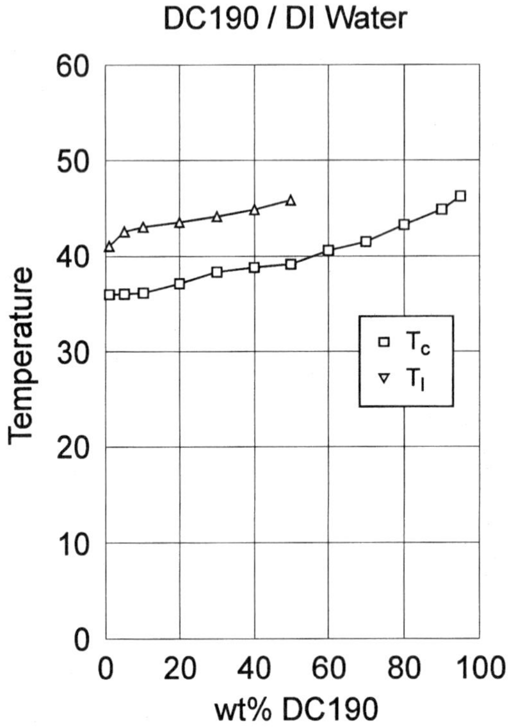

FIG. 18 Phase map of Dow Corning 190 in DI water: T_c, the cloud temperature; T_l, the temperature at which the cloudy dispersion begins to rapidly separate into two layers.

cloud temperature, this surfactant forms clear isotropic solutions with water at all concentrations. Salt shifts the cloud temperature downward but does not otherwise change the behavior.

C. Ionic Siloxane Surfactants

Gradzielski et al. [2] reported a binary phase diagram for an anionic trisiloxane surfactant. This surfactant formed a wide region of lamellar phase that was stable to temperature in the range they worked. Snow et al. [43,185,186] reported the aqueous aggregation behavior of a series of trisiloxane cationic and zwitterionic surfactants. They found that these surfactants self-associate to form micelles and vesicles and that at high surfactant concentration they form lamellar liquid crystal, and a hydrated bilayer structure. The same workers also investigated the effect of several electrolytes on the phase behavior of these cationic trisiloxane surfactants and found that salt promotes a transition from micelles to vesicles at low con-

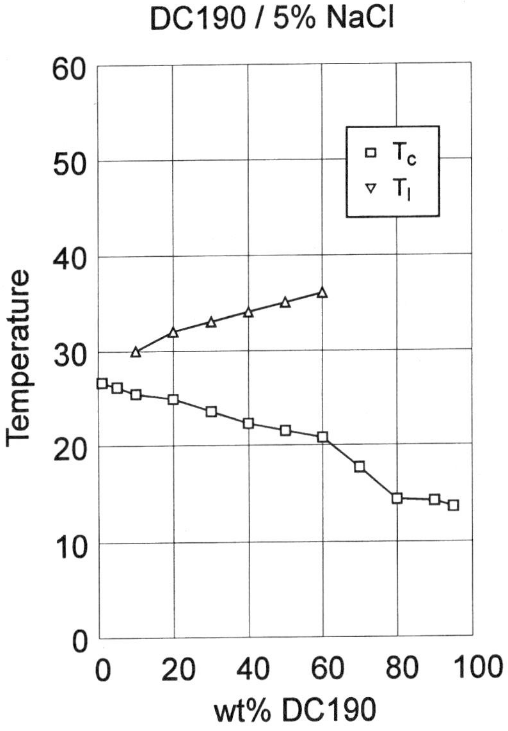

FIG. 19 Phase map of Dow Corning 190 in 5% NaCl solution.

centrations. Hill [187] found that a cationic ABA siloxane surfactant, RD_6R, where R = $CH_2CH_2CH_2(CH_3)_3N^+ Cl^-$, formed a wide region of lamellar phase, and a dispersion of unilamellar and multilamellar vesicles at low concentrations.

VI. TERNARY PHASE BEHAVIOR

Siloxane surfactants containing EO and EO/PO groups are typically soluble in ethers, alcohols, esters, ketones, polyoxypropylene diols, and aromatic and halogenated solvents. Unlike conventional nonionic surfactants, they are relatively immiscible with alkanes and with low and high molecular weight silicone oils [4].

The ternary phase behavior of mixtures of a rake-type polymeric siloxane surfactant with decane and decanol was reported by Stürmer et al. [3]. They found a ringing gel and regions of hexagonal and lamellar phase. These authors did not discuss the relationship between the behavior of this system and that of ternary systems based on hydrocarbon surfactants. Steytler et al. [188] found lamellar microstructures in the surfactant-rich isotropic phase for ternary mixtures of water, L-77, and cyclohexane. Hill et al. [189] and Li et al. [179,190] recently reported the ternary phase behavior of a series of nonionic trisiloxane surfactants with several low molecular weight cyclic and linear siloxane oils. These results are reviewed by Li et al. elsewhere in this volume [191]. These ternary systems were found to behave much like analogous systems containing hydrocarbon surfactants and oils—forming liquid crystal phases and microemulsions.

VII. APPLICATIONS

There are a large number of diverse applications for siloxane surfactants encompassing such seemingly unrelated problems as polyurethane foam manufacture, controlling adhesion of photographic film to itself [192,193], and cosmetic formulation. All these applications of siloxane surfactants exploit in one way or another the unusual features of this class of surfactants—surface activity in nonaqueous media, wetting/spreading, lubricity, and the ability to impart a certain feel to skin and hair. This section does not attempt to discuss these applications in detail, since many of them are covered elsewhere in this volume. Rather, we briefly survey some of the more important applications.

A. Polyurethane Foam Manufacture

The first and still the largest commercial application of silicone surfactants is their use as additives for the production of polyurethane foam. Worldwide volume for siloxane surfactants in polyurethane foam manufacture was estimated to be about 30,000 metric tons/year in 1994 [194]. Polyurethane foams were first commercialized in the 1950s in Germany—Fritz Hostettler of Union Carbide was granted the first patent on the use of siloxane polyether copolymers in the manufacture of polyurethane foam [195]. Literature references describing this technology began to appear in 1960 [196].

A number of excellent reviews of the chemistry of polyurethane foam exist [197]. Polyurethane foams are made starting with a mixture of polyol, isocyanate, water, blowing agents, and catalysts. The two primary reactions involve the formation of urethane (the "gelling" reaction) and the formation of urea (the "blowing" reaction), which generates CO_2. During the course of these reactions the temperature in the foam rises rapidly, the viscosity increases dramatically, urea begins to phase-separate, and bubbles nucleate and rapidly expand. This is obviously a complex and dynamic system, and there continue to be fundamental questions about the relationship between the composition and molecular structure of siloxane surfactants, and their function in polyurethane foam manufacture. Polyurethane foam can range from a flexible porous elastomer to a rigid resin, and siloxane surfactants are available that have been optimized for each type. The functions of the siloxane surfactant include stabilizing the foam, controlling film drainage, emulsifying the relatively immiscible polyol and isocyanate, and dispersing water and blowing agent. The relationship between the function and performance of the surfactant and its structure is unusually complex [198]. The reader interested in further details is referred to the chapter in this volume by Snow and Stevens [7].

B. Textile and Fiber Industry

Nonionic surfactants are used in many aspects of fabric manufacture including the wet-out bath, desizing, scouring, bleaching, and fabric finishing operations [96]. Siloxane surfactants are used to facilitate wetting and dispersion of water-insoluble substances, and are used as spinning and sewing lubricants. The siloxane surfactants are unique in being thermally stable lubricants with good wetting and low coefficients of friction at high speeds. Schmidt [6] briefly describes four applications of siloxane surfactants relevant to textile and fiber processing: silicone oil-in-water emulsifers, improving the adhesion of adhesives to silicone-treated fabrics, fiber lubricants, and release agents. During fiber production, siloxane surfactants enable the lubricant to spread quickly and completely even at very low pickup amounts [77]. These functions are related to the ability of the siloxane surfactant to lower the surface tension of the organic oil [77]. The types of siloxane surfactant useful for this application are tabulated and discussed by Schmidt [77]. A wide variety of uses exist, some of which do not seem obviously related to the surfactancy of these materials. Use of siloxane surfactants as wetting agents in aqueous textile treatments has been discussed only in terms of laboratory results; little published information is available about actual use.

Organophilic siloxanes of various structures have been shown to restore adhesion of hot-melt adhesives to silicone-coated rainwear. Siloxane surfactants improve spinnability of certain thermoplastic fibers. Siloxane surfactants can, in certain instances, migrate to the surfaces of fibers during and after extrusion to modify their properties [77].

C. Personal Care and Cosmetic Applications

Schaefer [47] reviewed the application of siloxane surfactants in personal care products. Oil-soluble siloxane surfactants improve the spreading ability of organic oils and waxes and as such are used in makeup and skin care formulations. More hydrophilic siloxane surfactants are used in aqueous applications such as shampoos and shower gels [199]. Cationic and zwitterionic functionality improve substantivity and are used in hair care products. Substantivity of nonionic siloxane surfactants is determined partly by their water solubility—low solubility, as evidenced by a low cloud point, leads to improved substantivity [47]. SPEs are claimed to reduce eye irritation of anionic surfactants [200]. They are known for their low physiological risk in cosmetic formulations [2]. In hair care preparations SPEs improve combing and gloss and impart a silky feel [47]. In lotions they impart smoothness and softness to the skin, and also minimize the whitening on rub-out, by means of their defoaming action. SPEs were found to help prevent cracking in soap and synthetic detergent bars [28,47]. Under nonneutral pH conditions, the trisiloxane surfactants show only limited hydrolytic stability, which has hampered their wide application [47]. Cationics and zwitterionics have found some application in hair care products [4,47].

Vick [27] discussed the structure–property relationships that are responsible for the behavior of siloxane surfactants in cosmetic applications. Vick analyzed the behavior of siloxane surfactants in terms of surface tension lowering, wetting, emulsification, and foam characteristics, all features that, he asserts, play a key role in their performance in personal care applications.

D. Paints and Coatings

Fink [95] has reviewed the use of siloxane surfactants in paint and coating applications. A wide range of structures are used in this area, including structures that incorporate hydrophilic, organophilic, and fluorocarbon functionality, and even combinations of these. These additives are used as wetting agents (substrate adhesion), flow promoters (prevent formation of blisters and craters), lubricants (glide and scratch control), and as antifoams and deaerators. Obviously, each of these uses involves distinct and not necessarily related surface phenomena. Each use has a group of siloxane surfactants that have been developed for that purpose and specifically tailored for the type of coating system involved (solvent or water based, acrylic, etc.). There is little insight available at the present time concerning the mode of action in these varied systems, but Fink's review is a useful starting point for the technologist seeking help with specific problems.

VIII. SUMMARY

Siloxane surfactants are a class of novel surfactants consisting of a permethylated siloxane backbone with one or more hydrophilic groups attached. Most siloxane surfactants are intermediate molecular weight copolymers. This class of surfactants is highly surface active in both aqueous and nonaqueous media, lowering surface tensions to nearly 20 dyn/cm. They are used in a wide range of applications in which their unique properties are required, including polyurethane foam and textile and fiber manufacture, in personal care products, in paints and coatings, and as wetting agents for pesticides. It remains difficult in most of these areas to correlate specific features of molecular architecture with performance. Recent work on the aqueous aggregation behavior of this class of surfactants shows that siloxane surfactants ranging from trisiloxanes to polymeric siloxanes form the usual micelles, vesicles, and normal and inverse liquid crystal phases, which is surprising since they would be expected to adopt rather disordered coiled configurations in solution. This information, which is useful for understanding the physical properties of concentrated surfactant solutions, may be particularly useful for understanding interactions between siloxane surfactants and conventional organic surfactants. Since the type of aggregate a surfactant forms is indicative of the intrinsic curvature it prefers at an interface, knowledge of aggregation behavior also helps one to understand emulsification and wetting. The unusual rapid wetting properties of dilute solutions of certain trisiloxane surfactants turn out to represent a special case of surfactant-enhanced wetting. Many possible siloxane surfactant structures remain to be characterized, and much work remains to be done, especially in specific application areas.

REFERENCES

1. Hill, R. M., in *Specialist Surfactants* (Robb, I. D., ed.), Blackie Academic and Professional, London, 1997.
2. Gradzielski, M.; Hoffmann, H.; Robisch, P.; Ulbricht, W., *Tenside Surf. Deterg.* 1990, *27*, 366.
3. Stürmer, A.; Thunig, C.; Hoffmann, H.; Gruening, B., *Tenside Surf. Deterg.* 1994, *31*, 90.
4. Gruning, B.; Koerner, G., *Tenside Surf. Deterg.* 1989, *26*, 312.
5. Plumb, J. B.; Atherton, J. H., in *Block Copolymers* (Allport, D. C.; Janes, W. H., eds.), Applied Science Publishers, London, 1973, p. 305.
6. Schmidt, G. L. F., in *Industrial Applications of Surfactants* (Karsa, D. R., ed.), Royal Society for Chemistry, London, 1987, p. 24.
7. Snow, S. A.; Stevens, R. G., this volume, Chapter 5.
8. Bass, R. L., *Chem. Ind.* July 18, 1959, p. 912.
9. Owen, M. J., in *Surfactants in Solution, Vol. 6* (Mittal, K. L.; Bothorel, P., eds.), Plenum Press, New York, 1986, p. 1557.
10. Owen, M. J.; Kendrick, T. C., *Macromolecules* 1970, *3*, 458.
11. Kendrick, T. C.; Kingston, B. M.; Lloyd, N. C.; Owen, M. J., *J. Colloid Interface Sci.* 1967, *24*, 135.
12. Kendrick, T. C.; Owen, M. J., *Chem. Phys. Appl. Pract. Surface (Spain) 5th* 1968, *2*, 571.
13. Hill, R. M.; He, M.; Lin, Z.; Davis, H. T.; Scriven, L. E., *Langmuir* 1993, *9*, 2789.
14. Hill, R. M.; He, M.; Davis, H. T.; Scriven, L. E., *Langmuir* 1994, *10*, 1724.
15. He, M.; Hill, R. M.; Lin, Z.; Scriven, L. E.; Davis, H. T., *J. Phys. Chem.* 1993, *97*, 8820.
16. Ross, S.; Nishioka, G., *Colloid Polym. Sci.* 1977, *255*, 560.
17. Noshay, A.; McGrath, J. E., *Block Copolymers: Overview and Critical Survey*, Academic Press, New York, 1977.
18. Yilgor, I.; Yilgor, E.; Venzmer, J.; Spiegler, R., *Polym. Mater. Sci. Eng.* 1996, *75*, 283.
19. Flaris, V.; Baker, W. E.; Lambla, M., *Polym. Networks Blends* 1996, *6*, 29.
20. Hu, W.; Koberstein, J. T.; Lingelser, J. P.; Gallot, Y., *Macromolecules* 1995, *28*, 5209.
21. Rossmy, G.; Spiegler, R.; Venzmer, J., German Patent 4,206,191 (1993).
22. Ward, B. J.; Williams, D. A.; Willey, P. R., U.S. Patent 4,992,512 (1991).

23. Macosko, C.; Sundararaj, U.; Rolando, R. J.; Chan, H. T., *Polym. Eng. Sci.* 1992, *32,* 1814.
24. Yilgor, I., this volume, Chapter 10.
25. Bailey, D. L., U.S. Patent 3,359,212 (1967).
26. Kobayashi, H.; Owen, M. J., *J. Colloid Interface Sci.* 1993, *156*, 415.
27. Vick, S. C., *Soap Cosmet. Chem. Spec.* May 1984, p. 36.
28. Gould, C., *Spec. Chem.* August 1991, p. 354.
29. Hsieh, C.-J.; Hsu, C.-S.; Hsiue, G.-H., *Polym. Mater. Sci. Eng.* 1988, *59*, 1014.
30. Finkelmann, H.; Rehage, G., *Adv. Polym. Sci.* 1984, *60/61*, 99.
31. Gray, G. W., in *Side Chain Liquid Crystal Polymers* (McArdle, C. B., ed.), Blackie, Glasgow, 1989, p. 106.
32. Gray, G. W.; Hill, J. S.; Lacey, D., *Mol. Cryst. Liq. Cryst.* 1991, *197*, 43.
33. Gray, G. W.; Hawthorne, W. D.; Hill, J. S.; Lacey, D.; Lee, M. S. K.; Nestor, G.; White, M. S., *Polymer* 1989, *30*, 964.
34. Lühmann, B.; Finkelmann, H., *Colloid Polym. Sci.* 1987, *265*, 506.
35. Lühmann, B.; Finkelmann, H., *Makromol. Chem.* 1985, *186*, 1059.
36. Noll, W., *The Chemistry and Technology of Silicones*, Academic Press, New York, 1968.
37. Clarson, S. J.; Semlyen, J. A., eds., *Siloxane Polymers*, PTR Prentice Hall, New York, 1993.
38. LeGrow, G. E.; Petroff, L. J., this volume, Chapter 2.
39. Bailey, D. L., U.S. Patent 3,299,112 (1967).
40. Kanner, B.; Reid, W. G.; Petersen, I. H., *Ind. Eng. Chem. Prod. Res. Dev.* 1967, *6*, 88.
41. Snow, S. A.; Fenton, W. N.; Owen, M. J., *Langmuir* 1990, *6*, 385.
42. Maki, H.; Murakami, Y.; Ikeda, I.; Komori, S., *Kogyo Kagaku Zasshi (J. Chem. Soc., Jpn., Ind. Chem. Sec.)* 1968, *71*, 1675.
43. Snow, S. A., *Langmuir* 1993, *9*, 424.
44. Morehouse, E. L., U.S. Patent 3,660,452 (1972).
45. Snow, S. A.; Fenton, W. N.; Owen, M. J., *Langmuir* 1991, *7*, 868.
46. Maki, H.; Horiguchi, Y.; Suga, T.; Komori, S., *Yukagaku* 1970, *19*, 1029.
47. Klein, K.-D.; Schaefer, D.; Lersch, P., *Tenside Surf. Deterg.* 1994, *31*, 115.
48. Schaefer, D., *Tenside Surg. Deterg.* 1990, *27*, 154.
49. Wagner, R.; Richter, L.; Wersig, R.; Schmaucks, G.; Weiland, B.; Weissmuller, J.; Reiners, J., *Appl. Organomet. Chem.* 1996, *10*, 421.
50. Wagner, R.; Richter, L.; Weiland, B.; Reiners, J.; Weissmuller, J., *Appl. Organomet. Chem.* 1996, *10*, 437.
51. Wagner, R.; Richter, L.; Wu, Y.; Weiland, B.; Weissmuller, J.; Reiners, J.; Hengge, E.; Kleewein, A., *Appl. Organomet. Chem.* 1998, *12*, 47.
52. Wagner, R.; Richter, L.; Weiland, B.; Weissmuller, J.; Reiners, J.; Kramer, W., *Appl. Organomet. Chem.* 1997, *11*, 523.
53. Jonas, G.; Stadler, R., *Acta Polym.* 1994, *45*, 14.
54. Wersig, R.; Sonnek, G.; Niemann, C., *Appl. Organomet. Chem.* 1992, *6*, 701.
55. Schmaucks, G. *J. Prakt. Chem.* 1994, *336*, 514.
56. Roidl, J., *Parfüm. Kosmet.* 1986, *67*, 232.
57. Wolfes, W., *Parfüm. Kosmet.* 1987, *68*, 195.
58. Rentsch, S. F., U.S. Patent 5,387,417 (1995).

59. Rigano, L.; Leporatti, R., *Cosmet. News*, 1996, *19*, 360.
60. Derian, P.-J., European Patent 738,511 (1996).
61. Legrow, G. E.; Glover, D. A., U.S. Patent 5,389,365 (1995).
62. Floyd, D. T., in *Cosmet. Pharm. Appl. Polym.* (Gebelein, C. G.; Cheng, T. C.; Yang, V. C.-M., eds.), Plenum Press, New York, 1991, p. 49.
63. Maki, H.; Saeki, S.; Ikeda, I.; Komori, S., *J. Am. Oil Chem. Soc.* 1969, *46*, 635.
64. Colas, A. R. L.; Renauld, F. A. D.; Sawicki, G. C., European Patent 367,381 (1990).
65. Colas, A. R. L.; Renauld, F. A. D.; Sawicki, G. C., British Patent 2,203,152 (1988).
66. Renauld, F. A. D.; Tonge, J. S., British Patent 2,234,511 (1991).
67. Klein, K.-D.; Knott, W.; Koerner, G., German Patent 4,313,130 (1994).
68. Klein, K.-D.; Knott, W.; Koerner, G., German Patent 4,320,920 (1994).
69. Klein, K.-D.; Knott, W.; Koerner, G.; Krakenberg, M., German Patent 4,330,059 (1994).
70. Klein, K.-D.; Wilkowski, S.; Selby, J., *4th Int. Symp. Adjuvants Agrochemicals*, Vol. 1, 1995, p. 27.
71. Klein, K.-D.; Knott, W.; Koerner, G., in *Organosilicon Chemistry*, Vol. II, *From Molecules to Materials* (Auner, N.; Weis, J., eds.), VCH, Weinheim, 1996, p. 613.
72. Knoche, M.; Tamura, H.; Bukovac, M. J., *J. Agric. Food Chem.* 1991, *39*, 202.
73. Knoche, M., *Weed Res.* 1994, *34*, 221.
74. Hoffmann, H.; Ulbricht, W., this volume, Chapter 4.
75. Stevens, M. P., *Polymer Chemistry, An Introduction*, 2nd ed., Oxford University Press, Oxford, 1990.
76. Owen, M. J., *Ind. Eng. Chem. Prod. Res. Dev.* 1980, *19*, 97.
77. Schmidt, G., *Tenside Surf. Deterg* 1990, *27*, 324.
78. Gentle, T. E.; Snow, S. A., *Langmuir* 1995, *11*, 2905.
79. Pandya, K. P.; Lad, K. N.; Bahadur, P., *Tenside Surf. Deterg* 1996, *33*, 374.
80. Hasan, F. B.; Huang, D. D., *J. Colloid Interface Sci.*, 1997, *190*, 161.
81. Zhang, L.; Somasundaran, P.; Maltesh, C., *J. Colloid Interface Sci.*, 1997, *191*, 202.
82. Schmaucks, G.; Sonnek, G.; Wustneck, R.; Herbst, M.; Ramm, M., *Langmuir* 1992, *8*, 1724.
83. Maki, H.; Komatsu, M.; Komori, S., *Kogyo Kagaku Zasshi (J. Chem. Soc., Jpn. Ind. chem. Sec.)* 1967, *70*, 1771.
84. Maki, H.; Murakami, Y.; Ikeda, I.; Komori, S., *Kogyo Kagaku Zasshi (J. Chem. Soc., Jpn, Jap. Ind. Chem. Sec.)* 1968, *71*, 1679.
85. Maki, H.; Horiguchi, Y.; Komori, S., *Kogyo Kagaku Zasshi (J. Chem. Soc., Jpn. Ind. Chem. Sec.)* 1970, *73*, 1142.
86. Maki, H.; Suga, T.; Ikeda, I.; Komori, S., *Yukagaku* 1970, *19*, 245.
87. Maki, H.; Horiguchi, Y.; Suga, T.; Komori, S., *Yukagaku* 1970, *19*, 1029.
88. Wagner R.; Richter L.; Weissmuller J.; Reiners J.; Klein K.-D.; Schaefer D.; Stadtmuller S., *Appl. Organomet. Chem.* 1997, *11*, 617.
89. Wagner, R.; Sonnek, G.; Wustneck, R.; Janicke, A.; Herbst, M.; Richter, L.; Engelbrecht, L., *Tenside Surf. Deterg. 31*, 1994, 344.
90. Schwarz, E. G.; Reid, W. G., *Ind. Eng. Chem.* 1964, *56*, 26.
91. Kanellopoulos, A. G.; Owen, M. J., *J. Colloid Interface Sci.* 1971, *35*, 120.
92. Bailey, D. L.; Peterson, I. H.; Reid, W. G., *Chem. Phys. Appl. Surface Active Substances Proc. 4th Int. Congr.*, 1967, *1*, 173.

93. Svitova, T.; Hoffmann, H.; Hill, R. M., *Langmuir* 1996, *12*, 1712.
94. Adams, J. W., in *Surface Phenomena and Additives in Water-Based Coatings and Printing Technology*, Sharma, M. K., ed.), Plenum Press, New York, 1991, p. 73.
95. Fink, H. F., *Tenside Surf. Deterg.*, 1991, *28*, 306.
96. Sabia, A. J., *Am. Dyest. Rep.* May (1982), p. 45.
97. Callaghan, I. C., in *Defoaming, Theory and Industrial Applications* (Garrett, P. R., ed.), Marcel Dekker, New York, 1993, p. 19.
98. Stevens, P. J. G., *Pestic. Sci.* 1993, *38*, 103.
99. Adamson, A. W., *Physical Chemistry of Surfaces*, 5th ed., John Wiley & Sons, New York, 1990, p. 399.
100. Murphy, G. J.; Policello, G. A.; Ruckle, R. E., *Brighton Crop Prot. Conf.—Weeds* 1991, *1*, 355.
101. Schiefer, H. M.; McClarnon, K. S., U.S. Patent 4,765,243 (1988).
102. Dayawon, M.; Bohn, J.; Striebel, S. M.; Rao, S.; Sandbrink, J. J., WO 8,912,394 (1989).
103. Petroff, L. J.; Romenesko, D. J.; Bahr, B. C., U.S. Patent 4,933,002 (1990).
104. Zhu, X.; Miller, W. G.; Scriven, L. E.; Davis, H. T., *Colloids Surf. A* 1994, *90*, 63.
105. Lin, Z.; Hill, R. M.; Davis, H. T.; Ward, M. D., *Langmuir* 1994, *10*, 4060.
106. Stoebe, T.; Lin, Z.; Hill, R. M.; Ward, M. D.; Davis, H. T., *Langmuir* 1996, *12*, 337.
107. Roggenbuck, F. C.; Rowe, L.; Penner, D.; Petroff, L.; Burow, R., *Weed Technol.* 1990, *4*, 576.
108. Penner, D.; Burow, R. F.; Roggenbuck, F., this volume, Chapter 9.
109. Stoebe, T.; Hill, R. M.; Ward, M. D.; Scriven, L. E.; Davis, H. T., this volume, Chapter 11.
110. Murphy, D. S.; Policello, G. A.; Goddard, E. D.; Stevens, P. J. G., in *Pesticide Formulations and Application Systems*, Vol. 12, ASTM STP 1146 (Devisetty, B. N.; Chasin, D. G.; Berger, P. D., eds.), American Society for Testing and Materials, Philadelphia, 1993, p. 45.
111. Gaskin, R. E.; Stevens, P. J. G., *Pestic. Sci.* 1993, *38*, 193.
112. Ananthapadmanabhan, K. P.; Goddard, E. D.; Chandar, P., *Colloids Surf.* 1990, *44*, 281.
113. Goddard, E. D.; Anantha Padmanabhan, K. P., in *Adjuvants Agrichemicals* (Foy, C. L., ed.), CRC Press, Boca Raton, FL, 1992, p. 373.
114. Zhu, X., Ph.D. thesis, University of Minnesota, 1992.
115. Lin, Z.; Hill, R. M.; Davis, H. T.; Ward, M. D., *Langmuir* 1996, *12*, 345.
116. Stoebe, T.; Lin, Z.; Hill, R. M.; Ward, M. D.; Davis, H. T., *Langmuir* 1997, *13*, 7270.
117. Stoebe, T.; Hill, R. M.; Ward, M. D.; Davis, H. T., *Langmuir* 1997, *13*, 7276.
118. Stoebe, T.; Lin, Z.; Hill, R. M.; Ward, M. D.; Davis, H. T., *Langmuir* 1997, *13*, 7282.
119. Ruckenstein, E., *J. Colloid Interface Sci.* 1996, *179*, 136.
120. Rosen, M. J.; Song, L. D., *Langmuir* 1996, *12*, 4945.
121. Bardon, S.; Cachile, M.; Cazabat, A.-M.; Fanton X.; Valignat, M.-P.; Villette, S., *Faraday Discuss.* 1996, *104*, 307.
122. Tiberg, F.; Cazabat, A.-M., *Europhys. Lett.*, 1994, *25*, 205.
123. Shanahan, M. E. R.; Houzelle, M.-C.; Carré, A., *Langmuir* 1998, *14*, 528.
124. Policello, G. A., U.S. Patent 5,104,647 (1992).
125. Policello, G. A.; Murphy, D. S., U.S. 5,558,806 (1996).

126. Porter, M. R., *Handbook of Surfactants*, 2nd ed., Blackie Academic & Professional, London, 1994, p. 110.
127. Ogino, K., Abe, M., eds., *Mixed Surfactant Systems*, Marcel Dekker, New York, 1993.
128. D. L. Bailey; Pater, A. S.; Morehouse, E. L., U.S. Patent 3,562,786 (1971).
129. Hill, R. M., in *Mixed Surfactant Systems, ACS Symp. Ser. Vol. 501* (Holland, P. M.; Rubingh, D. N., eds.), American Chemical Society, Washington, DC, 1992, p. 278.
130. Ohno, M.; Esumi, K.; Meguro, K., *J. Am. Oil Chem. Soc.* 1992, *69*, 80.
131. L. J. Petroff, U.S. Patent 4,784,799 (1988).
132. Kilgour, J. A.; Policello, G. A., U.S. Patent 5,360,571 (1994).
133. Policello, G. A., U.S. Patent 5,104,647 (1992).
134. Policello, G. A.; Stevens, P. J. G.; Forster, W. A.; Gaskin, R. E., in *Pesticide Formulations and Application Systems*, Vol. 15, ASTM STP 1268 (Collins; H. M.; Hall; F. R.; Hopkinson, M., eds.), American Society for Testing and Materials, Philadelphia, 1995, p. 59.
135. Policello, G. A.; Murphy, D. S., U.S. Patent 5,558,806 (1996).
136. Gruning, B.; Bungard, A., this volume, Chapter 8.
137. Bailey, D. L., German Patent 1,241,551 (1967).
138. Gasperlin, M., Kristl, J., Smid-Korbar, J., *Int. J. Pharm.* 1994, *107*, 51.
139. Keil, J. W., U.S. Patent 4,268,499 (1981).
140. Keil, J. W., U.S. Patent 4,265,878 (1981).
141. Dahms, G. H.; Zombeck, A., *Cosme. Toiletries* 1995, *110*, 91.
142. Grüning, B; Hameyer, P; Weitemeyer, C., *Tenside Surf. Deterg.* 1992, *29*, 78.
143. Gee, R. P.; Keil, J. W., U.S. Patent 4,122,029 (1978).
144. Gasperlin, M.; Smid-Korbar, J.; Kristl, J., *Pharm. J. Slov.* 1992, *43*, 3.
145. Rentsch, S. F., U.S. Patent 5,387,417 (1995).
146. Gasperlin, M.; Kristl, J.; Smid-Korbar, J., *S.T.P. Pharma Sci.* 1997, *7*, 158.
147. Smid-Korbar, J.; Kristl, J.; Stare, M., *Int. J. Cosmet. Sci.* 1990, *12*, 135.
148. Guthauser, B., U.S. Patent 5,162,378 (1992).
149. Starch, M. S., U.S. 4,311,695 (1982).
150. Gum, M. L., U.S. 4,782,095 (1988).
151. Gum, M. L. U.S. 4,801,447 (1989).
152. Zotto, A. A.; Thimineur, R. J.; Raleigh, W. J., U.S. Patent 4,988,504 (1991).
153. Raleigh, W. J.; Thimineur, R. J., U.S. Patent 5,292,503 (1994).
154. Pereira, M. C.; Spiegel, U., U.S. Patent 5,216,033 (1993).
155. Hill, R. M., European Patent 774,482 (1997).
156. Hill, R. M., U.S. Patent 5,623,017 (1997).
157. Hill, R. M., U.S. Patent 5,705,562 (1998).
158. Hill, R. M., U.S. Patent 5,707,613 (1998).
159. Schick, M. J.; Schmolka, I. R., in *Nonionic Surfactants, Physical Chemistry* (Schick, M. J., ed.), Marcel Dekker, New York, 1987, p. 835.
160. Lo, S. J.; Snow, S. A., U.S. Patent 4,879,051 (1989).
161. Crossin; M. C., U.S. Patent 5,132,053 (1992).
162. Fey, K. C., U.S. Patent 5,620,485 (1997).
163. Rehrer, D. H., U.S. Patent 4,460,380 (1984).
164. Callaghan, I. C.; Gould, C. M., U.S. Patent 4,711,714 (1987).

165. Spiegler, R.; Keup, M.; Kugel, K.; Lersch, P.; Silber; S., U.S. Patent 5,613,988 (1997).
166. Owen, M. J., *Chem., Phys. Chem. Anwendungstech. Grenzflaechenaktiven Stoffe, Ber. 6th Int. Kongr.*, September 1972, Vol. 3, Carl Hanser, Munich, 1973, pp. 623–630.
167. Blease, T. G.; Evans, J. G.; Hughes, L.; Loll, P., in *Defoaming, Theory and Industrial Applications* (Garrett, P. R., ed.), Marcel Dekker, New York, 1993, p. 299.
168. Gritskova, I. A.; Chirikova, O. V.; Shchegolikhina, O. I.; Zhdanov, A. A., *Colloid J.*, 1995, *57*, 25.
169. Laughlin, R. G., *The Aqueous Phase Behavior of Surfactants*, Academic Press, New York, 1994.
170. Tiddy, G. J. T., *Phys. Rep.* 1980, *57*, 1.
171. Israelachvili, J. N.; Mitchell, D. J.; Ninham, B. W., *J. Chem. Soc., Faraday Trans.* 2, 1976, *72*, 1525.
172. Mitchell, J. D.; Tiddy, G. J. T.; Waring, L.; Bostock, T.; McDonald, M. P., *J. Chem. Soc., Faraday Trans. 1* 1983, *79*, 975.
173. See Herb, C.; Prudhomme, R. K., eds., *Structure and Flow in Surfactant Solutions* ACS Symp. Ser., Vol. 578, American Chemical Society, Washington, DC., 1994.
174. Rosen, M. J., *Surfactants and Interfacial Phenomena*, 2nd ed., John Wiley & Sons, New York, 1989, pp. 191–195.
175. Nakagawa, T., in *Nonionic Surfactants* (Schick, M. J., ed.), Marcel Dekker, New York, 1966, p. 572.
176. Hill, R. M., unpublished results, Dow Corning Corp.
177. Lasic, D. D., *Liposomes: From Physics to Applications*, Elsevier, Amsterdam, 1993.
178. He, M.; Hill, R. M.; Doumaux, H. A.; Bates, F. S.; Davis, H. T.; Evans, D. F.; Scriven, L. E., in *Structure and Flow in Surfactant Solutions* (Herb, C.; Prudhomme, R. K., eds.), ACS Symp. Ser., Vol. 578, American Chemical Society, Washington, DC, 1994, p. 192.
179. Li, X.; Washenberger, R. M.; Scriven, L. E.; Davis, H. T.; Hill, R. M., submitted to *Langmuir.*
180. Doumaux, H., Ph.D. thesis, University of Minnesota, 1995.
181. Yang, J.; Wegner, G., *Macromolecules* 1992, *25*, 1786.
182. Yang, J.; Wegner, G., *Macromolecules* 1992, *25*, 1791.
183. Lin, Z.; Hill, R. M.; Davis, H. T.; Scriven, L. E.; Talmon, Y., *Langmuir* 1994, *10*, 1008.
184. The use of cryo-transmission electron microscopy (TEM) to image surfactant microstructures is described by Talmon, Y., *Colloids Surf.* 1986, *19*, 237.
185. Lin, Z.; He, M.; Scriven, L. E.; Davis, H. T.; Snow, S. A., *J. Phys. Chem.* 1993, *97*, 3571.
186. He, M.; Lin, Z.; Scriven, L. E.; Davis, H. T.; Snow, S. A., *J. Phys. Chem.* 1994, *98*, 6148.
187. Hill, R. M.; Snow S. A., U.S. Patent 5,235,082 (1993).
188. Steytler, D. C.; Sargeant, D. L.; Robinson, B. H.; Eastoe J.; Heenan R. K., *Langmuir* 1994, *10*, 2213.
189. Hill, R. M.; Li, X.; Scriven, L. E.; Davis, H. T., manuscript in preparation.

190. Li, X.; Washenberger, R. M.; Scriven, L. E.; Davis, H. T.; Hill, R. M., submitted to *Langmuir.*
191. Li, X.; Hill, R. M.; Davis, H. T., this volume, Chapter.
192. Yoneyama, M.; Usami, T.; Ino, S.; Sata, Y.; Yamamoto, N., U.S. Patent 4,047,958 (1977).
193. Japanese Patent Publication No. 34.230 (1970).
194. Reed, D., *Urethanes Technol.* Jan./Feb. 1995, p. 22.
195. Hostettler, F., German Patent 1,091,324 (1960).
196. Sandridge, R. L.; Morecroft, A. S.; Hardy, E. E.; Saunders, J. H., *J. Chem. Eng. Data* 1960, *5*, 495.
197. Oertel, G., ed., *Polyurethane Handbook*, Carl Hanser, Munich, 1985.
198. Zhang, X. D.; Macosko, C. W.; Davis, H. T., *ACS Symp. Ser.* 1997, *669*, 130.
199. Starch, M. S., *Drug Cosmet. Ind.* 1984, *134*, 38.
200. Starch, M.; De Vries, C., *Parfüm. Kosme.* 1986, *67*, 148.

2

Silicone Polyether Copolymers: Synthetic Methods and Chemical Compositions

GARY E. LEGROW* and LENIN J. PETROFF Designed Materials Development, Dow Corning Corporation, Midland, Michigan

*Retired.

I. INTRODUCTION

Silicone (or siloxane) polyether copolymers are commonly composed of a silicone polymer component chemically combined with one or more organic polyether components. The focus of this chapter is the synthesis of such silicone polyether copolymers. Other polar organic groups such as carbohydrate, quaternary ammonium salts, and betaines have been chemically bound to the polysiloxane backbone to form siloxane surfactants and are described by Schmaucks elsewhere in this volume [1]. The chemical combination of a polysiloxane and a polyether may occur either with the formation of hydrolytically stable silicon–carbon bonds (Si—C) or with hydrolytically unstable silicon–oxygen–carbon bonds (Si—O—C). Which of these types of linkage is desired in the silicone polyether copolymer determines the choices of reagents necessary for the chemical coupling reaction as well as the physical properties of the resulting copolymer.

II. SILICONE POLYETHER COPOLYMERS

A. Si—C Linked Copolymers

The formation of a Si—C bond is accomplished by hydrosilylation of a silicon hydride functional polysiloxane to an allyloxy polyether as described by Haluska [2]. The rake, or pendant, types of silicone polyether are the most common structures and are represented by the formula $Me_3SiO(Me_2SiO)_x(RMeSiO)_ySiMe_3$, where Me represents a methyl group and R represents the polyether functionality. The levels of x and y and the ratio x/y, which can vary widely, determine the physical, chemical, and surface properties of the resulting copolymer. The form of R can vary significantly as well. It can be represented by an $(EO)_m(PO)_n(BO)_0$ notation, where EO represents an ethylene oxide (CH_2CH_2O) unit, PO represents a propylene oxide (CH_2CHCH_3O) unit, and BO represents a butylene oxide [$CH_2CH(CH_2CH_3)O$] unit. Numerous literature and patent references, as well as other chapters of this volume, describe the structure–activity relationships for these compositions.

A wide variety of structures are possible, ranging from high molecular weight siloxane polymers, where x is large, to those containing no dimethylsiloxane units. The simplest cases of structures of these types are the discrete compounds shown in Figs. 1–4, where Me represents a methyl group and R represents a polyether substitution. The compound in Fig. 1 is a trisiloxane-based surfactant. Figure 2 shows multiple polyether rake or pendant substitutions, whereas Fig. 3 shows both dimethylsiloxane and monomethylsiloxane–polyether rake substitutions. The compound in Fig. 4 is an ABA-type block copolymer. Compounds 2, 3, and 4 are the simplest cases of a particular class of materials. More complex analogs can be found with multiple repeat units of dimethylsiloxane or monomethylsiloxane–polyether. Branched polymers can be obtained by the incorporation of siloxane T

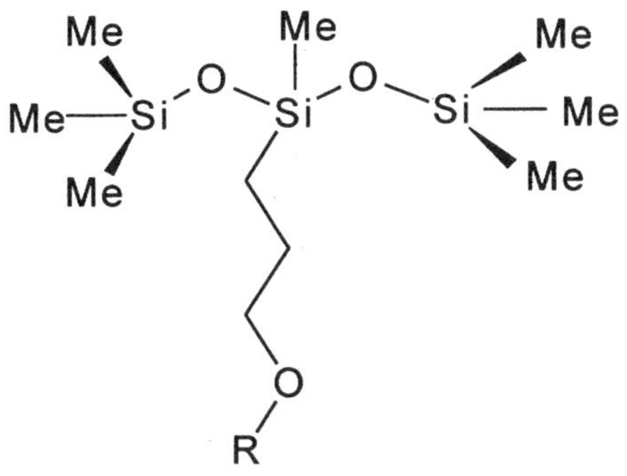

FIG. 1 Trisiloxane-based surfactant.

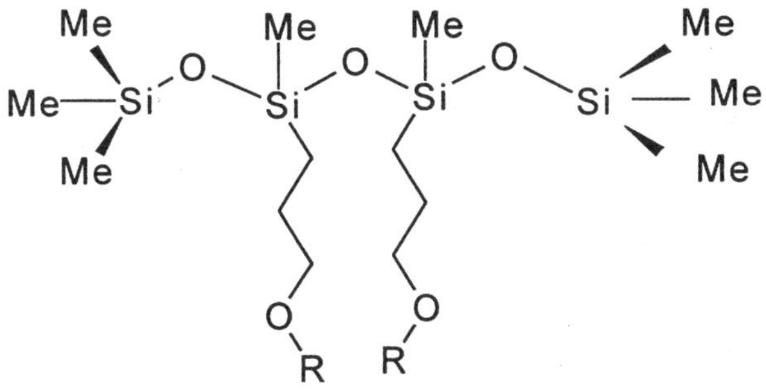

FIG. 2 Multiple polyether rake surfactant.

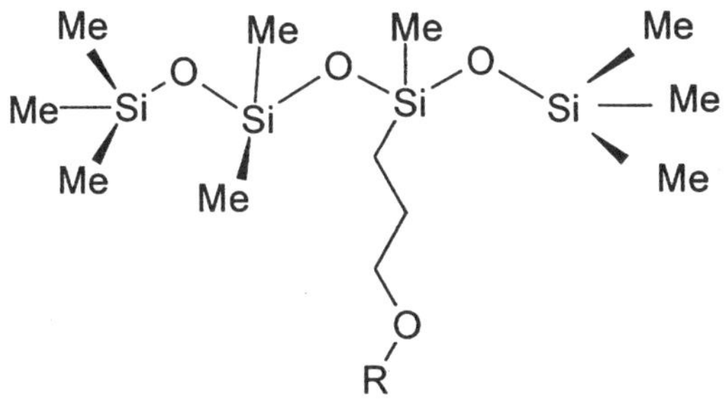

FIG. 3 Dimethylsiloxane/monoethylpolyether rake surfactant.

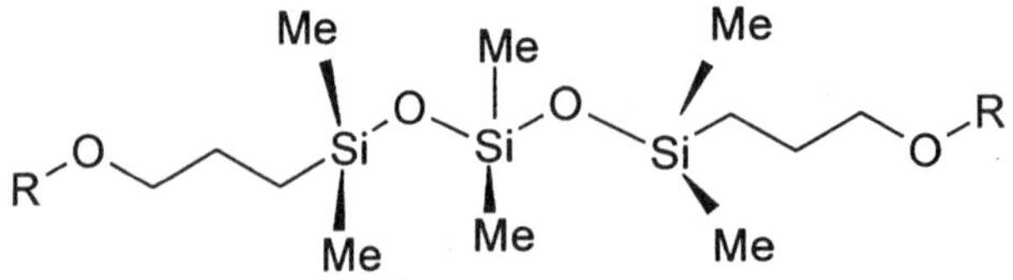

FIG. 4 ABA-type surfactant.

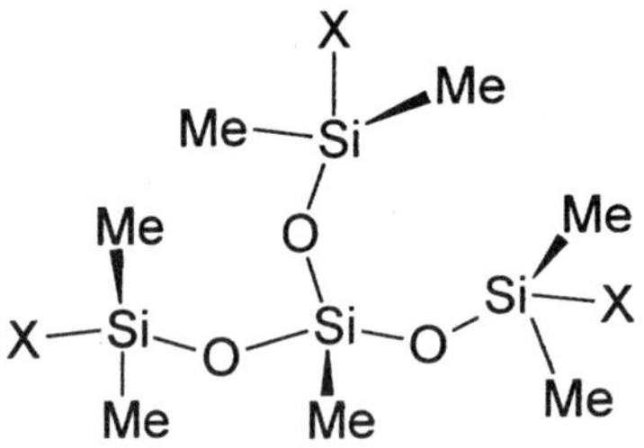

FIG. 5 T-branched surfactant.

units into the structure. The simplest case of such a T structure is shown in Fig. 5, where X represents the continuation of the polysiloxane chain.

B. Si—O—C Linked Copolymers

The second general class of silicone polyether copolymers comprises those in which the polyether group is linked to the siloxane backbone by an Si—O—C linkage. Structural families similar to those shown in Figs. 1–5 can be obtained. This linkage makes this class of materials less hydrolytically stable than their Si—C linked counterparts [3]. Rossmy and Wassermeyer describe the initial development of this form of copolymer via an esterification process and the resulting formation of an A–B–A silicone polyether, where A is a methoxy end-capped polyether and B is a polydimethylsiloxane chain as shown in Fig. 4 [4]. Branching can be incorporated into the silicone component of these copolymers, resulting in the types of material described in Fig. 5, or it can be introduced via polyethers containing multiple hydroxyl groups as shown in Fig. 6 [5].

III. SYNTHESIS OF Si—C LINKED COPOLYMERS

The most common commercial preparation of silicone polyether copolymers with hydrolytically stable silicon–carbon bonds involves a three-step process: (1) preparation of a siloxane hydride intermediate, (2) preparation of an allyloxy

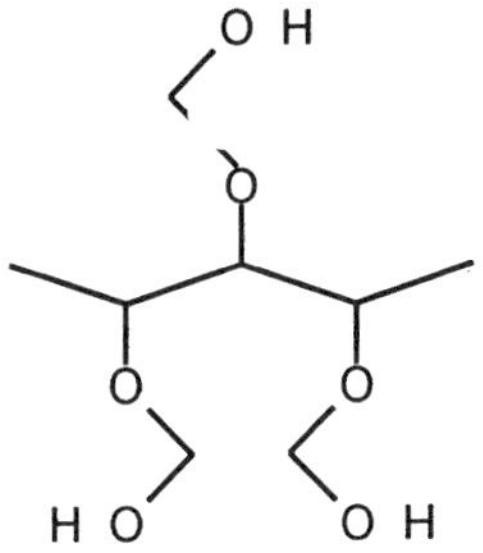

FIG. 6 Branched Si—O—C surfactant.

polyether intermediate, and (3) hydrosilylation of the second compound with the first to form the final copolymer. Kanner et al. describe a range of siloxane–polyether surfactants produced via this method [6]. Stewart and Bryant describe the evolution of this class of copolymers and the synthetic approaches that were explored in their development [7]. The next section develops the chemistry behind the three-step process.

A. Siloxane Hydrides for the Hydrosilylation Addition Reaction

Silicon hydride containing siloxane intermediates can be synthesized via several different process routes. The most direct route involves the preparation of basic hydridochlorosilanes. These SiH-containing silane monomers are produced by a series of chemical reactions known as the Rochow synthesis, or the "direct process" [8,9]. The Rochow process combines organic halides with silicon metal to form chlorosilane intermediates, including $HMeSiCl_2$, HMe_2SiCl, $MeSiCl_3$, and Me_3SiCl. These silanes can then be hydrolyzed to prepare a silicone hydride containing siloxane intermediate. For example, polymeric hydridomethylsiloxanes are produced commercially by the hydrolysis of $HMeSiCl_2$ in the presence of Me_3SiCl [Eq. (1)]. The trimethylchlorosilane acts as an end-blocking group to control molecular weight. Hydrochloric acid is formed as a by-product and is removed by water washing. This polymer can be depicted as $Me_3SiO(HMeSiO)_xSiMe_3$, where the average value of x may range from 10 to 100 [10]. Cyclosiloxanes, with y ranging from 4 to 6, are also formed in this reaction. They may or may not be removed by vacuum distillation from the linear polymer, depending on the intended use of this type of product.

$$2Me_3SiCl + (w + x)\ HMeSiCl_2 \xrightarrow{\text{excess } H_2O} Me_3SiO(HMeSiO)_xSiMe_3 + w(HMeSiO)_y + 2(w + x + 1)\ HCl \qquad (1)$$

Methyldichlorosilane, $HMeSiCl_2$, can also be hydrolyzed alone to form a mixture of cyclosiloxanes, $(HMeSiO)_y$, as shown in Eq. (2).

$$z\mathrm{HMeSiCl_2} \xrightarrow{\text{excess } H_2O} z/y\ (\mathrm{HMeSiO})_y + 2z\ \mathrm{HCl} \quad (2)$$

Homopolymers of the $Me_3SiO(HMeSiO)_xSiMe_3$ type can then again be produced by acid-catalyzed, ring-opening polymerization of the $(HMeSiO)_y$ cyclosiloxanes in the presence of hexamethyldisiloxane, $Me_3SiOSiMe_3$, for molecular weight control as shown in Eq. (3). A minimal amount of cyclosiloxanes is re-formed in this type of reaction [11].

$$(\mathrm{Me_3Si})_2\mathrm{O} + z(\mathrm{HMeSiO})_y \overset{H+}{\rightarrow} \mathrm{Me_3SiO(HMeSiO)}_{y,z}\mathrm{SiMe_3} + (\mathrm{HMeSiO})_y \quad (3)$$

The hydridosiloxanes above can be directly utilized as intermediates in the preparation of silicone polyether copolymers. More commonly they are used in the preparation of silicon hydride/dimethyl siloxane containing copolymers, which are subsequently further processed into the final silicone polyether. The silicon hydride/dimethyl siloxane copolymers can be prepared by acid-catalyzed equilibration using either linear [Eq. (1)] or cyclic [Eq. (2)] siloxane hydride containing materials along with a combination of hexamethyldisiloxane, $(Me_3Si)_2O$, for chain length control, and dimethylcyclosiloxanes, $(Me_2SiO)_v$, (v = 4–6), for molecular weight growth, as shown in Eqs. (4) and (5) [12]. Typical acid catalysts include sulfuric acid, trifluoroacetic acid, and trifluormethanesulfonic acid [13]. Heterogeneous acid-supported clays or ion exchange resins may also be employed as a catalyst source. Kricheldorf summarizes this catalysis and describes the mechanism of the polymerization [14].

$$(\mathrm{Me_3Si})_2\mathrm{O} + \mathrm{Me_3Si(HMeSiO)}_x\mathrm{SiMe_3} + (\mathrm{Me_2SiO})_v \overset{H+}{\rightarrow}$$
$$\mathrm{Me_3SiO(HMeSiO)}_y\mathrm{(Me2SiO)}_z\mathrm{SiMe_3} + (\mathrm{HMeSiO})_a(\mathrm{Me2SiO})_b \quad (4)$$

$$(\mathrm{Me_3Si})_2\mathrm{O} + (\mathrm{HMeSiO})_y + (\mathrm{Me_2SiO})_v \overset{H+}{\rightarrow} \mathrm{Me_3SiO(HMeSiO)}_y(\mathrm{Me_2SiO})_z\mathrm{SiMe_3}$$
$$+ (\mathrm{HMeSiO})_a(\mathrm{Me_2SiO})_b \quad (5)$$

The final product is a combination of linear and cyclosiloxanes. The ratio of dimethylsiloxane to monomethyl hydridosiloxanes can be varied in infinite combinations. This allows for the vast number of possibilities in terms of final surfactant or copolymer combinations and their resultant properties. The ratios of linear siloxanes to cyclosiloxanes range from 85:15 (at high Me_2SiO content) to 90:10 (at high HMeSiO content) [13]. The cyclosiloxanes will have an average Me_2SiO/HMeSiO ratio identical to that of the linear polymer. The linear formulas shown in Eqs. (4) and (5) are depicted as block formulas; however, this is only for convenience of writing. These SiH segments are randomly distributed throughout the polymer backbone. The polymer has a normal molecular weight distribution centered around the target molecular weight.

The cyclosiloxanes may or may not be removed from the mixture, depending on the end use of the intermediate. If the cyclosiloxanes are to be removed, the crude product mixture must be fully neutralized with base prior to separation. Typical bases are sodium bicarbonate and ammonia gas. The resulting salt is insoluble in the siloxane polymer and may be removed by filtration. The final neutral polymer can then be stripped under vacuum to remove the cyclosiloxanes [9]. Attempting to remove the cyclosiloxanes prior to neutralization would result in reequilibration or cracking of the polysiloxane as the equilibrium cyclosiloxanes were removed. This would continue until the polymer had been completely depolymerized into the starting cyclosiloxane species.

A structurally different type of siloxane hydride containing polymer is obtained from the use of HMe_2SiCl as the source for end blocking or chain termination. The simplest siloxane hydride end-blocked material is formed from the hydrolysis of dimethylchlorosilane, HMe_2SiCl, to form tetramethyldisiloxane, as shown in Eq. (6).

$$HMe_2SiCl \xrightarrow{\text{excess } H_2O} HMe_2SiOSiMe_2H + 2HCl \tag{6}$$

This disiloxane can be used as the end-blocking agent in an acid-catalyzed, ring-opening polymerization of dimethylsiloxane cyclics, $(Me2SiO)_y$ (y = 4–6), as described earlier and as shown in Eq. (7).

$$(HMe_2Si)_2O + x\,(Me_2SiO)_4 \xrightarrow{H+} HMe_2SiO\text{—}(4x - z)Me_2SiO\text{—}SiMe_2H + z/4\,(Me_2SiO)_z \tag{7}$$

As previously described, the ratio of linear to cyclosiloxanes formed is approximately 85:15; however, in this case the cyclosiloxanes formed are nonfunctional in that they lack any silicon hydride containing species. As a consequence they do not react in subsequent hydrosilylation reactions of the siloxane hydride terminated polymer. Because of the limited availability of HMe_2SiCl in the organosilicon industry, intermediates of the $HMe_2SiO(Me_2SiO)_xSiMe_2H$ type (ABA) have seen less development activity than their rake-type counterparts.

B. Polyether Chemistry

The second step in production of a siloxane polyether copolymer is the preparation of the unsaturated allyloxy polyether. This is accomplished by the base-catalyzed ring opening of an oxirane such as ethylene, propylene, or butylene oxide. Depending on the target application, one or a combination of these oxides may be required for optimal performance. Ethylene oxide polymers are readily water soluble and allow for the preparation of water-soluble or dispersible silicone polymers. Propylene oxide derivatives allow for more organic compatibility. Butylene oxide derivatives are even less polar and can allow for improved dispersibility in organic oils.

The polymerization is conducted as a three-step process as described by Whitmarsh [15]. This reaction schematic is shown in Eqs. (8)–(10). The polymerization is initiated by the formation of an alkoxide anion of the starting alcohol. Allyl alcohol is the most common means of introducing unsaturation into the final polymer, although Bennett has described other unsaturated alcohols useful as initiators [16,17].

$$H_2C{=}CHCH_2OH + NaOH \rightarrow H_2C{=}CHCH_2O^-Na^+ + H_2O \qquad (8)$$

The appropriate oxide(s) is then added to the alkoxide initiator, which ring-opens the oxide and propagates polymer chain growth. Oxides are fed into the reactor until the target molecular weight and ratio of oxides have been achieved. Equation (9) represents this process for ethylene and propylene oxides.

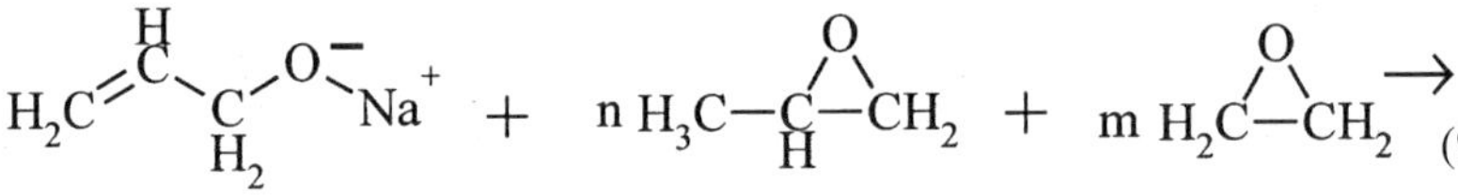

(9)

$$H_2C{=}CH_2CH_2O(CH_2CH_2O)_m(CH_2CHCH_3O)_n^-Na^+$$

The final step of the process is the neutralization of the alkoxide anion with acid to terminate the polymerization. An alkali metal salt is formed as a by-product, which can be left in the product or removed by filtration. Equation (10) shows this neutralization scheme for acetic acid.

$$H_2C{=}CH_2CH_2O(CH_2CH_2O)_m(CH_2CHCH_3O)_n^-Na^+ + H_3CCOOH \rightarrow$$

$$H_2C{=}CH_2CH_2O(CH_2CH_2O)_m(CH_2CHCH_3O)_nH + H_3CCOO^-Na^+ \qquad (10)$$

During this polymerization process, side reactions occur that have an impact on the final polymer properties and lead to limitations, hence must be considered in any assessment of the activity of the resulting silicone polyether copolymer. A feature of this polymerization chemistry is the formation of water during the initation step, as shown in Eq. (8). This water, as well as that introduced into the polymerization as a result of reactant contamination, leads to the formation of diol functional polyethers [Eq. 11]. The carbinol functional alkoxide formed during the first portion of this side reaction is then available for reaction with additional base to form a dialkoxide, which can propagate in both directions, giving rise to the diol functional polyether [15]. The final molecular weight of the diol is roughly twice that of the target allyl-initiated counterpart. These components are present in the final composition at levels ranging from 6 to 12%, depending on reaction conditions and the ratio of ethylene to propylene oxide in the target composition.

$$H_2O + NaOH\ (excess) + H_2\overset{O}{C\!-\!C}H_2 \rightarrow HOCH_2CH_2O^-Na^+ \rightarrow Na^{+-}OCH_2CH_2O^-Na^+ \qquad (11)$$

These polymeric diols contain no unsaturation and thus cannot be hydrosilylated by a silicon hydride functional siloxane. The diols are difficult or impossible to separate from the polymer mixture and will carry through as a component of the final product.

It is also known that propylene oxide undergoes rearrangement in the presence of base to form allyl alcohol ($H_2C{=}CHCH_2OH$) [18]. This has the effect of forming initiator in situ during the process. If not adjusted for, this side reaction will decrease the molecular weight of the polyether product. This effect is very sensitive to the level of PO introduced into the polymerization [19].

C. The Hydrosilylation Addition Reaction: Formation of the Si—C Linkage

The third and final step in this reaction scheme is the hydrosilylation of allyloxy polyether intermediates by a siloxane hydride forming the siloxane polyether copolymer. A simple model of this reaction is depicted in Eq. (12), where heptamethyltrisiloxane is used to hydrosilylate allyl alcohol.

$$(Me_3SiO)_2MeSiH + H_2C{=}CH{-}CH_2{-}OH \xrightarrow{Pt} (Me_3SiO)_2MeSi(CH_2)_3{-}OH \qquad (12)$$

A platinum compound is generally used to catalyze the hydrosilylation reaction. This compound may be acidic, such as chloroplatinic acid (Speier's catalyst) or it may be a neutral complex such as Karsted's catalyst, $[(ViMe_2Si)_2O]PtCl_2$ [20]. Compounds of other transition metals, such as those containing rhodium, nickel, and cobalt, have been investigated, as have various supported catalysts [21]. Numerous reports on hydrosilylation and various forms of catalysis have been published and are summarized in reference articles by Patai [22], and Marciniec [21,23]. Chalk and Harrod also describe the mechanism for the hydrosilylation of unsaturated polyethers [24]. The reaction is quite exothermic, with a heat of reaction of +28 kcal/mol. Various solvents, such as toluene and isopropanol, are used to help render the components compatible, as well as to act as a heat sink for the energy liberated from the reaction.

Although the hydrosilylation reaction is usually depicted as above, it can be, and usually is complicated by competing side reactions. Internal isomerization [25] of the terminal olefinic group of the polyether can occur, resulting in retardation and, in the extreme, termination of the hydrosilylation reaction, as depicted in Eq. (13). High temperature promotes the isomerization reaction; thus keeping the temperature of a hydrosilylation reaction below 80°C is preferred to maximize product yield.

$$H_2C{=}CH{-}CH_2{-}OH \xrightarrow{Pt} [H_3C{-}HC{=}CH{-}OH] \rightarrow H_3C{-}H_2C{-}HC{=}O \qquad (13)$$

In the example of isomerization of allyl alcohol, propionaldehyde is formed, while in the case of polyether polymers initiated by allyl alcohol, a vinylic ether is formed, as depicted in Eq. (14). The vinylic ether is hydrolytically unstable in the presence of acid, resulting again in formation of propionaldehyde and a diol functional polyether as shown in Eq. (15).

$$\begin{aligned}&H_2C{=}CH{-}CH_2{-}O{-}(CHR{-}CH_2{-}O)_x{-}H \xrightarrow{Pt}\\&H_3C{-}HC{=}CH{-}O{-}(CHR{-}CH_2{-}O)_x{-}H\end{aligned} \tag{14}$$

$$\begin{aligned}&H_3C{-}HC{=}CH{-}O{-}(CHR{-}CH_2{-}O)_x{-}H + H_2O \xrightarrow{H+}\\&[H_3C{-}HC{=}CH{-}OH] + HO{-}(CHR{-}CH_2{-}O)_x{-}H \rightarrow\\&H_3C{-}H_2C{-}HC{=}O + HO{-}(CHR{-}CH_2{-}O)_x{-}H\end{aligned} \tag{15}$$

To compensate for the isomerization that occurs, an excess of allyl functional polyether must be utilized to allow the hydrosilylation to go to completion. A 10–40 mol% excess of the allyl functional polyether per mole of hydridosiloxane is typical [2]. The excess unsaturated polyether is not readily removed and remains in the product. It may be in the form of excess allyl functional polyether, the isomerized analog as shown in Eq. (13), or the hydrolysis products propionaldehyde and diol polyether as discussed above.

Another side reaction that can occur is silylation, wherein the siloxane hydride intermediate can react with an active hydrogen in the polyether to form an alkoxysilane as depicted in Eq. (16) [26]. The active hydrogen typically is the terminal carbinol of the polyether chain. This problem can be averted by protection of the active hydrogen in the unsaturated alcohol or polyether. One way to do this is via silylation of the alcohol [27]. This is a commonly used technique in the pharmaceutical industry, wherein hexamethyldisilazane, $(Me_3Si)_2NH$, often serves as the silylating agent. The protecting Me_3Si group can be removed readily by hydrolysis, or by alcoholysis after the polyether has been attached to the siloxane. Another route is the conversion of the carbinol to a methyl ether via a Williamson synthesis utilizing methylene chloride and sodium hydroxide as the catalyst [28].

$$\begin{aligned}&(Me_3SiO)_2MeSiH + H{-}O{-}CH_2{-}CH{=}CH_2 \xrightarrow{H+}\\&(Me_3SiO)_2MeSi{-}O{-}CH_2{-}CH{=}CH_2 + H_2\end{aligned} \tag{16}$$

Even with the use of an excess of allyl polyether, residual silicon hydride functionality may remain in the final copolymer. This can be measured as ppm Si—H. Retardation of the hydrosilylation reaction, leading to incomplete reaction, can be a problem if the silicone polyether is highly substituted with polyether grafts. The inclusion of more dimethylsiloxane spacers in the polymer chain minimizes this problem. Residual silicon hydride functionality leads to hydrolysis of the Si—H on contact with moisture, forming silanol functionality and hydrogen gas. The silanol groups may further condense and cross-link to form siloxanes. When this complication occurs, the initial effect is an increase in viscosity of the product. Eventually, gelation of the product can occur. Formation of hydrogen gas may also present a packaging hazard. For these reasons, residual Si—H should be avoided by careful control of the molecular composition and stoichiometry of the raw materials.

Terminal Si—H functionality in a siloxane polymer can be as much as 10 times more reactive toward hydrosilylation and other reactions than pendant Si—H functionality along the chain. This difference, which is due to a combination of electronic and steric effects, must be taken into account when these Si—H intermediates are used in the synthesis reactions from which silicone polyether copolymers are made. These intermediates will react with terminally unsaturated polyethers to form hydrolytically stable A—B—A block silicone polyether copolymers as shown in Eq. (17).

$$HMe_2SiO(Me_2SiO)_xSiMe_2H + 2\ H_2C{=}CH{-}CH_2{-}O{-}(CH_2CHRO)_yH \xrightarrow{Pt} H(OCHRCH_2)_y{-}O(CH_2)_3{-}(Me_2SiO)_{x+2}{-}(CH_2)_3{-}O{-}(CH_2CHRO)_yH \quad (17)$$

IV. SYNTHESIS OF Si—O—C LINKED COPOLYMERS

The earliest examples of silicone polyethers were those produced in a manner that yielded Si—O—C linkages between the silicone and polyether. Many of these early synthetic routes described in the literature referenced earlier are still commercially viable today. These routes also involve a multistep process, as do their Si—C linked counterparts. The key difference is in the types of reactive chemical bond in the target silicone intermediate and in the absence of hydrosilylation processes in the formation of the final silicone polyether copolymer. Plumb and Atherton give an excellent review of many of the routes and composition of these copolymers [29]. These all are based on the reaction between a hydroxyl group on the polyether and a siloxane containing an Si—X group, where X can be a chloride, hydrogen, a primary or secondary amine, an alkoxy (OR) group, or an acyloxy (COOR). This condensation reaction is shown in Eq. (18).

$$X{-}(Me_2SiO)_x{-}Me_2Si{-}X + 2HO{-}R \rightarrow RO(Me_2SiO)_x{-}Me_2SiOR + 2H{-}X \quad (18)$$

A. Synthesis from Chloride Functional Polydimethylsiloxanes

The first step in the preparation of all the Si—X containing intermediates described above is the preparation of a chloride functional siloxane as described in Eq. (19). This type of intermediate is produced by hydrolyzing Me_2SiCl_2 with insufficient H_2O, yielding a chloride end-blocked, linear polydimethylsiloxane intermediate. Other chlorosilanes can be incorporated into the reaction to impart branching ($MeSiCl_3$), trimethylsiloxy end-blocking (Me_3SiCl), or silicon hydride functionality (Me_2SiHCl or $MeSiHCl_2$).

$$x + y + 1\ (Me_2SiCl_2) \xrightarrow{\text{deficient } H_2O} Cl(Me2SiO)_x\text{—}Me_2\ SiCl + y/n\ (Me_2SiO)_n + 2HCl \quad (19)$$

A silicone polyether can be prepared directly from this chloride functional polydimethylsiloxane intermediate. A polyether of the general structure $HO(CH_2CHRO)_yR$ will react with the chloride functional polydimethylsiloxane, resulting in a Si—O—C linked silicone polyether and HCl, as described in Eq. (20).

$$Cl(Me_2SiO)_x\text{—}Me_2SiCl + 2OH\text{—}R \rightarrow RO(Me_2SiO)_x\text{—}Me_2SiOR + HCl \quad (20)$$

One drawback of attempting to directly form silicone polyether copolymers by this route is the difficulty of removing the by-product hydrochloric acid from the product. The residual acid can promote hydrolysis of the Si—O—C linkage causing degradation of the polymer. This is particularly problematic when the silicone polyether is to be applied in aqueous media. Therefore, this route is not the preferred means to prepare these copolymers.

B. Synthesis from Derivatized Chloride Functional Polydimethylsiloxanes

A common means of avoiding unwanted effects of residual hydrochloric acid is by derivatization of the chloride end-blocked siloxane prior to the formation of the silicone polyether. The chloride functional polydimethylsiloxane may be reacted with alcohols (in the presence of an acid acceptor such as pyridine), secondary amines, or anhydrides to form alkoxy-terminated, dialkylamino-terminated, or acyloxy-terminated polydimethylsiloxanes, respectively. Equations (21)–(23) show examples of these reactions with methanol, dimethylamine, or acetic anhydride [27,30,31]. All three reactions produce as by-products various salts, which are easily removed prior to the formation of the silicone polyether copolymer.

$$Cl(Me_2SiO)_x\text{—}Me_2SiCl + ROH/R_3N \rightarrow RO(Me_2SiO)_x\text{—}Me_2SiOR + R_3NH^+Cl^- \quad (21)$$

$$Cl(Me_2SiO)_x\text{—}Me_2SiCl + R_2NH \rightarrow R_2N(Me_2SiO)_x\text{—}Me_2SiNR_2 + R_2NH_2^+Cl^- \quad (22)$$

$$Cl(Me_2SiO)_x\text{—}Me_2SiCl + AcOAc \rightarrow AcO(Me_2SiO)_x\text{—}Me_2SiOAc + 2AcCl \quad (23)$$

The intermediates produced in Eqs. (21)–(23) can then be reacted with hydroxyl-terminated polyethers in accordance with Eq. (24), producing A–B–A block silicone polyether copolymers. The by-products of these coupling reactions are volatile nonpolar materials (e.g., methanol, dimethylamine, acetic acid), which can be removed from the mixture, driving the reaction to completion.

$$RO(Me_2SiO)_x\text{—}Me_2SiOR + HO(CH_2CHR'O)_yR'' \rightarrow$$
$$R''(OCHR'CH_2)_y\text{—}O(Me_2SiO)_{x+1}\text{—}(CH_2CHR'O)_yR'' + 2ROH \quad (24)$$

C. Synthesis from Silicon Hydride Functional Polydimethylsiloxanes

Another means of forming Si—O—C linked silicone polyethers uses as intermediates silicon hydride functional polydimethylsiloxanes, such as those produced by reactions (3)–(5) [32]. This route affords a keen advantage over the routes just described in that the by-product of the condensation is convenient-to-remove hydrogen gas. Equation (25) represents this reaction for a dimethylhydrogen endblocked polydimethylsiloxane. Rake and branched polymers can be prepared in an analogous fashion.

$$H(Me_2SiO)_x\text{—}Me_2SiH + HO(CH_2CHR'O)_yR'' \rightarrow$$
$$R''(OCHR'CH_2)_y\text{—}O(Me_2SiO)_{x+1}\text{—}(CH_2CHR'O)_yR'' + 2H_2 \quad (25)$$

These reactions are generally slow unless catalyzed with base. Piperidine, alkali metal oxides, or silanolates have been shown to be effective, as have metal acylates and dialkylhydroxylamines [29].

V. ANALYSIS OF SILICONE POLYETHER COPOLYMERS

Silicone polyether copolymers are prepared by combining silicones with polyethers via either a hydrosilylation addition reaction or a reactive siloxane plus alcohol condensation reaction as described above. The synthetic routes used to prepare these materials yield copolymers that are distribution products of both the siloxane and polyether polymers. Analysis of the products of these reaction schemes must take into account the polymer distributions present, as well as the various side reactions and product impurities present in the complex reaction product mixture.

The many species in this complex mixture can include the silicone polyether copolymer, residual unsaturated polyether, diol polyether, and residual unreacted cyclosiloxanes. Detailed characterization of this mixture requires a combination of analytical techniques. These can include Fourier transform infrared (FTIR), spectroscopy, gel permeation chromatography (GPC), and nuclear magnetic resonance (NMR) spectroscopy. The application of these techniques to silicone compounds is discussed in great detail by Smith [33] and is summarized below only for key points pertaining to silicone polyether materials.

A. Infrared Spectroscopy

Infrared spectroscopy can be used to obtain both qualitative and quantitative information about the reaction mixture. Lipp and Smith provide a table showing many of the characteristic stretching bands for organosilicon compounds [34]. One of the most useful bands for characterization of Si—C based silicone polyethers is the Si—H stretch between 2100 and 2300 cm^{-1}. This stretch is useful because it can provide information on the completeness of the hydrosilylation reaction by quantifying residual SiH [35].

B. Gel Permeation Chromatography

Gel permeation chromatography is useful in separating polymer mixtures and in determining molecular weights and distributions of individual fractions. Steinmeyer and Becker give a useful review of the details of GPC techniques as they relate to silicone polymers [36]. Key considerations in using this technique include the solubility of the silicone polyether in the appropriate solvent, the type of detector utilized, and the use of appropriate calibration standards. The average molecular weight of the components can be determined, as well as their polydispersity. GPC can usually separate the silicone polyether copolymer from unreacted polyethers and determine their respective concentrations in the product mixture.

C. Nuclear Magnetic Resonance Spectroscopy

Nuclear magnetic resonance spectroscopy, using a combination of ^{1}H, ^{13}C, and ^{29}Si NMR spectra, is a powerful technique in the characterization of silicone polyether materials. Taylor, Parbhoo, and Fillmore provide a thorough treatment of the analysis of silicone compounds via NMR [37].

Carbon-13 NMR is useful for characterization of the polyether portion of the copolymer. It can be used to determine the length of the polyether, the ratio of ethylene oxide to propylene oxide, the type of linkage between the polyether and silicone backbone (Si—C vs. Si—O—C), and the capping group on the polyether. Ethylene oxide groups resonate at about 70 ppm, while propylene oxide groups resonate at 16.5, 73, and 75 ppm. At 13 ppm, the carbon adjacent to Si will have a unique chemical shift that is absent from Si—O—C based copolymers.

Silicon-29 offers an excellent means of characterizing the siloxane portion of the copolymer. It can determine the linear or branched nature of the siloxane, the number of dimethylsiloxane repeat units, the number of polyether-substituted siloxane units, and the sequencing of these units. Taylor et al. [37] describe numerous chemical shifts and interpretation techniques involved in the analysis.

TABLE 1 Chemical Shift Data Required for Silicone–Polyether Analysis

Material	NMR type	Unique nuclei
Siloxane	^{1}H	Si$\underline{H}$, SiO$\underline{R}$ (e.g., C$\underline{H}_3$), Si$\underline{Me}$ (C$\underline{H}_3$) of the Si bearing the reactive group
Siloxane	^{29}Si	$\underline{Si}$H, $\underline{Si}$OR, $\underline{Si}$Me of the Si bearing the reactive group
Polyether	^{1}H	Terminal—C$\underline{H}_2$OH, internal —OC$\underline{H}_2$C$\underline{H}_2$O—, terminal OC$\underline{H}_3$ or OC$\underline{H}_2$C$\underline{H}_3$
Unsaturated polyether	^{1}H	Terminal $\underline{H}_2$C=C$\underline{H}$—C$\underline{H}_2$—O—
Silicone polyether	^{1}H	SiC$\underline{H}_2$C$\underline{H}_2$C$\underline{H}_2$O—, SiOC$\underline{H}_2$CH$_2$O, Si$\underline{Me}$ (C$\underline{H}_3$)
Silicone polyether	^{29}Si	$\underline{Si}$CH$_2$CH$_2$CH$_2$O—, $\underline{Si}$Me (CH$_3$)

As an introduction only, Table 1 presents the information necessary to characterize and interpret one of these spectra based on proton and ^{29}Si nuclei. It shows the unique Si nuclei and protons present in the chemical structure.

VI. SUMMARY

A wide range of synthetic routes are available to prepare silicone polyether materials. The particular route used can affect the physical, chemical, and surface properties of the resulting material. As a consequence, the end application must be considered when one is choosing a synthetic technique by which to prepare the silicone polyether material. This chapter provides an overview of such techniques and can be used in combination with other chapters in this volume to determine the optimal combination of process route and application.

REFERENCES

1 G. Schmaucks, this volume, Chapter 3.
2. L. A. Haluska, U.S. Patent 2,868,824 (Aug. 9, 1956).
3. M. Gradzielski, H. Hoffmann, P. Robisch, and W. Ulbricht, *Tenside Surf. Deterg.* 1990, *27*, 366.
4. G. Rossmy and J. Wassermeyer, U.S. Patent 3,115,512 (Dec. 24, 1963).
5. B. Gruning and G. Koerner, *Tenside Surf. Deterg.* 1989, *26*, 312.
6. B. Kanner, W. Reid, and I. Peterson, *Ind. Eng. Chem., Prod. Res. Dev.* 1967, *6*(2) 88.
7. H. Stewart, and R. Bryant, *J. Cell. Plast.* March/April, 1973.
8. E. G. Rochow, U.S. Patent 2,380,995, 1945.
9. W. Noll, *Chemistry and Technology of Silicones*, Academic Press, New York, 1968.
10. P. V. Wright and J. A. Semlyen, *Polymer* 1970, *11*, 462.
11. T. N. Biggs and G. E. LeGrow, U.S. Patent 5,516,870 (May 14, 1996).
12. C. J. Litteral and D. L. Mullins, U.S. Patent 3,980,688 (Sept. 14, 1976).

13. S. Clarson and J. A. Semlyen, *Siloxane Polymers*, PTR Prentice Hall, Englewood Cliffs, NJ, 1993.
14. H. R. Kricheldorf, *Silicon in Polymer Synthesis*, Springer-Verlag, New York. 1996
15. R. Whitmarsh, in *Nonionic Surfactants—Polyoxyalkylene Block Copolymers* (V. Nace, ed.), Marcel Dekker, New York, 1996.
16. E. Bennett, U.S. Patent 4,059,605, 1977.
17. E. Bennett, U.S. Patent 3,957,843, 1976.
18. J. March, *Advanced Organic Chemistry: Reactions, Mechanisms, and Structure*, 3rd ed., John Wiley & Sons, New York, 1985, pp. 329–332.
19. D. M. Simons and J. J. Verbanc, *J. Polym. Sci.*, *44*, 303, 1960.
20. J. L. Speier, J. A. Webster, and G. H. Barnes, *J. Am. Chem. Soc.* 1957, *79*, 974.
21. B. Marciniec, *Comprehensive Handbook on Hydrosilylation*, Pergamon Press, Oxford, 1992.
22. S. Patai and Z. Rappoport, eds., *The Chemistry of Organosilicone Compounds*, John Wiley & Sons, New York, 1989.
23. B. Marciniec and J. Gulinski, *J. Organometal. Chem.* 1993, *446*, 15.
24. A. J. Chalk and J. F. Harrod, *J. Am. Chem. Soc.* (1965), *87*, 1.
25. J. C. Saam and J. L. Speier, *J. Am. Chem. Soc.* 1958, *80*, 4104.
26. J. E. Baines and C. Eaborn, *J. Chem. Soc.* 1956, 1436.
27. A. H. Ward, U.S. Patent 5,194,452 (Mar. 16, 1993).
28. R. Morrison, and R. Boyd, Ethers and epoxides, in *Organic Chemistry*, 4th ed. Allyn & Bacon, Boston, 1983.
29. J. B. Plumb and J. H. Atherton, Copolymers containing polysiloxane blocks, in *Block Copolymers* (D. C. Allport and W. H. Janes, eds.), John Wiley & Sons, New York, 1973.
30. D. L. Bailey and F. M. O'Conner, U.S. Patent 2834748 (Mar. 22, 1954).
31. H. Niederprum and W. Simmler, British Patent 1,073,156 (Apr. 9, 1965).
32. L. Bailey and F. M. O'Conner, British Patent 892,819 (May 23, 1957).
33. A. L. Smith, *The Analytical Chemistry of Silicones*, John Wiley & Sons, New York, 1991.
34. E. D. Lipp and A. L. Smith Infrared, Raman, near-infrared and ultraviolet spectroscopy, in *The Analytical Chemistry of Silicones* (A. L. Smith, ed.), John Wiley & Sons, New York, 1991, p. 305.
35. F. A. Zhokhova and V. V. Zharkov, *Plast. Massy (Plastics)* 1983, *6*, 54.
36. R. D. Steinmeyer and M. A. Becker, Chromatographic methods, in *The Analytical Chemistry of Silicones* (A. L. Smith, ed.), John Wiley & Sons, New York, 1991, p. 255.
37. R. B. Taylor, B. Parbhoo, and D. M. Fillmore, Nuclear magnetic spectroscopy in *The Analytical Chemistry of Silicones* (A. L. Smith, ed.), John Wiley & Sons, New York, 1991, p. 347.

3

Novel Siloxane Surfactant Structures

GERD SCHMAUCKS Research and Development, Schill & Seilacher GmbH, Hamburg, Germany

I. INTRODUCTION

In nature three types of self-organizing system are known: proteins, liquid crystals, and surfactants.

A compound can be a member of all three classes; this is true, for example, of some proteins. The self-organization of molecules in three dimensions forms crystals; their arrangement in two dimensions at an interface causes surface activity.

In principle crystallization starts with two molecules and goes on with the first layer. This first layer may have the same arrangement as the surfactant molecules build up at an interface. To supply evidence of such an arrangement, it was necessary to synthesize crystalline surfactants.

Most of the siloxane surfactants used in different fields of industrial application are mixtures of polydimethylsiloxanes with or without functional groups, which consist of compounds with a distribution of molecular weight and/or of isomers. Those mixtures do not crystallize. To carry out fundamental research it is very important to study pure compounds because small differences in the structure cause significant changes in x-ray crystal structure and may disturb the arrangement at the interface.

This chapter, which is focused on investigations carried out at the Max Planck Institute for Colloidal and Interfacial Studies in Berlin, begins with the synthesis and discusses the interfacial behavior of quaternary ammonium groups containing siloxane surfactants. New directions of the development of siloxane surfactants are described in the final sections.

II. *N,N,N*-TRIALKYL-3-(SILOXANYLPROPYL) AMMONIUM HALIDES

A. Synthesis

The general formula of *N,N,N*-trialkyl-3-(siloxanylpropyl)ammonium halides is shown in Scheme 1. This class of siloxane surfactants was first mentioned by Maki et al. in 1970 [1]. A number of compounds having the structural unit SiCCCN were investigated by Speier et al. [2].

One step in the synthesis of siloxane surfactants is in most cases the hydrosilylation of unsaturated compounds: that is, the addition of a SiH group to a double or triple bond, catalyzed by hexachloroplatinic acid in isopropanol (Speier's catalyst [3]) or other metal compounds. This reaction leads to a mixture of α- and β-addition products. The ratio of the products depends on the siloxane used [4], the substrate, and the catalyst. When special Rh catalysts are used, the yield of γ-isomer in the reaction of triethoxysilane with allylamine can be increased up to 99.8% [5].

In the synthesis of quaternary siloxanylpropylammonium salts, ^{13}C-NMR measurements were used to show that the quaternization of the mixture of β- and

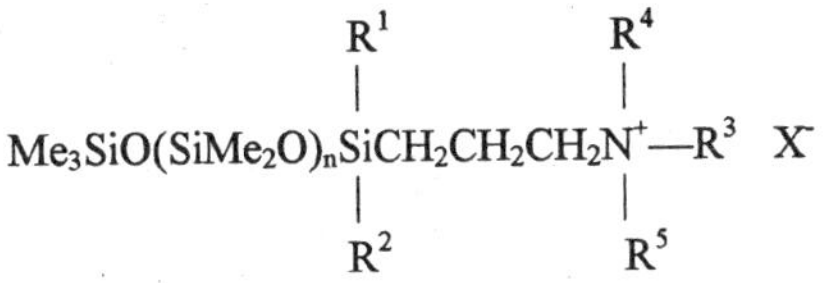

3a-d R^1=R^2= Me; R^3= Me, Et, Pr, n-Bu; R^4=R^5= Me; X= Br; n= 0

4b-d R^1=R^2= Me; R^3= Et, Pr, n-Bu; R^4=R^5= Me; X= Br; n= 1

5a-h R^1= Me; R^2= $OSiMe_3$; R^3= Me, Et, Pr, n-Bu, i-Bu, allyl; R^4=R^5= Me; X= Cl, Br, I; n= 0

6a-d, f R^1=R^2= $OSiMe_3$; R^3= Me, Et, Pr, n-Bu, allyl; R^4=R^5= Me; X=Br; n= 0

7b-d R^1= Me; R^2= $OSiMe_3$; R^3= Et, Pr, n-Bu; R^4=R^5=Et; X= Br; n= 0

SCHEME 1 General formula of *N,N,N*-trialkyl-3-(siloxanylpropyl)ammonium halides: **3a–3d,**

γ-addition products [Eq. (2)] received from the hydrosilylation of dimethylallylamine with H-siloxanes [Eq. (1)] is kinetically controlled, and the quaternization of the γ-isomer with alkyl halides is faster than the reaction of the β-isomer [6]. So it was possible to synthesize isomerically pure compounds for the investigation of the adsorption behavior and for the determination of their x-ray crystal structure.

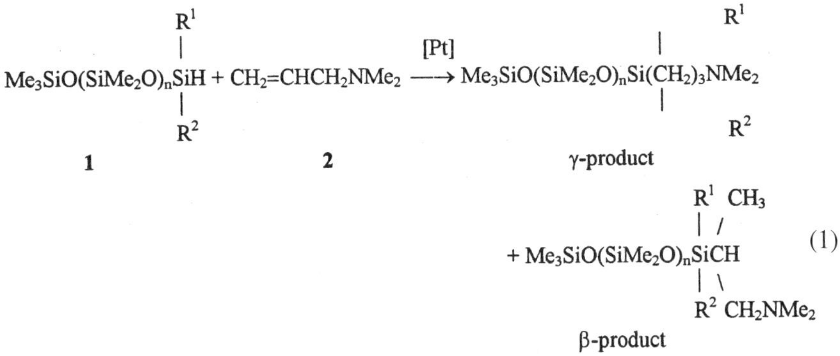

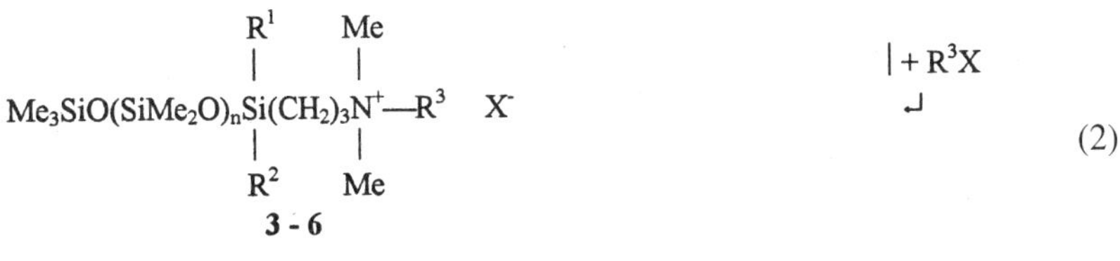

TABLE 1 *N,N,N*-Trialkyl-(3-siloxanylpropyl)ammonium Halides

Compound	R^1	R^2	R^3	X	*n*
3a	Me	Me	Me	Br	0
3b	Me	Me	Et	Br	0
3c	Me	Me	Pr	Br	0
3d	Me	Me	*n*-Bu	Br	0
4b	Me	Me	Et	Br	1
4c	Me	Me	Pr	Br	1
4d	Me	Me	*n*-Bu	Br	1
5a	Me	$OSiMe_3$	Me	Br	0
5b	Me	$OSiMe_3$	Et	Br	0
5c	Me	$OSiMe_3$	Pr	Br	0
5d	Me	$OSiMe_3$	*n*-Bu	Br	0
5e	Me	$OSiMe_3$	*i*-Bu	Br	0
5f	Me	$OSiMe_3$	Allyl	Br	0
5g	Me	$OSiMe_3$	Me	I	0
5h	Me	$OSiMe_3$	Allyl	Cl	0
6a	$OSiMe_3$	$OSiMe_3$	Me	Br	0
6b	$OSiMe_3$	$OSiMe_3$	Et	Br	0
6c	$OSiMe_3$	$OSiMe_3$	Pr	Br	0
6d	$OSiMe_3$	$OSiMe_3$	*n*-Bu	Br	0
6f	$OSiMe_3$	$OSiMe_3$	Allyl	Br	0
7b	Me	$OSiMe_3$	Et	Br	0
7c	Me	$OSiMe_3$	Pr	Br	0
7d	Me	$OSiMe_3$	*n*-Bu	Br	0

Table 1 lists the compounds represented in Scheme 1. Each one is a hygroscopic crystalline white solid with good solubility in water and in many organic solvents. The solubility in water decreases with the increasing number of silicon atoms in the molecule.

B. X-Ray Crystal Structure

The x-ray crystal structures of selected compounds were determined at room temperature [7] and in some cases also at low temperature (**3d, 4d, 5f**) [8]. Examples of the different types of structure are given in Figs. 1–3. Surprisingly, the siloxane groups of the disiloxane **3d** and the straight chain trisiloxane **4d** do not extend the chain N–C–C–C but form an angle of about 108.2° at Si1.

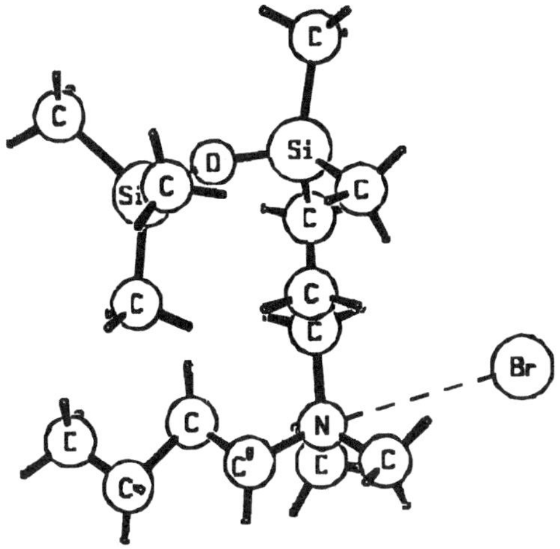

FIG. 1 X-ray crystal structure of *N*-butyl-*N*,*N*-dimethyl-[3-(tetramethyldisiloxanyl)-propyl]ammonium bromide, **3d.**

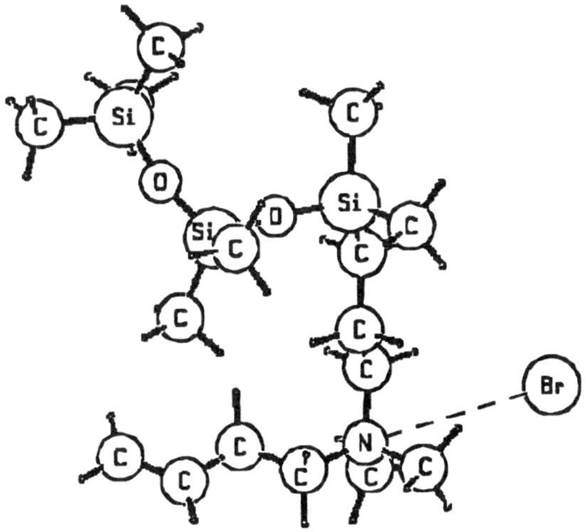

FIG. 2 X-ray crystal structure of *N*-butyl-*N*,*N*-dimethyl-[3-(1,1,3,3,5,5,5-heptamethyltrisiloxan-1-yl)propyl]ammonium bromide, **4d.**

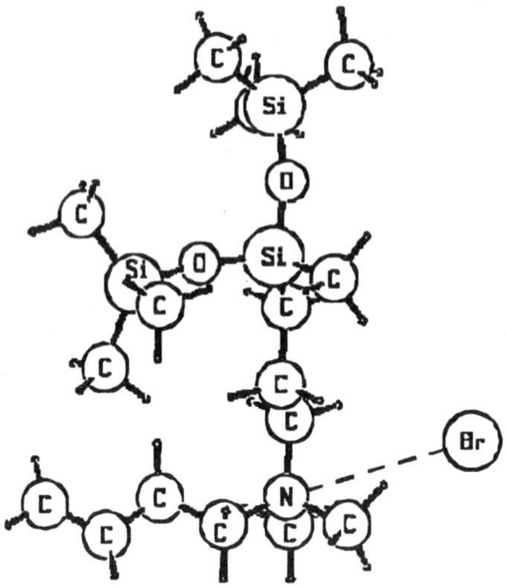

FIG. 3 X-ray crystal structure of *N*-butyl-*N*,*N*-dimethyl-[3-(1,1,1,3,5,5,5-heptamethyltrisiloxan-3-yl)propyl]ammonium bromide, **5d.**

The temperature dependence of the Si–O–Si angle was determined by investigating compound **4d.** At room temperature the Si–O–Si angles are 163.1 and 164.5°; at 173 K they are 147.0 and 159.5°. The molecules are rather flexible. The $SiMe_3$ parts of all compounds are able to freely rotate around the Si—O bond, and the whole siloxane group can rotate around the Si1—C bond in steps of 120° with respect to the rest of the molecule.

In the literature [9] the high flexibility of the siloxane groups is often considered to be a major reason for their extraordinary surface activity. This flexibility of the molecules facilitates their adaptability to different surroundings, but it complicates the prediction of their arrangement in the crystal and at the air–water interface.

All the compounds of this type investigated so far by x-ray crystal structure analysis arrange in the crystal to form a bilayer structure. Examples are shown in Figs. 4–6.

C. Adsorption Behavior at the Air–Water Interface

The adsorption behavior was characterized by surface tension measurements using the du Noüy ring method followed by calculations with the Frumkin

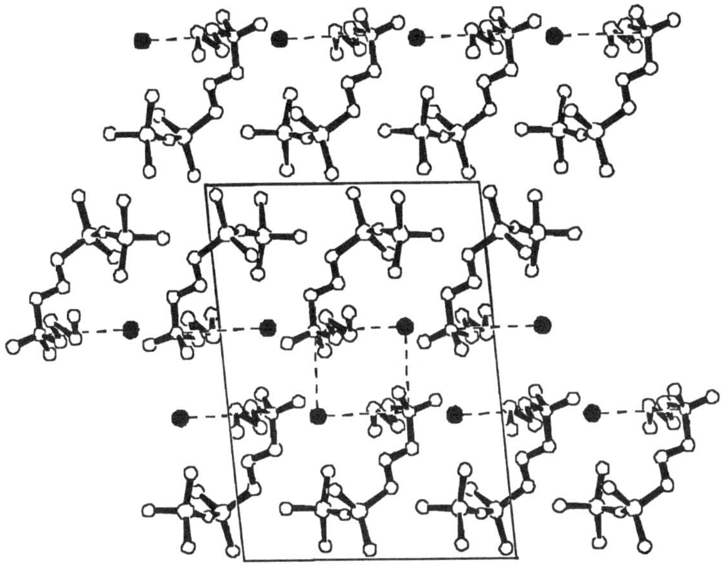

FIG. 4 Packing of the molecules of **3d** in the crystal lattice.

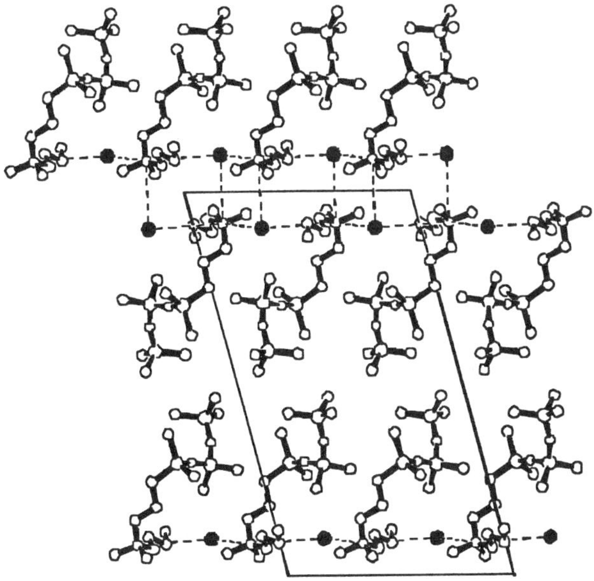

FIG. 5 Packing of the molecules of **4d** in the crystal lattice.

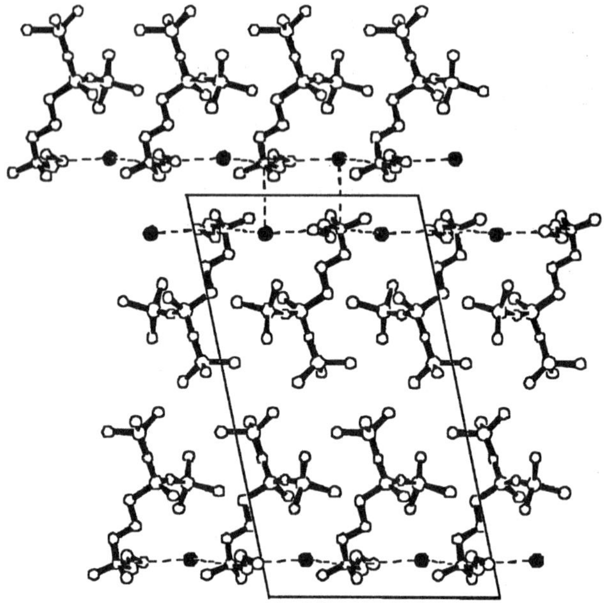

FIG. 6 Packing of the molecules of **5d** in the crystal lattice.

isotherm [10]. The results are listed in Table 2. The strong dependence of the adsorption behavior on the siloxane group is obvious. The maximum surface tension depression of the tetrasiloxanes is in the range of the theoretical minimum. In addition, the wetting properties of all compounds excluding the disiloxanes are excellent.

Comparison of the required surface area per molecule calculated for the adsorption layer with the value determined in the x-ray crystal structure can give information on the arrangement of the molecules at the interface. Whereas the surface areas per molecule of the branched trisiloxanes **5c** and **5d** in the crystal structure and at the interface are nearly the same, which can mean the same arrangement, the surface areas per molecule of the disiloxane **3d** and the straight-chain trisiloxane **4d** are determined to be 0.597 (**3d**) and 0.605 (**4d**) in the crystal lattice but significantly smaller at the interface. The reason seems to be a change of the position of the siloxane group related to the alkyl chain at the nitrogen atom by turning around the Si1—C3 bond.

TABLE 2 Interfacial Activity of Alkyl-*N*,*N*-dimethyl-(3-siloxanylpropyl) ammonium Halides

Compound	cmc (mol/dm^3)	Maximum surface tension depression (mN/m)	Adsorption energy (kJ/mol)	Surface activity (mol/cm^3)	Surface area per molecule (nm^2)
3b	6.9×10^{-2}	26.5	−5.29	4.98×10^{-7}	0.342
3c	2.6×10^{-2}	23.5	−6.78	2.69×10^{-7}	0.431
3d	5.7×10^{-2}	24.6	−5.69	2.36×10^{-7}	0.474
4b	2.3×10^{-2}	20.4	−7.69	1.87×10^{-7}	0.442
4c	9.7×10^{-3}	19.0	−14.49	2.70×10^{-8}	0.431
4d	1.0×10^{-2}	20.0	−10.18	6.57×10^{-8}	0.451
5a	1.4×10^{-2}	22.5	−15.30	7.88×10^{-9}	0.646
5b	2.2×10^{-2}	20.6	−13.30	1.81×10^{-8}	0.589
5c	2.0×10^{-2}	20.2	−14.91	9.24×10^{-9}	0.646
5d	9.6×10^{-3}	18.5	−15.45	8.32×10^{-9}	0.628
5e	1.3×10^{-2}	21.6	−14.63	1.04×10^{-8}	0.612
5f	2.4×10^{-2}	19.5	−14.18	1.25×10^{-8}	0.632
5g	2.2×10^{-3}	30.7	−14.11	1.29×10^{-8}	0.590
5h	2.1×10^{-2}	24.0	−13.01	2.04×10^{-8}	0.605
6a	4.5×10^{-3}	22.4	−20.90	7.67×10^{-10}	0.699
6b	2.4×10^{-3}	17.9	−22.83	3.48×10^{-10}	0.694
6c	2.8×10^{-3}	18.8	−22.46	4.00×10^{-10}	0.779
6d	2.6×10^{-3}	19.3	−23.70	2.43×10^{-10}	0.766
6f	2.9×10^{-3}	18.4	−23.80	2.30×10^{-10}	0.745

D. Adsorption at the Mercury–Electrolyte Interface

The adsorption of some selected *N*,*N*,*N*-trialkyl-(3-siloxanylpropyl)ammonium halides was investigated by means of a slowly dropping mercury electrode and a 0.5 M solution of sodium fluoride at pH 7 and a temperature of 25°C. The amplitude of the measuring alternating voltage was 5 mV, and the measuring frequency was 520 Hz. Details of the method used and the calculations are published elsewhere [11].

The most interesting result of these measurements is the strong dependence of the adsorption energy on the variation of the alkyl chain at the nitrogen atom. The

TABLE 3 Adsorption Energy of *N,N,N*-trialkyl-(3-siloxanylpropyl)ammonium Bromides at the Mercury Electrode, Related to $E = -0.8$ V

Compound	Adsorption energy ΔG_A (kJ/mol × 10^3)
3a	–32.03
3b	–33.14
3c	–34.66
3d	–36.27
4b	–34.19
4c	–27.07
4d	–36.44
5a	–27.37
5b	–30.96
5c	–32.38
5d	–34.98
5f	–31.29
6a	–26.58
6c	–29.06
6d	–30.22
7b	–31.73
7c	–33.32
7d	–36.07

different siloxane groups determine the ground level of the adsorption energy, and every additional methylene group at the nitrogen atom causes a stepwise increase. From the results, given in Table 3, it can be concluded that at the mercury electrode the ammonium group is adsorbed exclusively.

III. *N,N*-DIALKYL-3-SILOXANYLPYRROLINIUM DERIVATIVES

A. Synthesis

1. *N,N*-Dialkyl-3-siloxanylpyrrolinium Halides

The *N,N*-dialkyl-3-siloxanylpyrrolinium halides represent a new kind of siloxane surfactants. The synthesis starts with the hydrosilylation of 1,4-dichlorobutyne with H-siloxanes [Eq. (3)] followed by the reaction of the received 1,4-dichloro-2-siloxanylbutenes with primary or secondary amines, respectively. ^{13}C-NMR studies of the hydrosilylation showed that the E products are received exclusively, as in the case of the hydrosilylation of but-2-yne-1,4-diole [12].

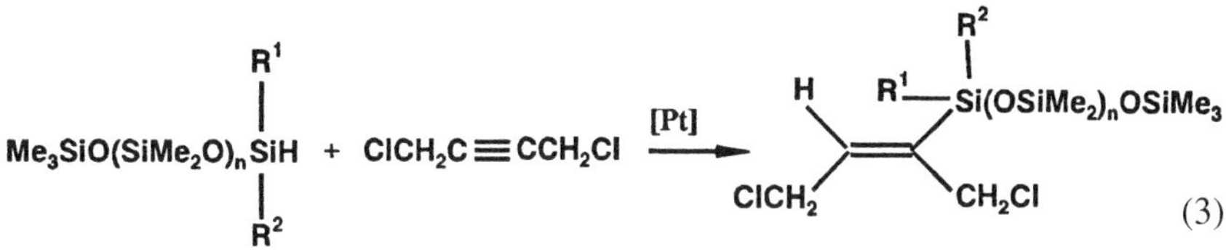

(3)

The reaction with primary amines gives, via an intramolecular ring linking *N*-alkyl-3-siloxanylpyrrolines [Eq. (4)], colorless liquids, which can be quaternized with alkyl halides [Eq. (5)].

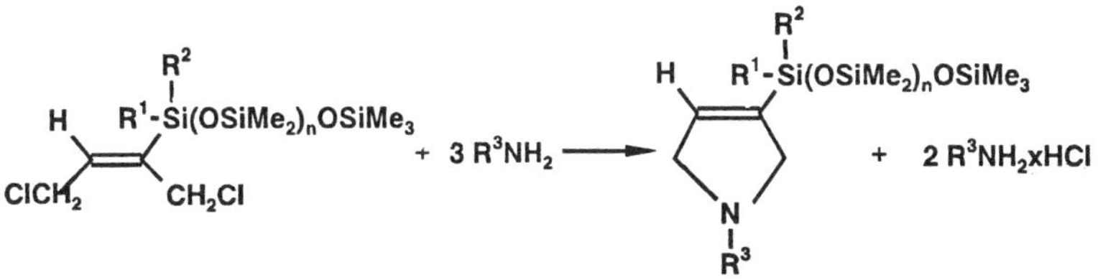

(4)

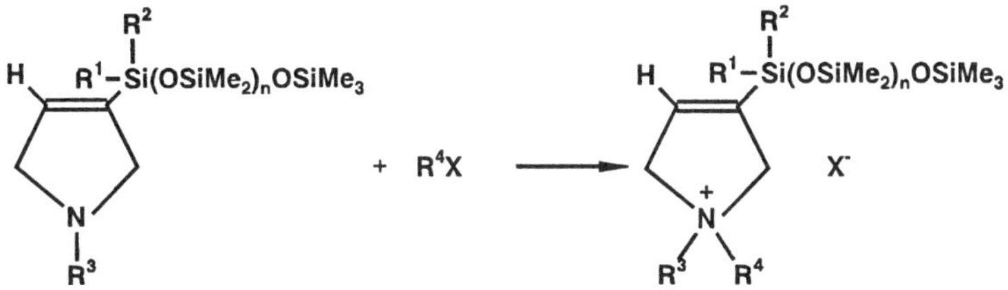

(5)

8a–d
9a–b
10a
11a–f
12a–d

The reaction with secondary amines leads to *N,N*-dialkyl-3-siloxanylpyrrolinium chlorides [Eq. (6)], but because of difficulties in the separation of the by-product, the first path is favored [13].

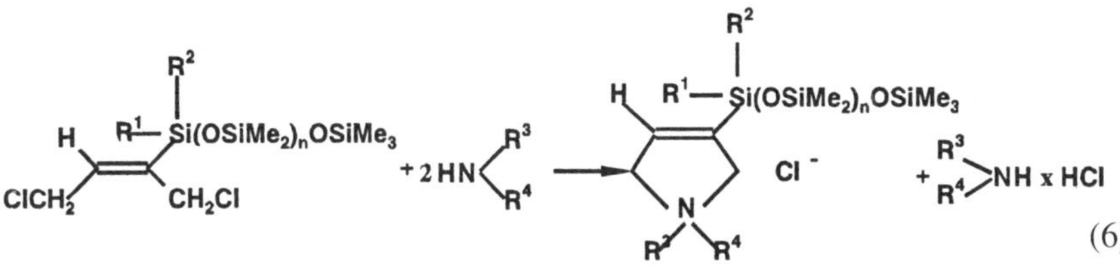

(6)

TABLE 4 *N,N*-Dialkyl-3-siloxanylpyrrolinium Halides

Compound	R^1	R^2	R^3	R^4	X
8a	Me	$OSiMe_3$	Me	Me	Cl
8b	Me	$OSiMe_3$	Et	Et	Cl
8c	Me	$OSiMe_3$	Me	Allyl	Cl
8d	Me	$OSiMe_3$	H	Et	Cl
9a	$OSiMe_3$	$OSiMe_3$	Me	Me	Cl
9b	$OSiMe_3$	$OSiMe_3$	H	Et	Cl
10a	Me	Me	Me	Et	Br
11a	Me	$OSiMe_3$	Me	Me	Br
11b	Me	$OSiMe_3$	Me	Et	Br
11c	Me	$OSiMe_3$	Et	Et	Br
11d	Me	$OSiMe_3$	Allyl	Et	Br
11e	Me	$OSiMe_3$	Allyl	Pr	Br
11f	Me	$OSiMe_3$	Pr	Pr	Br
12a	$OSiMe_3$	$OSiMe_3$	Me	Me	Br
12b	$OSiMe_3$	$OSiMe_3$	Me	Et	Br
12c	$OSiMe_3$	$OSiMe_3$	Et	Et	Br
12d	$OSiMe_3$	$OSiMe_3$	Allyl	Et	Br

The list of compounds is given in Table 4. All pyrrolinium salts are good soluble in water and in some organic solvents. They form white crystals. The characterization was carried out by elemental analysis and ^{13}C-NMR spectroscopy.

2. *N*-Alkyl-*N*-sulfonatoalkyl-3-siloxanylpyrrolinium Sulfobetaines

Straight-chain siloxanylsulfo- and siloxanylcarbobetains were synthesized and investigated by Snow et al. [14,15]. The synthesis can be carried out using two methods: the addition of sulfite or hydrogen sulfite to siloxanes with unsaturated functional groups [16] or the reaction of tertiary amines with sultones. To get pure compounds without extensive separations the second path is more convenient.

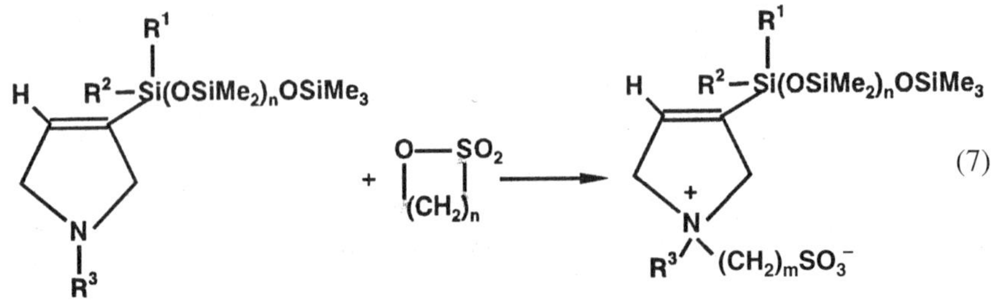

(7)

13a–e
14a–c

TABLE 5 *N*-Alkyl-*N*-sulfonatoalkyl-3-siloxanylpyrrolinium Sulfobetaines

Compound	R^1	R^2	R^3	*m*
13a	Me	$OSiMe_3$	Me	3
13b	Me	$OSiMe_3$	Me	4
13c	Me	$OSiMe_3$	Et	4
13d	Me	$OSiMe_3$	Allyl	4
13e	Me	$OSiMe_3$	Pr	4
14a	$OSiMe_3$	$OSiMe_3$	Me	3
14b	$OSiMe_3$	$OSiMe_3$	Me	4
14c	$OSiMe_3$	$OSiMe_3$	Allyl	4

In accordance with Eq. (7), the reaction of *N*-alkyl-3-siloxanylpyrrolines with propane- or butanesultone, respectively, gives a number of new siloxanylsulfobetaines listed in Table 5 [13].

The products are white solids with good solubility in water, and some organic solvents, and were characterized by elemental analysis and ^{13}C-NMR spectroscopy.

B. X-Ray Crystal Structure of 3-Siloxanylpyrrolinium Derivatives

1. *N,N*-Dialkyl-3-siloxanylpyrrolinium Halides

The investigation of the crystal structure was carried out at room temperature using CuK_α radiation. The x-ray crystal structures of **8d, 9a, 9b, 11a,** and **12a** could be determined. All of them are different. The molecular structures are shown in Figs. 7–10.

The packing of the molecules in the crystal lattice depends on the ratio of the siloxane group to the polar group. If the polar part of the molecule dominates, the molecules arrange in double layers; if the siloxane part is larger, they build up columns. The flexibility of the Si–O–Si angle and the influence of neighboring molecules can be demonstrated by comparing branched tetrasiloxane compounds that have three chemically equivalent—$OSiMe_3$ groups. Despite their equivalency, the Si–O–Si angles range from 148° to 162° in **9a** and from 148° to 158° in **9b.** The variations of the packing principles are shown in the Figs. 11–15. In this structure water molecules are integrated into the layer to fill the space between the polar groups because the large tetrasiloxane groups hinder a closer packing.

2. *N*-Alkyl-*N*-sulfonatoalkyl-3-siloxanylpyrrolinium Sulfobetaines

The determination of the x-ray crystal structure of siloxanyl groups containing sulfobetaines was very difficult because it is not easy to get crystals of the desired

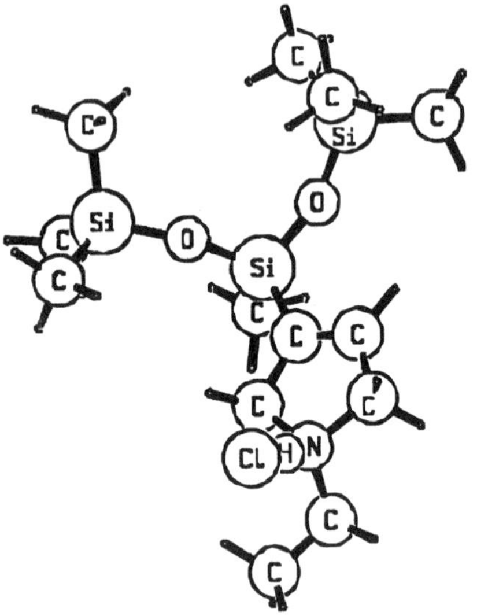

FIG. 7 Molecular structure of *N*-ethyl-3-(1,1,1,3,5,5,5-heptamethyltrisiloxan-3-yl)pyrroline hydrochloride, **8d.**

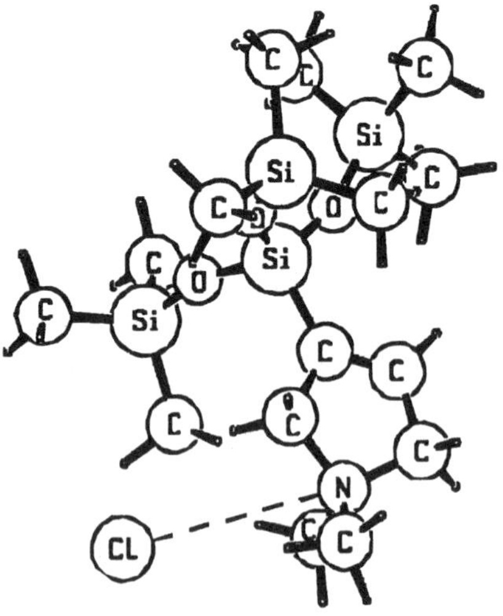

FIG. 8 Molecular structure of *N,N*-dimethyl-3-[tris(trimethylsiloxanyl)silyl]pyrrolinium chloride, **9a.**

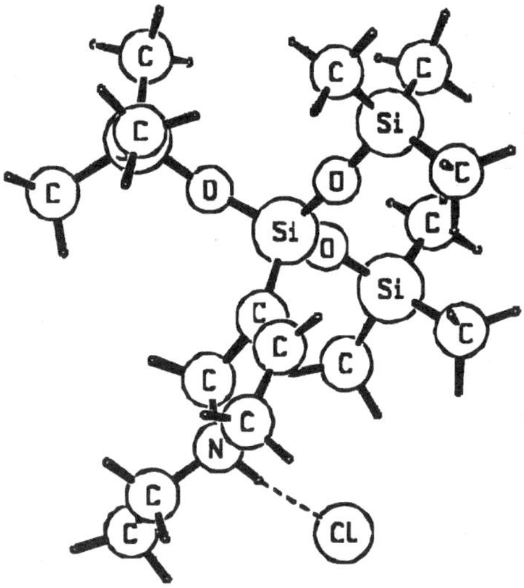

FIG. 9 Molecular structure of *N*-ethyl-3-[tris(trimethylsiloxanyl)silyl]pyrroline hydrochloride, **9b.**

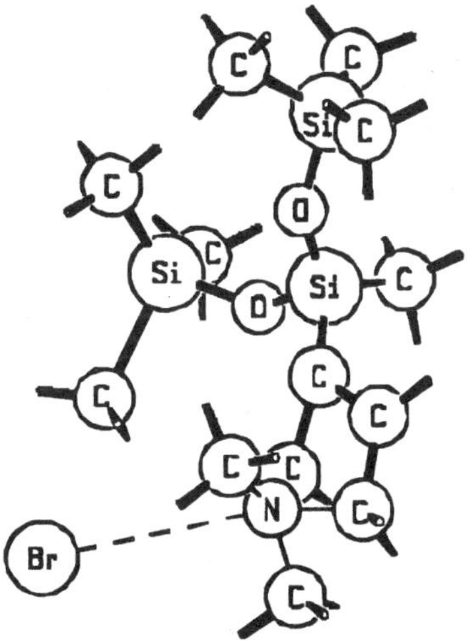

FIG. 10 Molecular structure of *N*,*N*-dimethyl-3-(1,1,1,3,5,5,5-heptamethyltrisiloxan-3-yl)pyrrolinium bromide, **11a.**

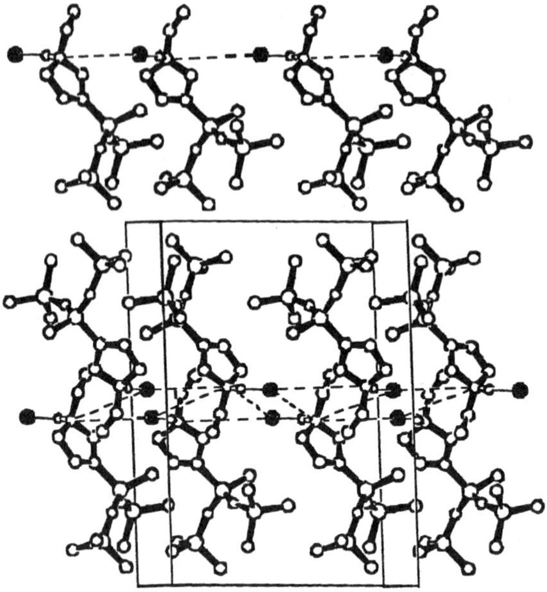

FIG. 11 Molecular packing of **8d** in the crystal lattice.

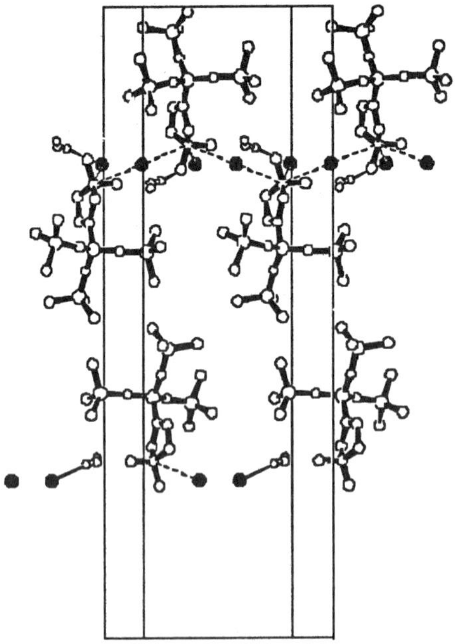

FIG. 12 Molecular packing of **9a** in the crystal lattice.

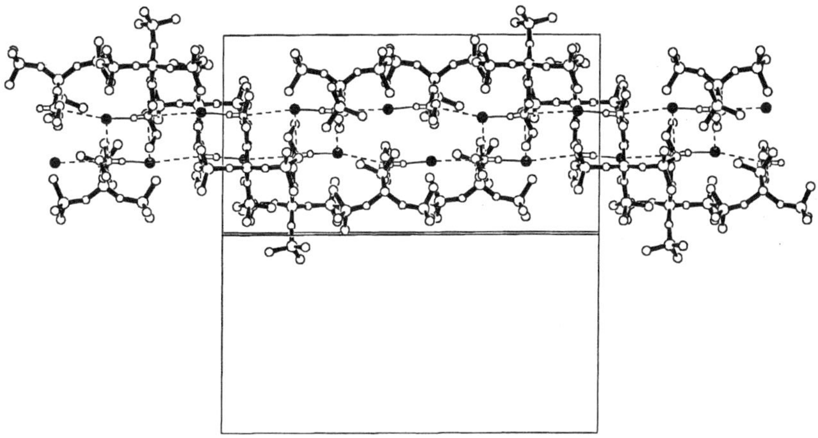

FIG. 13 Molecular packing of **9b** in the crystal lattice.

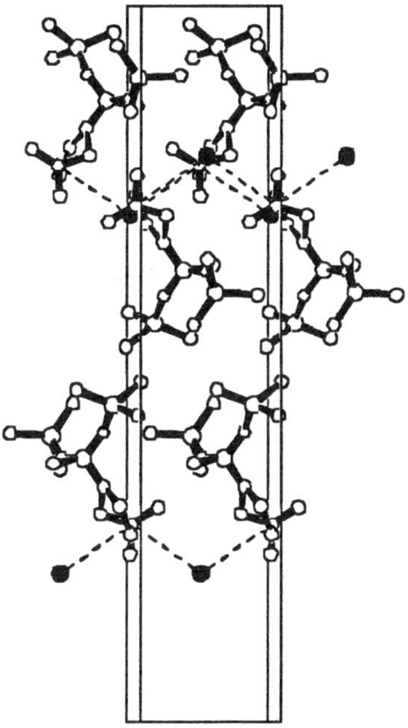

FIG. 14 Molecular packing of **11a** in the crystal lattice.

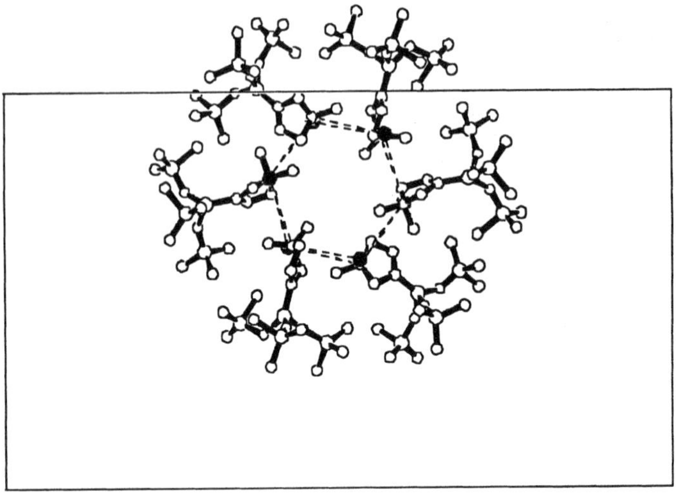

FIG. 15 Molecular packing of **12a** in the crystal lattice.

quality. So it is a great success to report the previously unpublished structure of *N*-methyl-*N*-3-(sulfonato)propyl-3-[tris(trimethylsiloxanyl)silyl]pyrrolinium sulfobetain, **14a** (Fig. 16). This structure contains one molecule of water between two sulfobetaine molecules. Figure 17 shows the crystal packing of the molecules.

C. Adsorption Behavior at the Air–Water Interface

The results of the determination of the adsorption behavior and the calculations using the Frumkin isotherm are listed in Table 6.

The interpretation of the results is not as easy as in the case of the siloxanylalkylammonium halides. This seems to evidence the important role of the spacer between the hydrophobic siloxane group and the hydrophilic polar group, but the decisive influence of the number of silicon atoms in the siloxane group on the surface activity can be seen here too. These results show also the possibility of varying adsorption behavior by the choice of functional group. Having the same number of siloxane units, the sulfobetaines possess a significant lower critical micelle concentration.

IV. BOLAFORM SURFACTANTS

A. Synthesis

The so-called bolaform surfactants are molecules with two hydrophilic and only one hydrophobic group. The literature describes many different types of this class

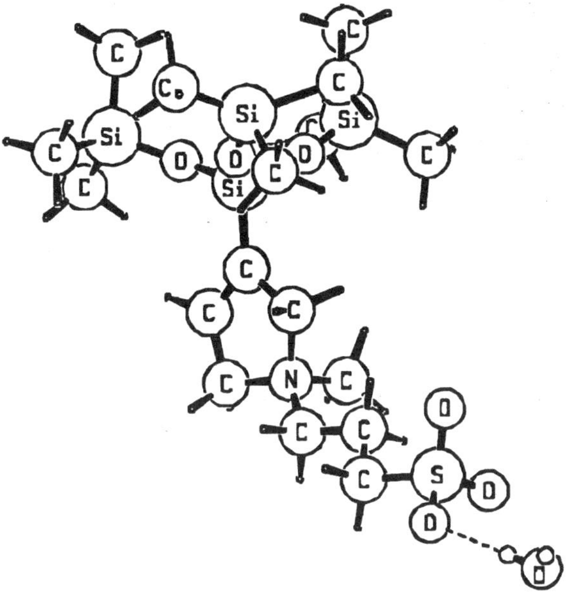

FIG. 16 Molecular structure of *N*-methyl-*N*-3-(sulfonato)propyl-3-[tris(trimethylsiloxanyl)silyl]pyrrolinium sulfobetaine, **14a.**

of surfactants. For example, Zara et al. [17] investigated quaternary ammonium salts of the general formula

$$C_nH_{2n}\text{-}\alpha,\omega\text{-bis}(C_{12}H_{25}N^+Me_2Br^-)$$

Nonionics [18,19], anionics [20], and cationics [13] were synthesized based on but-2-yne-1,4-diole. The general formula is given in Scheme 2.

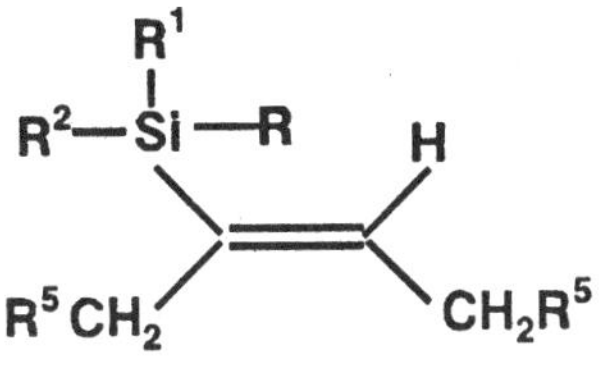

SCHEME 2 General formula of bolaform surfactants based on butynediole: R = Me, pOSiMe$_3$; R^1 = Me, OSiMe$_3$; R^2 = OSiMe$_3$; R^5 = $(OCH_2CH_2)_nOH$, OCOCH = CHCOOR^6, N^+Me_3X; R^6 = H, Na; X = halogen; n = degree of ethoxylation.

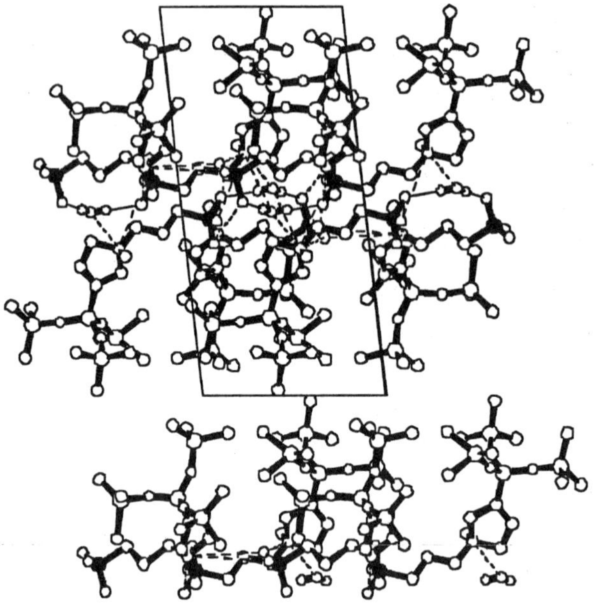

FIG. 17 Molecular packing of **14a** in the crystal lattice; the black circles are the sulfur atoms.

In general the synthesis is carried out in two steps—the hydrosilylation of the triple bond of butynediole and the functionalization of the OH groups. Sometimes additional steps are necessary (e.g., the use of protection groups).

1. 1,4-Bis(trimethylammonium)-2-[tris(trimethylsiloxanyl)silyl]but-2-ene Dichloride, **16**

This synthesis shall serve as example. Starting with butynediol the conversion into 1,4-dichlorobut-2-yne can be carried out by standard methods, for instance, via thionylchloride. The hydrosilylation with H-siloxanes [Eq. (3)] is followed by the reaction of the product with tertiary amines [Eq. (8)]:

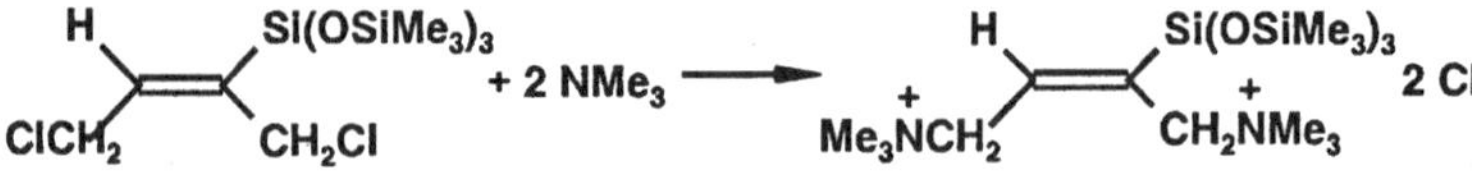

(8)

2. 3,3′-(1,1,3,3-Tetramethyldisiloxane-1,3-diyl)bis(*N*-ethyl-*N*,*N*-dimethyl-1-propanaminium) Diperchlorate, **17**

This compound represents another type of bolaform surfactant. From a 0.01 M solution of *N*-ethyl-*N*,*N*-dimethyl-3-(pentamethyldisiloxanyl)-1-propanaminium

TABLE 6 Interfacial Activity of *N,N*-Dialkyl-3-siloxanylpyrrolinium Derivatives

Compound	cmc (mol/dm^3)	Maximum surface tension depression (mN/m)	Adsorption energy (kJ/mol)	Surface activity (mol/cm^3)	Surface area per molecule (nm^2)
9a	1.1×10^{-2}	21.0	–22.5	4.22×10^{-10}	0.889
10a	1.7×10^{-2}	26.0	–5.82	4.07×10^{-7}	0.320
11a	2.7×10^{-3}	25.0	–8.33	1.45×10^{-7}	0.547
11b	2.8×10^{-2}	22.0	–8.71	1.24×10^{-7}	0.430
11c	4.6×10^{-3}	26.2	—	—	—
11f	1.0×10^{-2}	20.8	–12.70	2.37×10^{-8}	0.521
12a	7.4×10^{-3}	20.8	–16.90	4.36×10^{-9}	0.599
12b	6.0×10^{-3}	19.5	–23.2	3.21×10^{-10}	0.644
12c	No minimum	—	–18.3	2.37×10^{-9}	0.613
13a	7.5×10^{-3}	20.0	–20.3	1.07×10^{-9}	0.761
13b	No minimum	—	–15.0	9.20×10^{-9}	0.745
13c	2.8×10^{-3}	21.5	–19.4	1.53×10^{-9}	0.685
13d	6.2×10^{-3}	24.5	–7.9	1.73×10^{-7}	0.312
13e	4.6×10^{-3}	21.0	–19.3	1.62×10^{-9}	0.639
14a	1.1×10^{-4}	30.4	–19.7	1.33×10^{-9}	0.607
14b	8.0×10^{-4}	21.6	–22.1	5.10×10^{-10}	0.607
14c	1.4×10^{-4}	22.0	–17.7	3.06×10^{-9}	0.612

bromide (**3b**) in 0.5 mol/dm^3 of sodium perchlorate in the presence of a phosphate buffer to keep the pH value at 7, white needles spontaneously began to crystallize. These needles were isolated and recrystallized from acetone. The analytical characterization evidences the formula given in Scheme 3.

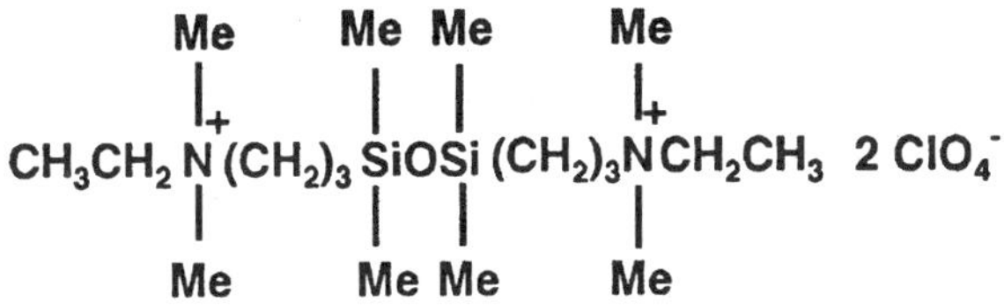

SCHEME 3 Formula of 3,3′-(1,1,3,3-tetramethyldisiloxane1,3-diyl)bis(*N*-ethyl-*N,N*-dimethyl-1-propanaminium) diperchlorate.

B. X-Ray Crystal Structures

1. 1,4-Bis(trimethylammonium)-2-[tris(trimethylsiloxanyl)silyl]but-2-ene Dichloride, **16**

The x-ray crystal structure was determined by means of CuK_{α} radiation at room temperature. The Si–O–Si angles range from 148.3 to 163.4°. The molecular structure is shown in Fig. 18.

The packing of the molecules in the crystal lattice is demonstrated in Fig. 19. The stabilization of the lattice is reached by N^{+}–water contacts additional to the N^{+}–Cl^{-} contacts. The required area per molecule in the crystal lattice was calculated to be 0.845 nm^2.

2. 3,3′-(1,1,3,3-Tetramethyldisiloxane-1,3-diyl)bis(*N*-ethyl-*N*,*N*-dimethyl-1-propanaminium) Diperchlorate, **17**

The x-ray crystal structure was determined at 293 and 163 K using MoK_{α} radiation. The Si–O–Si angle is relatively small (141.8–144.4°) and depends on the measuring temperature [21]. The molecular structure of **17** is given in Fig. 20. The contacts between N^{+} and ClO_4^{-} build up a three-dimensional net, as shown in Fig. 21.

C. Adsorption Behavior at the Air–Water Interface

The adsorption behavior of 1,4-bis(trimethylammonium)-2-(1,1,1,3,5,5-heptamethyltrisiloxan-3-yl)but-2-ene dichloride (**15**) and 1,4-bis(trimethylammo-

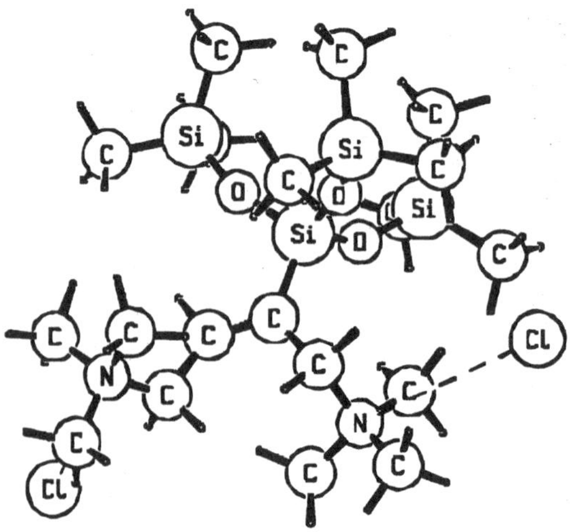

FIG. 18 Molecular structure of 1,4-bis(trimethylammonium)-2-[tris(trimethylsiloxanyl)silyl]but-2-ene dichloride, **16.**

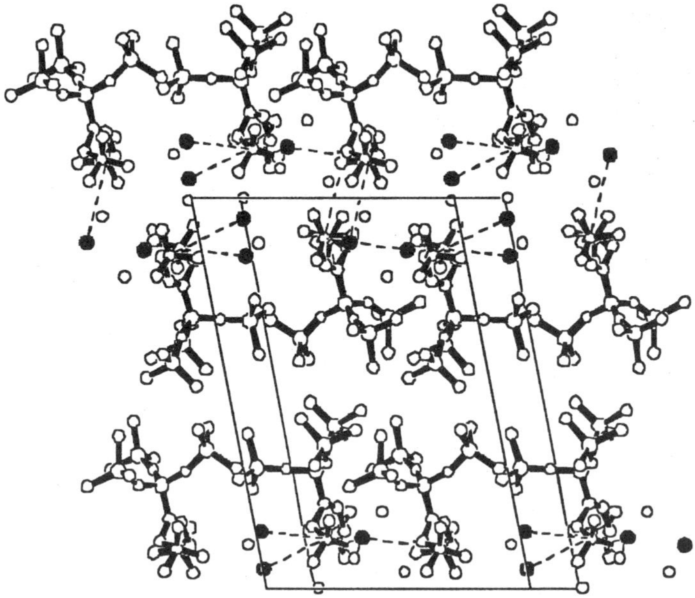

FIG. 19 Packing of **16** in the crystal lattice.

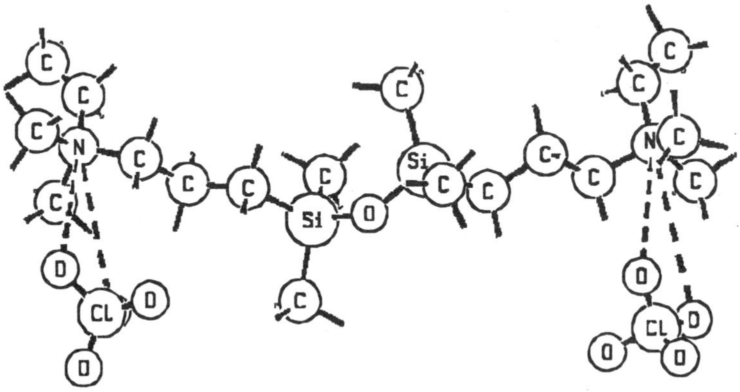

FIG. 20 Molecular structure of 3,3′-(1,1,3,3-tetramethyldisiloxane-1,3-diyl)bis(*N*-ethyl-*N*,*N*-dimethyl-1-propanaminium) diperchlorate, **17.**

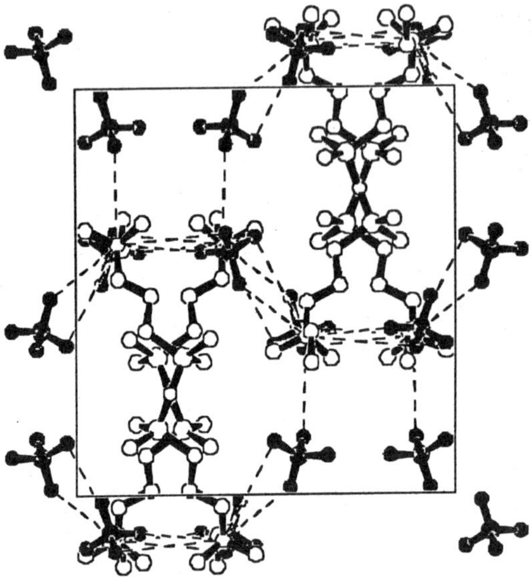

FIG. 21 Packing of **17** in the crystal lattice.

TABLE 7 Interfacial Activity of Bolaform Surfactants

Compound	cmc (mol/dm^3)	Maximum surface tension depression (mN/m)	Adsorption energy (kJ/mol)	Surface activity (mol/dm^3)	Surface area per molecule (nm^2)
15	9.5×10^{-3}	23.8	–9.75	8.08×10^{-8}	0.393
16	8.5×10^{-3}	24.3	–11.50	3.96×10^{-8}	0.416

nium)-2-[tris(trimethylsiloxanyl)silyl]but-2-ene dichloride (16) was determined and the Frumkin isotherm was calculated. The results are given in Table 7.

The data of **15** and **16** are nearly equal, although the siloxane groups are different and all results reported before showed that the surface activity is influenced significantly by the number of silicon atoms in the hydrophobic group. Comparison of the areas per molecule determined from the crystal structure and calculated from the adsorption isotherm shows that the value at the interface is smaller. One possible conclusion is that at the air–water interface the hydrophobic siloxane backbone prevents the second ammonium group from reaching the surface. On the other hand, the positive charge on the ammonium group minimizes the influence of the number of silicon atoms.

V. SILOXANYLPHOSPHINOXIDES

A. Synthesis

Siloxanylethyldiethylphosphinoxides were synthesized by addition of diethylphosphinoxide to vinylsiloxanes at 70°C in a chain reaction started with azobisisobutyronitrile [eq. (9)] [22,23]. The useful reaction time is about 2 h; longer times decreased the yield of the addition product in favor of polymerization products.

The corresponding vinylsiloxanes can be received by equilibration of chlorovinylsilanes [24], reaction of trimethylsilanole with vinylsilanes, or reaction of sodium trimethylsilanolate with chlorovinylsilanes.

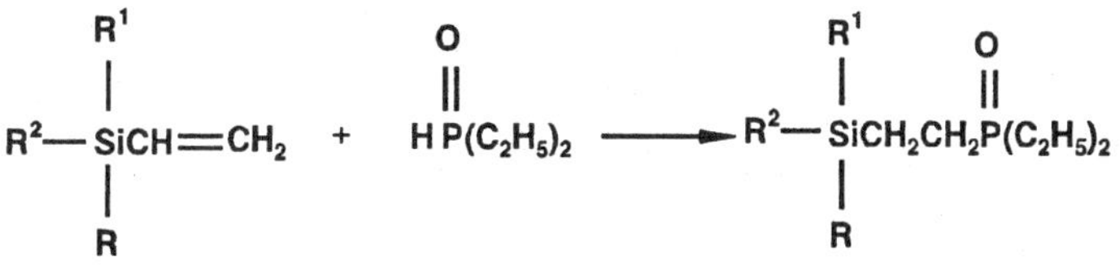
(9)

where R = Me, $OSiMe_3$; R^1 = Me, $OSiMe_3$; and R^2 = Me, $OSiMe_3$.

The following compounds were synthesized and characterized by elemental analysis and NMR spectroscopy:

$Me_3SiCH_2CH_2P(O)(C_2H_5)_2$
18

$Me_3SiOSiMe_2CH_2CH_2P(O)(C_2H_5)_2$
19

$(Me_3SiO)_2SiMeCH_2CH_2CH_2P(O)(C_2H_5)_2$
20

$(Me_3SiO)_3SiCH_2CH_2P(O)(C_2H_5)_2$
21

All these compounds are viscous colorless liquids with good solubility in water.

B. Adsorption Behavior at the Air–Water Interface

The adsorption behavior of compounds **18–21** at the air–water interface was characterized by means of the du Noüy ring method. The results are summarized in Table 8.

TABLE 8 Adsorption Behavior of Siloxanylethyldiethylphosphinoxides

Compound	Molecular weight (g/mol)	cmc (mol/dm^3)	Maximum surface tension depression (mN/m)
18	206	6.3×10^{-2}	26.6
19	280	7.0×10^{-4}	21.5
20	354.6	1.5×10^{-4}	24.5
21	428.7	4.2×10^{-5}	23.4

Comparing the results of Table 8 with literature data on alkylphosphinoxides [25], one sees that the siloxane surfactants possess nearly the same critical micelle concentration but decrease the surface tension of water by about 10 mN/m.

Especially remarkable is the hydrophobic effect of the trimethylsilyl group in compound **18**. It is a functionalized silane having the surface activity of a siloxane. This characteristic is difficult to explain because here are no flexible Si–O–Si angles.

VI. CARBOHYDRATE-MODIFIED SILOXANES

A. Synthetic Principles and Some Examples

An increasing number of recent publications, especially patents, describe carbohydrate-modified siloxanes [26,27]. The idea is to combine the extraordinary surface activity and wetting properties of siloxanes with the biodegradability of carbohydrates. Some years ago alkyl polyglycosides were first synthesized and now they are common surfactants for different applications. The combination of long alkyl chains with carbohydrate molecules is achieved by means of glycosyl bonds.

The connection of siloxanes and carbohydrates can be achieved by Si—C or Si—O—C bonds, with or without a spacer. Since the stability of carbohydrates is a problem under the conditions of hydrosilylation, it is more convenient to react the carbohydrates with molecules that have been hydrosilylated with siloxanes before. The compounds resulting from this approach include a spacer, which can influence the important properties.

The stability, and so the surface activity of carbohydrate-modified siloxanes, depends strongly on the pH value. Depending on the application field, it is possible to synthesize degradable surfactants that lose their surface activity when the pH value is changed in a desired range. The following examples give only a small view of this new but widespread field of functionalized siloxanes.

Nonionics—obtained, for instance, by combination of 2-siloxanyl-but-2-ene-1, 4-diol with one or two carbohydrate molecules, cationics, and anionics—were synthesized [28]. All these compounds are mixtures of stereoisomers.

The hydrosilylation of allyl glycidyl ether with H-siloxanes followed by the reaction of the product with the sodium salt of glucose maleate gives compound **22**, a nearly white solid with good solubility in water and some alcohols. The compound was characterized by NMR spectroscopy.

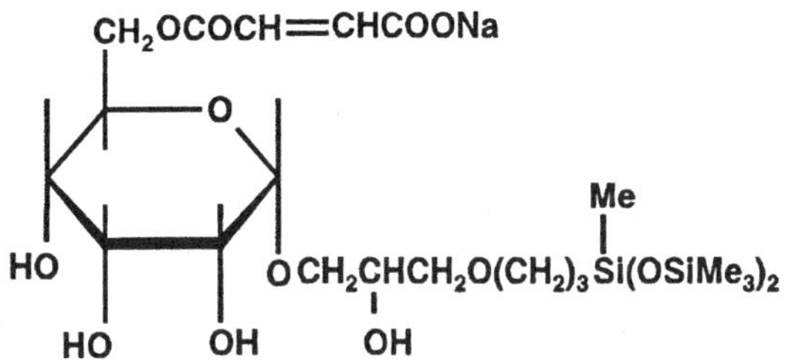

22

Using the same reaction path, compound **23** was synthesized, as well as a number of compounds with polymeric siloxane groups. The latter, however, were not soluble in water.

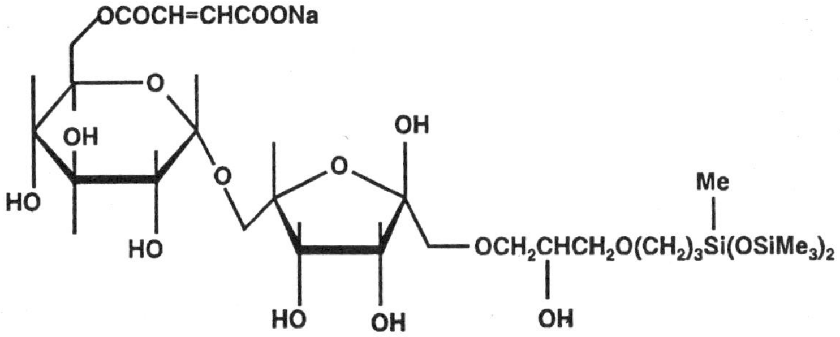

23

Sulfobetain compounds were synthesized by means of the reaction of siloxanylpropyl-2,3-epoxypropylether with glucamin or *N*-methylglucamine, respectively, followed by the addition of butanesultone. Some examples are given in Scheme 4. The biodegradability of compound **24,** as determined by the OECD

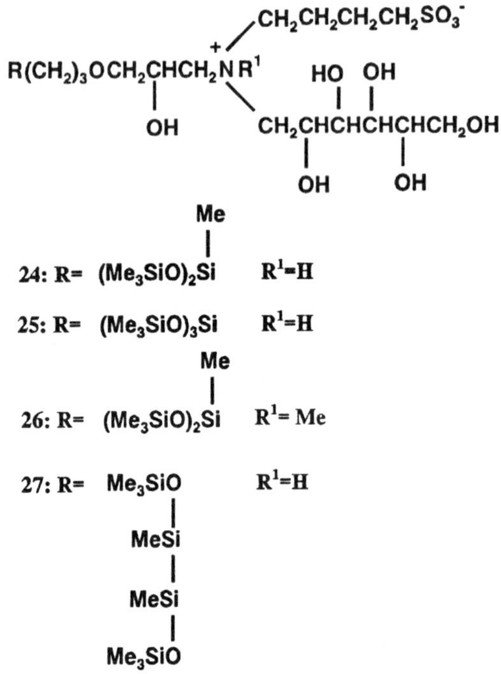

SCHEME 4 Siloxanyl-modified glucamine derivatives.

test, is 60% higher than of a common polydimethylsiloxane. This shows that the aim of the development of carbohydrate-modified siloxanes was reached, but the value of degradability is much lower than it is for hydrocarbon surfactants.

B. Adsorption Behavior at the Air–Water Interface

Surface tension measurements were carried out by means of the du Noüy ring method. The solubility of the compounds in water depends on the ratio of siloxane groups to carbohydrate molecules. Because of the complicated structure of the surfactant molecules, the arrangement at the interface is difficult to imagine.

The results of the measurements are given in Table 9. The very high surface activity of this type of siloxane is expressed by the low critical micelle concentration and the maximum surface tension depression. Experiments showed that all compounds are very good emulsifiers.

TABLE 9 Adsorption Behavior of Carbohydrate-Modified Siloxanes

Compound	cmc (mol/dm^3)	Maximum surface tension depression (mN/m)	Interfacial tension $c >$ cmc (mN/m)
22	2.45×10^{-4}	22.4	2.6
23	1.35×10^{-5}	20.7	0
24	2.80×10^{-3}	19.8	0
25	5.20×10^{-5}	19.4	0
26	4.10×10^{-4}	19.6	—
27	6.50×10^{-5}	23.4	0.6

VII. SUMMARY AND FURTHER WORK

The investigation of crystalline siloxane surfactants offers the opportunity to compare x-ray crystal structure data with surface activity measurements. In some cases it is possible to make proposals for the arrangement of the surfactant molecules at the interface. However the high flexibility of the siloxane groups allows the molecules to enter into arrangements at the interface that are not possible in the crystal lattice for different reasons.

The synthesis of crystalline siloxane surfactants with other functional groups and their investigation should give validate the present ideas for the siloxane arrangement at the interface.

Another interesting question is the origin of the surface activity of silicone compounds. The flexibility of the siloxane groups, expressed by the wide variation in the range of the Si–O–Si angle, is well known but carbosilanes in which the oxygen atom is replaced by a CH_2 group also possess a remarkable surface activity [29]. The synthesis and investigation of surface active silanes could contribute to answering this question if compounds stable enough to carry out surface activity measurements can be prepared.

The modification of siloxanes with carbohydrate moieties led to a new class of surface active silicone compounds with improved biodegradability. This advance can expand the application of siloxane surfactants into new fields.

ACKNOWLEDGMENTS

I thank all colleagues who contributed to this work, especially Rainer Rudert and U. Retter, and the Deutsche Forschungsgemeinschaft, the Max-Planck-Gesellschaft, and the Bayer AG for financial support.

All x-ray crystal structures were presented with the aid of SCHAKAL92, a FORTRAN program for the graphic representation of molecular and crystallographic models, by E. Keller, of the University of Freiburg, Freiburg, Germany.

REFERENCES

1. H. Maki, T. Suga, I. Ikeda, and S. Komori, *Yukagaku 19*: 245 (1970); H. Maki, Y. Horiguchi, T. Suga, and S. Komori, *Yukagaku 19*: 1029 (1970).
2. J. L. Speier, C. A. Roth, and J. W. Ryan, *J. Org. Chem. 36*: 3120 (1971).
3. J. C. Suam, and J. L. Speier, *J. Am. Chem. Soc. 80*: 4104 (1958).
4. V. B. Pukhnarevich, L. I. Kopylova, B. A. Trofimov, and M. G. Voronkov, *Zh. Obshch. Khim. 45*: 89, 2638 (1975).
5. K. Takatsuna, M. Nakajima, M. Tachikawa, A. Shinohara, K. Shinozawa, and Y. Okumura European Patent *321*, 174 (1989).
6. G. Schmaucks, G. Sonnek, R. Wüstneck, M. Herbst, and M. Ramm, *Langmuir 8*: 1724 (1992).
7. M. Ramm, B. Schulz, G. Sonnek, and G. Schmaucks, *Cryst. Res. Technol. 25*: 763 (1990).
8. R. Rudert and G. Schmaucks, *Acta Crystollogr. C 50*: 631 (1994).
9. See, for instance: A. Messier, G. Schorsch, J. Rouviere, and L. Tenebre, *Prog. Colloid Polym. Sci. 79*: 249 (1989).
10. A. N. Frumkin, *Z. Phys. Chem. 116*: 466 (1925).
11. U. Retter, R. Klinger, G. Sonnek, and G. Schmaucks, *J. Colloid Interface Sci. 156*: 85 (1993).
12. G. Sonnek, E. Drahs, H. Jancke, and H. Hammann, *J. Organomet. Chem. 386*: 29 (1990).
13. G. Schmaucks, *J. Prakt. Chem. 336*: 514 (1994).
14. S. A. Snow, W. N. Fenton, and M. J. Owen, *Langmuir 6*: 385 (1990).
15. S. A. Snow, W. N. Fenton, and M. J. Owen, *Langmuir 7*: 868 (1991).
16. B. Grüning and G. Koerner, *Tenside Surf. Deterg, 26*: 5 (1989).
17. E. Alami, G. Beinert, P. Marie, and R. Zara, *Langmuir 9*: 1465 (1993).
18. G. Sonnek, C. Raab, G. Müller, and H. Hamann, GDR Patent *255* 737 (1986).
19. R. Wersig, G. Sonnek, and C. Niemann, *Appl. Organomet. Chem. 6*: 701 (1992).
20. G. Sonnek and E. Drahs, GDR Patent *255* 346 (1986).
21. R. Rudert, G. Schmaucks, D. Zobel, and M. Strumpel, *Acta Crystallogr., C 51*: 763 (1995).
22. G. Schmaucks, G. Sonnek, M. Herbst, A. A. Zhdanov N. A. Kurasheva, L. I. Kutejnikova, and T. W. Strelkova, in *Gaussig-Report* Nünchritz, 1990, pp. 50–52.
23. A. A. Zhdanov, N. A. Kurasheva, L. I. Kutejnikova, and Le Ngog Khan, *Isv. Akad. Nauk SSSR, Ser. Khim. 1*: 183 (1985).
24. K. A. Andrianov, N. M. Petrovnina, T. V. Vasileva, V. E. Shklovev, and B. I. Dyachenko, *Zh. Obshch. Khim. 48*: 2692 (1978).
25. K. Lunkenheimer, K. Haage, and R. Miller, *Colloids Surf. 22*: 215 (1987).
26. G. Jonas and R. Stadler, *Acta Polym. 45*: 14 (1994).

27. J. Sejpka and F. Wimmer, *European Patent* 0612 759 A1, to Wacker Chemie GmbH (1994).
28. R. Wagner, R. Wersig, G. Schmaucks, B. Weiland, L. Richter, A. Hennig, A. Jänicke, J. Reiners, W. Krämer, J. Weissmüller, and W. Wirth, German Patents 4318, 536 A1, 4318, 537 A1, 4318, 539 A1 to Bayer AG (1994).
29. A. R. L. Colas, F. A. D. Renauld, and G. C. Sawicki, British Patent 2,203, 152 A (1988); U.S. Patent 5,026, 891 (1991) to Dow Corning Ltd.

4

Surface Activity and Aggregation Behavior of Siloxane Surfactants

H. HOFFMANN and W. ULBRICHT Physical Chemistry I, University of Bayreuth, Bayreuth, Germany

I. INTRODUCTION

A. Definition and Nomenclature of Siloxane Surfactants

Siloxane surfactants consist of normal surfactants having a hydrophilic and a hydrophobic group. The hydrophobic group is a siloxane backbone with a various number of dimethylsiloxane units and usually end-capped with a trimethylsiloxane group at each end. The hydrophilic groups can be ionic, zwitterionic, or nonionic and are usually attached by a short alkyl chain to the siloxane backbone. The constitution of the siloxane surfactants permits them to have a large variety of different structures. The number of dimethylsiloxane units can vary from 1 to a higher number. The number of siloxane groups determines the length of the hydrophobic group, hence the hydrophobicity of the whole surfactant. The hydrophilic unit can contain all the hydrophilic groups that are used in normal surfactants; that is, it can be cationic or anionic with counterions of all types, zwitterionic or nonionic. The hydrophilic moiety is attached to the backbone by an alkyl chain with different numbers of CH_2 groups, and the hydrophilic groups can be attached to the dimethylsiloxane units at any position and/or to the terminal siloxane units. Finally the hydrophilic groups can be replaced by purely hydrophobic alkyl or perfluoroalkyl groups, which leads to completely hydrophobic siloxane surfactants that show surface activity in nonpolar solvents.

From this large variety of possible structures it is obvious that many more siloxane surfactants with different constitutions and properties can exist in comparison to normal surfactants. It is therefore useful to introduce a nomenclature capable of characterizing the siloxane surfactants in an unambiguous way. According to a proposal of Noll [1], the hydrophilic groups are denoted with the symbol R and their structure is given in the usual way. The trimethylsiloxane units with one O atom to which other siloxane groups are attached are represented by the symbol M, while the dimethylsiloxane units are symbolized by D. Siloxane units with attached hydrophilic groups are characterized by the symbols D′(R) or M′(R). Figure 1 gives examples of the nomenclature of the siloxane surfactants.

From the large variety of siloxane surfactants mainly three types have been studied and used for various applications:

1. *Trimethylsiloxane surfactants.* Members of this class of siloxane surfactants have the general structure M—D′(R)—M; that is, they consist of the two

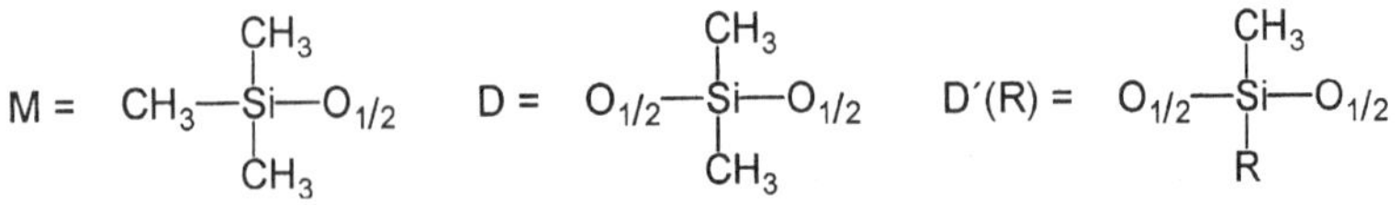

FIG. 1 Schematic graph of the nomenclature of the siloxane surfactants.

terminal M units and one D unit to which the hydrophilic group R is attached. The hydrophilic group R can be anionic (mostly sulfate with alkali counterions), cationic (mostly alkylated ammonium with various monovalent counterions), zwitterionic (betaine or sulfobetaine), or nonionic (in most cases polyglycol ethers with different numbers of EO units); usually it is attached to the silicon atom through a—$(CH_2)_3$—group, but there may also be hydrophilic—CHOH—or ether groups.

2. *Graft- or comblike polymers.* Members of this class of siloxane surfactants have the general structure M—D_x—D'_y(R)—M; that is, they consist of two terminal M units and a various number of D units to some of which R groups are attached.
3. *ABA triblock copolymers.* These siloxane surfactants have the general structure M′(R)—D_x—M′(R); that is, they consist of two terminal blocks R, which are attached to the M groups of the middle block, which is a siloxane backbone with various numbers of D units. The R blocks can be hydrophilic groups as in the case of the other two types of siloxane surfactant; they can, however, also be hydrophobic alkyl or perfluoroalkyl chains of various lengths.

Regarding this large variety of possible structures, the formula of most siloxane surfactants represents an average composition, and each sample consists of a mixture of many isomers. This is especially the case for the polymers, where the isomers can differ in the length of the siloxane backbone, the numbers and positions of the hydrophilic side groups, and for nonionic samples, also in the number of EO groups per side chain. On the other hand, the trisiloxane surfactants other than the nonionic ones are usually rather well-defined compounds. This situation must be kept in mind if literature data (e.g., surface tensions, cmc values, sizes and shapes of aggregates or phase diagrams) for a specific siloxane surfactant in solution are used or compared.

B. Properties of the Siloxane Surfactants

The hydrophilicity or the hydrophobicity of siloxane surfactants can be adjusted in very small steps by changing the length of the siloxane backbone, as well as the number and/or the constitution of the hydrophilic side chain.

Siloxane surfactants usually have a high thermal stability and a strong surface activity. Furthermore they are nontoxic and do not irritate the skin [2]. They are thus suitable for pharmaceutical and cosmetic applications. The biggest disadvantage of siloxane surfactants, however, is the ease of hydrolysis in aqueous solutions. Trisiloxane compounds can be completely hydrolyzed in water after they have been allowed to stand for a few hours. This hydrolysis is especially fast in acid or basic solutions, while the samples are stable for a few days at neutral pH values [2–4]. Such hydrolysis leads to the formation of hydrated SiO_2 and silicone

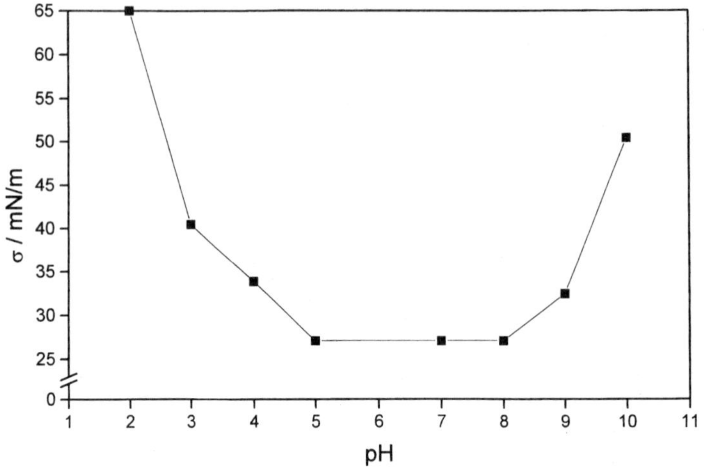

FIG. 2 The equilibrium surface tension of a siloxane surfactant (M—D′(R)—M with R = —$(CH_2)_3$—$N^+(CH_3)_2$—$(CH_2)_3$—SO_3^-; 0.1 wt%) in aqueous solutions as a function of pH (measured after solutions had stood for 30 min).

oils like hexamethyldisiloxane, and the samples lose their surface activity. The hydrolysis can thus be easily monitored by measuring the surface tension of the solutions with time after the preparation of the samples (Fig. 2). Samples with longer siloxane backbones as a whole are much more stable; longer graft copolymers are stable in neutral aqueous solutions for years [2,5,6]. This increased stability is explained by the decreasing water solubility and lower cmc values with increasing length of the siloxane backbone. Hydrolysis seems to take place preferentially in unassociated siloxane surfactant molecules [2,7,8]. A stabilizing effect can also come from the hydrophilic side chains, which can shield the folded siloxane backbone from the surrounding water [9], but the reason for the stability of the comblike polymers to hydrolysis is not completely clear at present.

II. THE SURFACE ACTIVITY OF SILOXANE SURFACTANTS

A. Adsorption at the Surface of Water and Other Polar Solvents Against Air

1. General Principles

If a surfactant is dissolved in a strongly polar solvent like water, the hydrophilic group is solvated by the water molecules, while the hydrophobic group is pushed out of the water. This hydrophobic effect [10] is the reason for a strong adsorption of surfactant molecules at the water surface and for the aggregation of surfactants to micelles; in both cases the hydrophilic groups remain solvated and in contact with the water phase, while the hydrophobic groups are either in the air or in the micellar interior, where they are more or less effectively shielded from unfavorable contact with the water molecules.

The adsorption of any molecule at a surface leads to a decrease of the surface tension. The quantitative relation between the amount of adsorption and the surface tension is given by the Gibbs adsorption isotherm [11].

$$-d\sigma = \Gamma\, RT\, d \ln c$$

where σ is the surface tension of the solution, Γ the surface concentration of adsorbed molecules, and c the concentration of the adsorbed sample in the bulk phase. Normally Γ, hence $d\sigma$, are small, but for surfactants in aqueous solutions, the hydrophobic effect produces a very strong adsorption of surfactant molecules at the water surface, and this enrichment of surfactant molecules leads to a corresponding strong reduction of the surface tension. According to the Gibbs equation, the adsorption increases with increasing bulk concentration of the surfactant, until a rather densely packed film of surfactant molecules is formed at the surface and the cmc of the surfactant is reached. At this concentration all further added surfactant molecules aggregate to micelles that are surface inactive; hence the surface tension remains constant above the cmc.

To characterize the surface activity of surfactants, Rosen [12,13] introduced the expressions "efficient" and "effective." The efficiency gives the surfactant concentration that is used for a defined reduction of the surface tension, while the effectiveness expresses the maximum reduction of the surface tension that can be obtained with a surfactant. Since the hydrophobic effect is the reason for the surface activity, it can be easily understood that the efficiency of a surfactant is mainly determined by the effective size of the hydrophobic group (i.e., the number of hydrophobic units in contact with water) and to a minor extent by the nature of the hydrophilic group. For example, nonionic surfactants with long *n*-alkyl chains are very efficient, while ionic surfactants having the same number of CH_2 groups but branched alkyl chains or also surfactants having shorter alkyl chains are much less efficient. The effectiveness of a surfactant cannot be estimated as easily from its structure. It is rather independent of the size of the hydrophobic group; it increases with decreasing polarity of the hydrophilic group and with increasing packing density of the adsorbed film, and it increases strongly with decreasing cohesion forces between the hydrophobic groups.

2. The Effectiveness of Siloxane Surfactants

Close to the cmc, siloxane surfactants in aqueous solutions behave much like normal hydrocarbon surfactants, as can be seen in Fig. 3 [14–16]. The surface tension decreases linearly with the logarithm of the surfactant concentration, and a densely packed film exists in this region at the surface. Above the cmc the surface

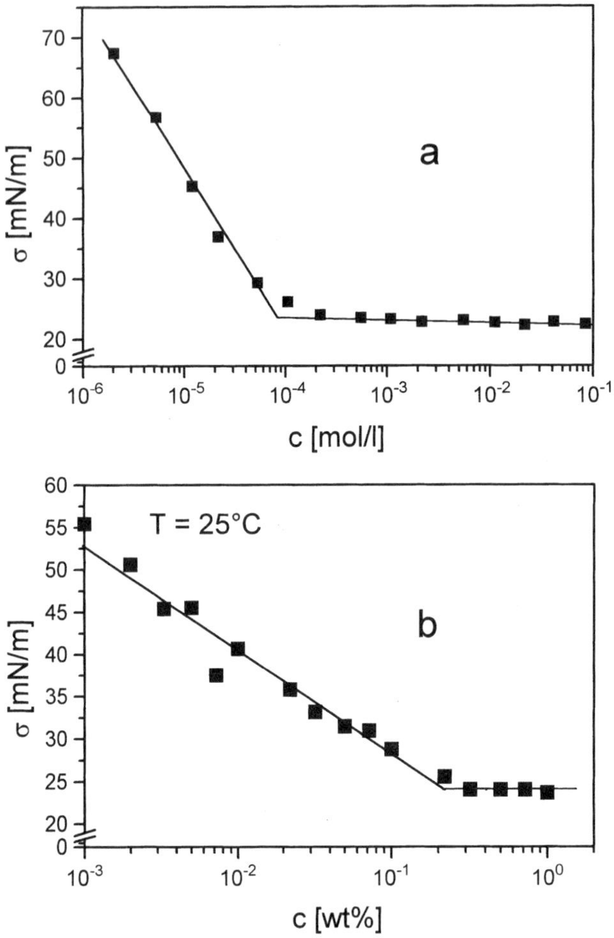

FIG. 3 The surface tension σ at 25°C for aqueous solutions of (a) M—D′ (R)—M with R = —$(CH_3)_2$—$(EO)_4$—OH and (b) M—$D_{13}D'_5$(R)—M with R = —$(CH_2)_3$—O—SO_3^- H_3N^+—$CH(CH_3)_2$ as a function of the logarithm of the concentration *c*.

tension remains constant. Similar curves are obtained for many other siloxane surfactants [3–6,17]. For the graft copolymers, whose molecules have a larger surface area, the slope of the curves is much smaller than for trisiloxane surfactants (see, e.g., Fig. 3b) [18]. Often the cmc does not show up as a sharp break in the curves, but as a more or less broad transition region; this effect is mainly due to the presence of impurities and numerous isomeric and homologous surfactant molecules. The surface tension curves usually do not show a minimum at the cmc in spite of the presence of surface active impurities, as has been found for hydrocarbon surfactants [19].

Figure 3 shows that siloxane surfactants are very effective surfactants [3–6,20]. While aqueous solutions of hydrocarbon surfactants above the cmc usually have surface tensions between 30 and 40 mN/m [21], the siloxane surfactant solutions have values between 20 and 30 mN/m. Only perfluoro surfactants have lower values: between 15 and 20 mN/m [22]. This can be explained by the cohesion forces between the hydrophobic groups of the different surfactants, which decrease from hydrocarbons to silicone oils to perfluoroalkanes. The cohesion forces are reflected in the surface tensions of the compounds against air. They are 21.7 mN/m for *n*-octane, 17.0 mN/m for octamethyltrisiloxane, and 14.3 mN/m for *n*-perfluorononane [23,24]. Siloxane surfactants can cover the surface very densely. It is assumed that the siloxane backbone lies flat on the surface, with the methyl groups exposed to the surrounding gas phase [3,4], as in Fig. 4. Since the siloxane chain is rather flexible, a very dense packing of the methyl groups at the surface can be achieved, and this dense packing, together with the low cohesion forces between siloxanes, results in the low surface tension values of the siloxane surfactant solutions.

As has been found for hydrocarbon surfactants [21], the effectiveness of siloxane surfactants is also influenced to a minor extent by the nature of the hydrophilic groups [3,4,6,18,25]. Most effective are nonionic and zwitterionic siloxane surfactants; with such samples minimum values for the surface tension around 20 mN/m can be obtained. Ionic siloxane surfactants are generally less effective, but these compounds also have surface tension values around 30 mN/m. Graft copolymers with a long siloxane backbone are less effective than trisiloxane surfactants [6,18]; this is certainly because these samples contain more hydrophilic side chains, a characteristic that generally lowers their effectiveness. The influence of electrolytes on the effectiveness of the siloxane surfactants is complicated; it can be either enhanced or reduced by increasing ionic strength [18].

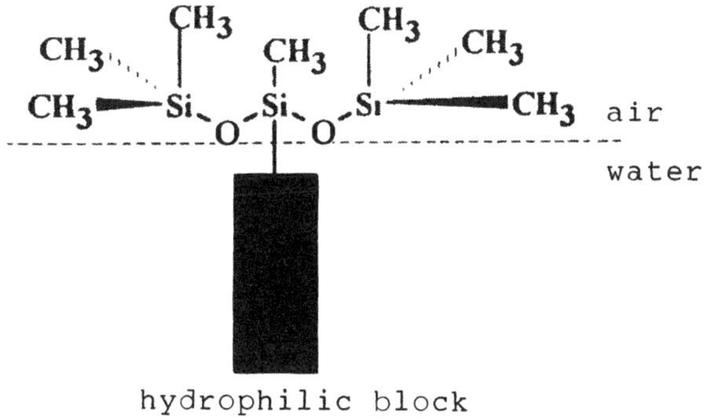

FIG. 4 Schematic drawing of the configuration of a trisiloxane surfactant at a phase boundary between water and air.

3. The Efficiency of Siloxane Surfactants

It is much more difficult to establish a correlation between the efficiency of siloxane surfactants and their structure. The principles for hydrocarbon surfactants are also valid. That is, the efficiency increases with increasing size of the hydrophobic siloxane backbone and decreases with increasing number and polarity of the hydrophilic groups. A quantitative relation like Traube's rule [26] has not been found for the siloxane surfactants. This limitation is partially due to the impurities, isomers, and homologs that are normally present in the available samples, but it also reflects the flexibility of the siloxane backbones, which allows folding or coiling of longer siloxane chains. Thus part of the siloxane units is shielded from contact with water, and such surfactants are much less efficient than the length of the hydrophobic chain would lead one to expect.

4. Experimental Results

Quantitative measurements of the surface activity of siloxane surfactants as a function of their structure have been described by Snow, Fenton, and Owen [3] for defined zwitterionic (sulfobetaine) short chain siloxane surfactants with the general formula M—D_x—$D'_y(R_1)$—M with R_1 = —$(CH_2)_3$—$N^+(R')_2$—$(CH_2)_z$—SO_3^-, where $x = 0$, 1, 2, or 3, $y = 1$ or 2, R′ = —CH_3, —CH_2—CH_3, or —CH_2—CH_2—CH_2OH, and $z = 3$ or 4. The results are shown in Table 1 and in Fig. 5 for several selected samples. All samples with one hydrophilic group are very effective and lower the surface tension above the cmc to values around 21 mN/m; this value is comparable to the surface tension of silicone oils against air. The efficiency of these surfactants is mainly determined by the length of the siloxane backbone; from Table 1 and Fig. 5 a relation similar to the Traube rule can be

TABLE 1 Values for the Surface Tension σ Against Air at 0.001, 0.01, and 0.1 wt% ($\sigma_{0.001}$, $\sigma_{0.01}$, $\sigma_{0.1}$), the cmc, and the Surface Area A[a]

				Surface tension (mN/m)			cmc		A
x	R′	R″	z	$\sigma_{0.001}$	$\sigma_{0.01}$	$\sigma_{0.1}$	wt%	mmol/L	(Å²/molecule)
0	CH_3	CH_3	3	52.5	40.5	28.2	0.3	7.0	75
0	CH_3	CH_3	4	53.1	41.2	28.0	0.4	9.0	75
0	C_2H_5	C_2H_5	3	53.6	39.7	23.8	0.2	4.4	65
0	CH_3	$(CH_2)_2OH$	3	58.9	48.7	33.9	1.1	24.0	75
1	CH_3	CH_3	3	38.3	25.7	21.1	0.03	0.6	70
2	CH_3	CH_3	3	23.0	22.0	21.2	0.08	1.4	>100
3	CH_3	CH_3	3	22.2	21.8	21.1	0.005	0.008	>100

[a]For different zwitterionic sulfobetaine siloxane surfactants with the general formula M—D_x—$D'(R_1)$—M with R_1 = —$(CH_2)_3$—$N^+(R'R'')$—$(CH_2)_z$—SO_3^- at 25°C ($\sigma_{cmc} \approx 21$ mN/m).

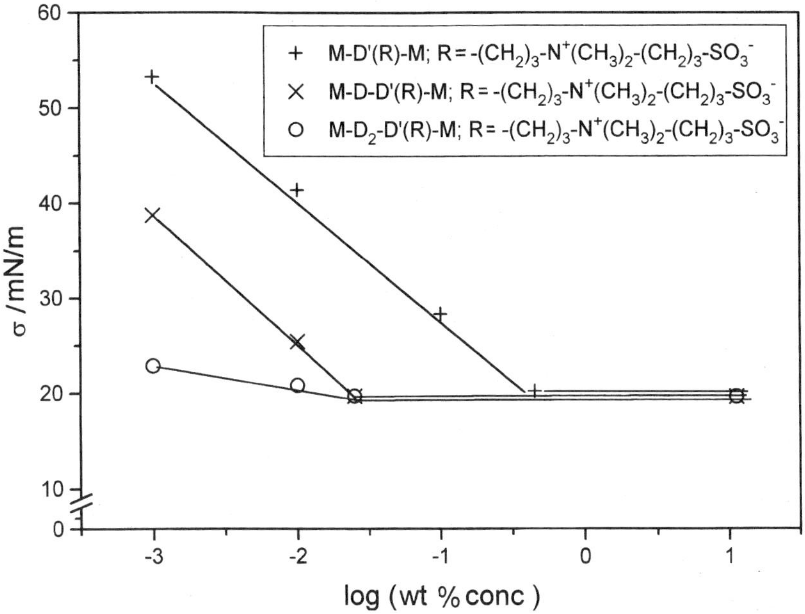

FIG. 5 The surface tension σ at 25°C for aqueous solutions of three zwitterionic sulfobetaine siloxane surfactants with different lengths of the siloxane backbone as a function of the logarithm of the concentration *c*.

assumed for short chain siloxane surfactants. Samples with two hydrophilic groups are less effective and efficient; unfortunately no quantitative data on these samples are given in the paper. These results agree with results [6] on commercially available siloxane surfactants.

In a second paper the same authors describe investigations on the surface activity of similar zwitterionic (betaine) short chain siloxane surfactants with the general formula $M—D_x—D'_y(R_1)—M$ with $R_1 = —(CH_2)_3—N^+(R')_2—(CH_2)_z—CO_2^-$, where $x = 0$, 1, or 2, $y = 1$, $R' = —CH_3$, and $z = 1$, 2, or 3 [4]. The results are represented in Table 2 and Fig. 6; it can be seen that neither the effectiveness nor the efficiency is influenced by replacing the sulfobetaine by betaine groups or by changing the value of *z*. Figure 6 shows furthermore that all betaine solutions show minimum surface tensions around 21 mN/m independent of the length of the siloxane backbone, but the efficiency increases again with the number of siloxane units in a pattern similar to that described by the Traube rule. All these short chain siloxane surfactants are very sensitive to hydrolysis in aqueous solutions; at room temperature and neutral pH they are stable for a few days, but they hydrolyze within minutes at acid or basic pH values and within hours at temperatures around 100°C.

TABLE 2 Values for the Surface Tension σ Against Air at 0.001, 0.01, 0.1, 0.5, and 1 wt% ($\sigma_{0.001}$, $\sigma_{0.01}$, $\sigma_{0.1}$, $\sigma_{0.5}$, $\sigma_{1.0}$), the cmc, and the Surface Area A[a]

		Surface tension (mN/m)						
x	z	$\sigma_{0.001}$	$\sigma_{0.01}$	$\sigma_{0.1}$	$\sigma_{0.5}$	$\sigma_{1.0}$	cmc (wt%)	A (Å²/molecule)
0	1	57.5	44.9	31.0	21.5	21.2	0.6	70
0	2	54.4	43.6	28.5	21.8	21.0	0.6	85
1	1	45.4	26.7	22.0	21.4	21.2	0.08	50

[a]For different zwitterionic betaine siloxane surfactants with the general formula M—D_x—D′(R_1)—M with R_1 = —$(CH_2)_3$—$N^+(CH_3)_2$—$(CH_2)_z$—CO_2^- at 25°C.

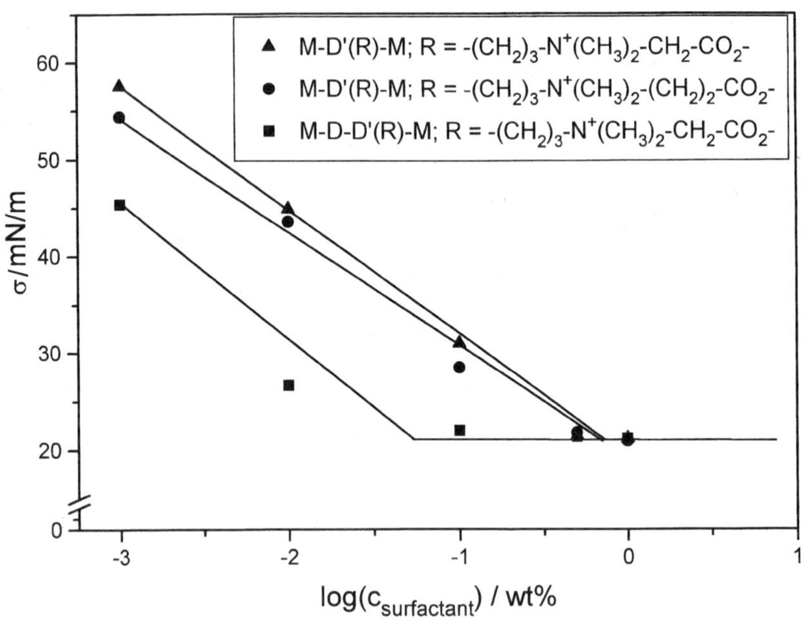

FIG. 6 The surface tension σ at 25°C for aqueous solutions of three zwitterionic betaine siloxane surfactants with different lengths either of the spacer between the ionic moieties of the betaines or of the siloxane backbone as a function of the logarithm of the concentration c.

A paper of Gentle and Snow [25] reports the surface activity of nonionic (polyoxyethylene) trisiloxane surfactants with various lengths of the EO chains. The surfactants have the general formula M—D′(R)—M with R = —$(CH_3)_2$—$(EO)_x$—OH, where x increases in steps of 2 EO units from 4 to 20. The results are given in Table 3, which shows that the surfactants are also very effective, although both effectiveness and efficiency decrease with increasing length of the EO chain. The trisiloxane surfactants thus behave in the same way as hydrocarbon surfactants. The surface area per surfactant molecule at the cmc is around 35 $Å^2$ per molecule up to *x* values of 16. For higher x values, the area per adsorbed molecule increases drastically by a factor of almost 2; this behavior is explained by the assumption that such long EO chains are not completely located in the bulk water but penetrate partially into the adsorbed surfactant film, which prevents the formation of a tightly packed film. It can be stated, however, that these siloxane surfactants are still more effective than hydrocarbon surfactants.

The surface activity of nonionic fluorinated short chain siloxane surfactants with the general formula $M'(R_F)$—D′(R′)—$M'(R_F)$, $M'(R_F)$—M′(R′), or $(R_F)_3M'(R')$ with R_F = CF_3 —$(CH_2)_2$— or CF_3—$(CF_2)_3$—$(CH_2)_2$—and R′ = —$(CH_2)_3$—$(EO)_x$—R with $x = 4, 7$, or 12, and R = —OH, —OCH_3, or CH_3CO_2— in aqueous solutions was studied by Kobayashi and Owen [27]. This class of siloxane surfactants is usually much more stable with respect to hydrolysis than other trisiloxane surfactants. The results, listed in Table 4, show that all samples are more effective than hydrocarbon surfactants, but the samples with the longer fluorinated alkyl chains have surface tensions values slightly below 21 mN/m and thus are similar in effectiveness to unfluorinated trisiloxane samples, while surfactants with a shorter fluoroalkyl chain are even significantly less effective. The effectiveness of pure perfluoro surfactants was not studied. The surfactants with

TABLE 3 Values for the Surface Tension σ Against Air at the cmc (σ_{cmc}), the cmc, the Surface Concentration Γ, and the Surface Area *A*[a]

x	σ_{cmc}	cmc (mmol/L)	Γ (× 10^{10} mol/cm^2)	*A* ($Å^2$/molecule)
4	22.6	0.079	5.0	33.5
8	22.8	0.1	5.1	32.6
10	23.2	0.16	4.4	38.2
12	24.9	0.63	4.2	39.2
14	24.5	0.63	4.7	35.3
16	25.8	1.0	4.2	39.3
18	26.9	3.2	2.6	64.6
20	27.8	1.6	2.3	72.2

[a]For nonionic siloxane surfactants with the general formula M—D′(R)—M with R = —$(CH_3)_2$—$(EO)_x$—OH at 25°C.

TABLE 4 Values for the Surface Tension σ Against Air at 0.001, 0.01, 0.1, and 1 wt% at the cmc ($\sigma_{0.001}$, $\sigma_{0.01}$, $\sigma_{0.1}$, $\sigma_{1.0}$, σ_{cmc}), the cmc and the surface area A[a]

								Surface tension (mN/m)					cmc	A
a	b	x	y	z	r	R	R″	$\sigma_{0.001}$	$\sigma_{0.01}$	$\sigma_{0.1}$		$\sigma_{1.0}$	(wt%)	(Å^2/molecule)
1	1	7	1	0	CH_3	OH	R_F	50.8	28.8	27.0	23.5	29	0.01	40
1	1	12	1	0	CH_3	OH	R_F	60.5	30.7	26.5	25.9	28	0.01	30
1	1	7	1	0	R_F	OH	R_F	46.6	46.0	31.0	28.3	31	0.1	50
1	1	12	1	0	R_F	OH	R_F	62.5	43.2	26.5	24.1	27	0.1	60
3	0	7	0	0	—	CH_3CO_2	—	46.6	36.6	28.1	26.3	28	0.1	100
1	1	7	1	3	CH_3	OCH_3	R_F	43.8	28.5	21.9	20.7	22	0.03	60
1	1	12	1	3	CH_3	OCH_3	R_F	54.7	41.8	27.9	20.0	20	0.3	70
1	1	4	0	3	—	OH	R′	41.8	29.9	22.1	20.5	22	0.1	100
1	1	7	0	3	—	OCH_3	R′	45.5	21.0	20.6	20.3	21	0.01	40
0	1	7	1	—	CH_3	OH	CH_3	39	22	21	21	21	0.01	60[b]
0	1	12	1	—	CH_3	OH	CH_3	36	22	21	21	21	0.01	70[b]

[a]For different fluorinated siloxane surfactants with the general formula $M'(R_F)_a$—D'_y (r, R′)—$M'_b(R'')$ with $R_F = CF_3$—$(CF_2)_z$—$(CH_2)_2$—, R′ = —$(CH_2)_3$—$(EO)_x$—R at 25°C.

[b]Nonfluorinated trisiloxane surfactant M—D′(R′)—M.

longer fluoroalkyl chains are also generally more efficient than the ones with shorter chains, but no strict correlation of efficiency with fluoroalkyl chain number can be derived from Table 4.

Similar studies of graft copolymers with long siloxane backbones are reported by Hoffmann et al. [6]. These samples are not available as pure compounds with a well-defined structure, but the data listed in Table 5 confirm nevertheless the conclusions that were drawn for the short chain surfactants. The surface tension

TABLE 5 Values for the Surface Tension σ at the cmc (σ_{cmc}), the Interfacial Tension γ Against *n*-Decane at the cmc (γ_{cmc}), the cmc, and the Dissociation Degree α of the Aggregates for Different Siloxane Surfactants at 25°C

Siloxane surfactant	γ_{cmc} (mN/m)	σ_{cmc} (mN/m)	cmc (wt%)	cmc (mmol/L)	α
M—D′(R)—M	—	—	0.35	8.0	0.7
R = —$(CH_2)_3$—O—CH_2—CHOH—CH_2—$SO_3^-Na^+$					
M—D′(R)—M	—	—	0.27	6.0	0.64
R = —$(CH_2)_3$—O—CH_2—CHOH—CH_2—$N^+(CH_3)_3CH_3CO_2^-$					
M—D′(R)—M	0.4	22.0	0.48	11.5	0.4
R = —$(CH_2)_3$—O—$SO_3^-H_3N^+$—$CH(CH_3)_2$					
M—D′(R)—M	—	—	0.03	0.7	—
R = —$(CH_2)_3$—$N^+(CH_3)_2$—$(CH_2)_3$—SO_3^-					
M—D′(R)—M	2.8	24.4	0.037	0.37	—
R = —$(CH_2)_3$—$(EO)_{16}$—O—CH_3					
M—D_{13}—D'_5(R)—M	8.0	29.9	$< 5 \times 10^{-4}$	$< 10^{-6}$	—
R = —$(CH_2)_3$—$(EO)_{12}$—OH					
M—D_{18}—D'_5(R)—M	9.4	30.8	$< 5 \times 10^{-4}$	$< 10^{-6}$	—
R = —$(CH_2)_3$—$(EO)_{12}$—OH					
M—D_{13}—D'_5(R)—M	7.3	24.5	0.3	0.96	—
R = —$(CH_2)_3$—O—$SO_3^-H_3N^+$—$CH(CH_3)_2$					
M—D_{13}—D'_5(R)—M + 50 mM NaCl	9.3	26.9	0.053	0.22	—
R = —$(CH_2)_3$—O—SO_3^- H_3N^+—$CH(CH_3)_2$					
M—D_{11}—D'_5(R)—M	9.7	30.0	0.4	1.5	—
R = —$(CH_2)_3$—O—CH_2—CHOH—CH_2—$N^+(CH_3)_2$—$CH(CH_3)_2$ $CH_3CO^-_2$					
M—D_{11}—D'_5(R)—M + 50 mM NaCl	6.1	28.2	0.185	0.72	—
R = —$(CH_2)_3$—O—CH_2—CHOH—CH_2—$N^+(CH_3)_2$—$CH(CH_3)_2CH_3CO_2^-$					
M—D_{13}—D'_5(R)—M	—	33.1	0.04	0.15	—
R = —$(CH_2)_3$—O—CH_2—CHOH—CH_2—$N^+(CH_3)_2$—$CH(CH_3)_2CH_3CO_2^-$					
M—D_{13}—D'_5(R)—M	4.3	27.1	0.006	0.024	—
R = —$(CH_2)_3$—O—CH_2—CHOH—CH_2—$N^+(CH_3)_2$—CH_2—CO_2^-					

for these solutions is in most cases around 30 mN/m, and the graft copolymers are more effective than most of the hydrocarbon detergents, but significantly less effective than the short chain siloxane surfactants. It is likely that this set of differences is due to the presence of more hydrophilic groups, which reduces the effectiveness of the short chain systems, too. The efficiency of the graft copolymers is high in comparison to most hydrocarbon surfactants, but it is many orders of magnitude lower than the efficiency that would be estimated from the short chain samples applying Traube's rule. This is partially also due to the presence of more hydrophilic groups, which reduces efficiency generally, but it must be mainly attributed to the flexibility of the siloxane chain, since this chain can be coiled, thus shielding part of the siloxane units from contact with water.

As was found in the case of hydrocarbon surfactants [28,29], siloxane surfactants show surface activity and aggregation in other strongly polar solvents with a similar high surface tension (e.g., water). The same was been found, for example, for glycol, glycerol, formamide and dimethyl sulfoxide [30]. This behavior can be explained by the attractive forces between the molecules of these solvents, which are in the same range as the forces between water molecules. Hence a similar solvophobic effect can be expected for surfactants of all types.

5. Dynamic Surface Tension Measurements

The dynamic surface tension has also been determined for short chain sulfobetaine siloxane surfactants [3]. This quantity, obtained by measuring the maximum bubble pressure at increasing bubble frequencies, gives information on the relative rates of surface tension reduction. It was found that the dynamic surface tension correlates with the size of the siloxane backbone such that the surface tension at higher bubble frequencies increases up to the value of pure water with increasing length of the siloxane backbone, while it remains low for trisiloxane samples. This correlation is attributed to the slower diffusion of the larger molecules and also to the lower monomer concentrations for the surfactants that have larger siloxane backbones as a result of their lower cmc values.

Svitova et al. [31] have carried out measurements of the dynamic surface (against air) and interfacial tension (against *n*-dodecane) of aqueous solutions of different nonionic trisiloxane and hydrocarbon (*n*-dodecyl) surfactants with increasing lengths of the hydrophilic EO chains. They determined the fall rate of the surface or interfacial tension with a drop volume tensiometer. They measured also the rate of spreading of these surfactant solutions on liquid hydrocarbons, and they found a good correlation between the surface/interfacial tension fall rate, the rate of spreading, and the dynamic spreading coefficient. For the "superwetter" M—D′(R)—M, with R = —$(CH_2)_3$—$(EO)_8$—OH, they calculated a diffusion coefficient that was about one order of magnitude higher than the value for the hydrocarbon surfactant with five EO groups. With increasing numbers of EO groups (i.e., with increasing hydrophilicity of the surfactant), decreasing diffusion coefficients and surface/interfacial tension fall rates for both types of surfactant were observed, which correlates with a suppression of the spreading ability of the samples.

Interestingly, the dynamic surface tension of the fluorinated siloxane surfactants is always worse than in normal siloxane surfactants [27]. These effects are attributed to the bulkiness of the fluoroalkyl groups, which in turn reduces the flexibility of the siloxane backbone, leading to a less dense packing of the molecules at the surface and to a slower diffusion and surface reorientation.

B. Adsorption at the Surface of Weakly Polar and Nonpolar Solvents Against Air

The large structural variety of siloxane surfactants allows the preparation of systems that are surface active also in weakly polar organic liquids like polyether polyols, polyester polyols, and isocyanates [2,14,32], while surface activity of hydrocarbon surfactants in such solvents has not been found. Nonionic graft copolymers are used for this purpose having the general formula R_1—O—D_x—$D'_y(R')$—D_x—O—R_1 with R′ = —D_x—R_1 and R_1 = —$(EO)_a$—$(PO)_b$—R with R = H or alkyl with various chain lengths. The polyethylene oxide and polypropylene oxide units can be arranged as blocks of various sizes, or they can be statistically distributed. By changing the values of a and b for the ethylene oxide and propylene oxide units, the hydrophilic or hydrophobic character of the surfactants can be adjusted very sensitively, hence the solubility and surface activity in the organic solvent. This advantage is especially important for the formulation of polyurethane foams for different applications [33,34].

Figure 7 shows an example of the surface activity of a nonionic siloxane surfactant with $x = 5$, $y = 2$, $a = 20$, $b = 19$, statistical distribution of EO and PO, and R = —C_4H_9 in a polyether polyol having a molar weight around 5000 g/mol [32]; principally the same curve for a similar system was observed by Rossmy et al. [14]. The figure shows a decrease of the surface tension with increasing surfactant concentration from about 35 mN/m for the pure solvent to a value close to the surface tension of the pure siloxane surfactant around 22 mN/m. These curves show a break at a characteristic concentration; it is not yet clear, however, whether this break indicates a micellar aggregation in the organic solvent. This point is discussed again in the next section.

If the siloxanes are substituted by hydrophobic long chain alkyl groups, one obtains strongly hydrophobic siloxane surfactants that are called siloxane waxes. Such samples, which can be prepared either from trisiloxanes or from graft copolymers, are soluble in hydrocarbons. They show surface activity in these nonpolar solvents and can reduce their surface tension from about 27 mN/m to 22 mN/m [35], which can improve the spreading ability of the solution significantly. The surface activity is due to adsorption of the surfactant on the phase boundary

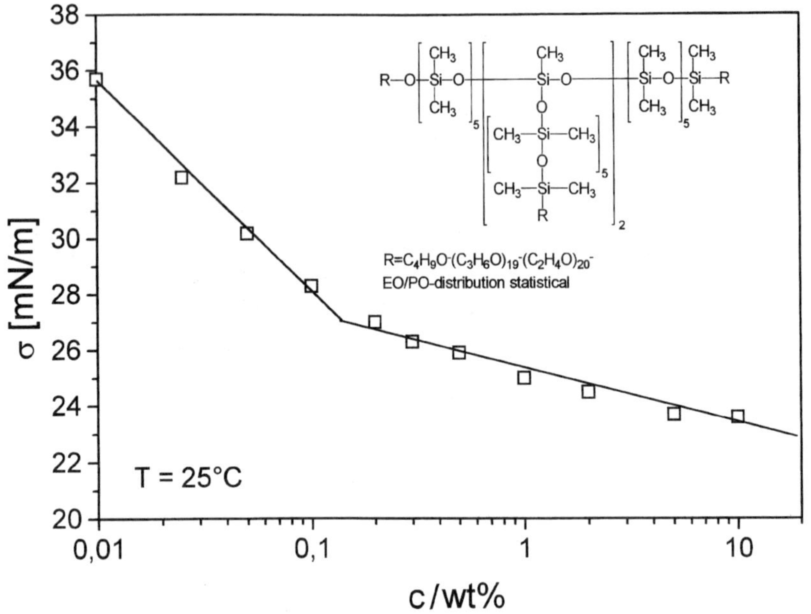

FIG. 7 The surface tension σ at 25°C for solutions of a nonionic siloxane surfactant in polyether polyol (M_W = 5000 g/mol) semilogarithmically plotted as a function of the concentration *c*.

by the partial incompatibility of the siloxane group and the hydrocarbon; it is likely that the siloxane groups are exposed to the surrounding air phase and thus reduce the surface tension. Normal hydrocarbon surfactants that can form reversed micelles in nonpolar solvents are not surface active in hydrocarbons. For the same reason the siloxane waxes do not show surface activity in silicone oils, although they can be soluble in such solvents.

C. Adsorption at the Interface Between Water and Nonpolar Solvents

The adsorption of siloxane surfactants at interfaces between water and nonpolar solvents follows the same principles described for the water surface against air. Figure 8 presents a typical example [18]. The interfacial tension γ decreases with increasing surfactant concentration according to the Gibbs equation and remains constant above the cmc. The cmc values from interfacial tension measurements agree very well with the values from surface tension measurements [18].

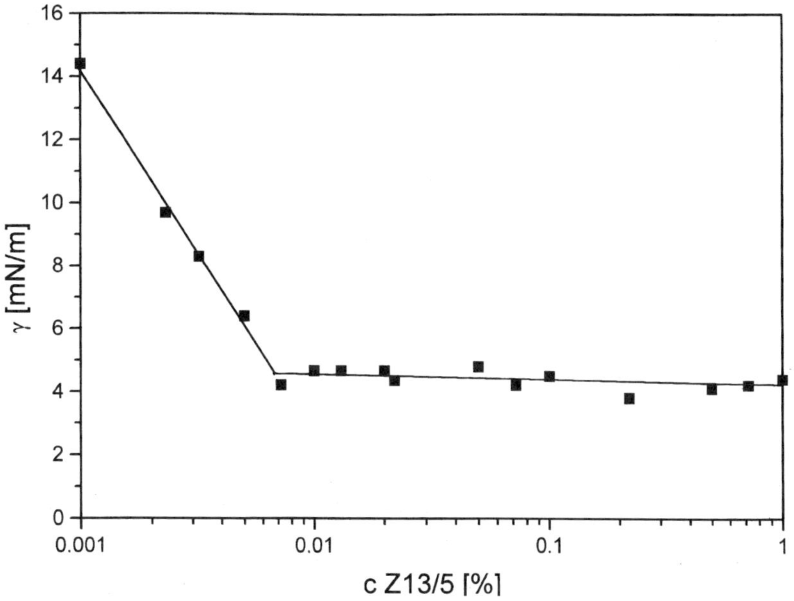

FIG. 8 The interfacial tension γ at 25°C against *n*-decane for aqueous solutions of $M—D_{13}—D'_5$ (R)—M, with R = $—(CH_2)_3—O—CH_2—CHOH—CH_2—N^+(CH_3)_2—CH_2—CO_2^-$ semilogarithmically plotted as a function of the concentration *c*.

Table 5 above also contains values of interfacial tensions against a hydrocarbon at the cmc. For hydrocarbon surfactants, the interfacial tensions above the cmc can be correlated with the shape of the micellar aggregates [36,37]. Globular micelles are present when γ exceeds 1 mN/m; rodlike micelles show values around or below 1 mN/m; and with disks or lamellar phases at low surfactant concentrations, γ can be several orders of magnitude below 1 mN/m ($\leq 10^{-3}$ mN/m). As Table 5 shows, such low values for the interfacial tension against a hydrocarbon have not been measured for the studied siloxane surfactant solutions. In most cases the γ values are between 5 and 10 mN/m, and the lowest values in the tables are still significantly higher than 1 mN/m. From these high values it can be supposed that in these solutions globular micelles are present; furthermore the siloxane surfactant solutions should have a poor solubilization capacity of around 1 molecule hydrocarbon per 10 surfactant molecules [38], which is indeed observed in solubilization measurements. Such siloxane surfactants should also not be very suitable for the formulation of thermodynamically stable microemulsions [38], although they have usually been found to be good emulsifiers for various types of oils [39]. It is mentioned, however, that siloxane surfactants synthesized recently lowered the surface tension against an oil phase to values below 0.1 mN/m [40,41]; also Svitova et al. [31] measured values for the interfacial tension for solutions of the nonionic siloxane surfactant M—D′(R)—M with R = —$(CH_2)_3$—$(EO)_8$—OH against aliphatic hydrocarbons below 0.5 mN/m. In this case a value of 0.03 mN/m was found against *n*-hexane, and the interfacial tension increases almost linearly with the chain length of the hydrocarbon up to a value of 0.4 mN/m for *n*-hexadecane. With such surfactants the formation of stable microemulsion phases in ternary systems of water, oil, and siloxane surfactant can be expected and has also been proven for the system M—D′(R)—M with R = —$(CH_2)_3$—$(EO)_8$—OH, *n*-hexane, and water [31].

The high interfacial tension values are somewhat surprising because one might expect the adsorption films at the water–oil interface to have the same structure and packing as the films at the water surface against air, which can be covered very densely and effectively by the siloxane surfactants. The high values are thus probably due to the partial incompatibility between silicone oils and hydrocarbons. In this connection it is interesting that at an interface has been found between water and a silicone oil an even higher interfacial tension at the cmc of a zwitterionic graft polymer around 15 mN/m [18]. In this case the measured cmc was more than one order of magnitude higher than the cmc measured in the solution against both air and *n*-decane. The higher cmc in the silicone oil is probably a result of the solubility of the surfactant in the silicone oil.

Many siloxane surfactants have a unique ability to remarkably improve the wettability of low energy solid surfaces such as waxes or polyethylene by aqueous solutions. This phenomenon is described as "superwetting" or "superspreading" in the literature [42,43]. Since it is described in another chapter, it shall not be discussed in this part in detail.

D. The Critical Micelle Concentration of Aqueous Siloxane Surfactant Solutions

The siloxane surfactants that are sufficiently soluble in water show an analogous aggregation behavior in strongly polar solvents like other surfactants as can be seen from Figs. 3–8. At low concentrations the surfactants are dissolved as monomers, and they start to form micelles at a characteristic concentration (cmc). In first approximation above the cmc the monomer concentration remains constant and all further surfactant molecules aggregate. Therefore the cmc of siloxane surfactants can be measured by the same methods that have been used for hydrocarbon surfactants [44]; Fig. 9 shows results of electric conductivity measurements that are very suitable for cmc measurements on ionic siloxane surfactants [6,18]. The cmc values for siloxane surfactant of different types are also listed in Tables 1–5.

According to the pseudophase model [45,46] there is equilibrium between the monomers in the micelles and in the aqueous medium; hence the chemical poten-

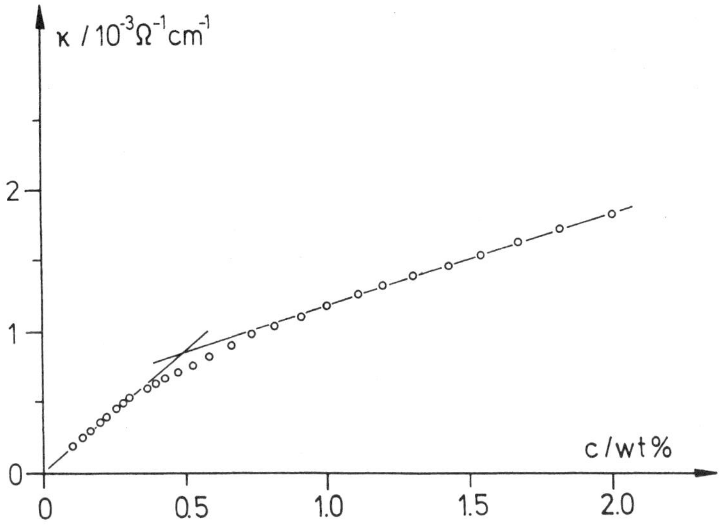

FIG. 9 The specific conductivity κ at 25°C for aqueous solutions of M—D′(R)—M with R = —$(CH_2)_3$—O—SO_3^- H_3N^+—$CH(CH_3)_2$ as a function of the concentration *c*.

tials of the monomers in the micelle (μ_M) and in the water (μ_1) must be equal. Because μ_1 is concentration dependent according to the formula $\mu_1 = \mu_{01} + RT \ln (X_1)$ (here X_1 represents the mole fraction of surfactant in water and is equal to the cmc for surfactant concentrations above the cmc), the energy $\Delta\varepsilon$ can be calculated by the equation $\Delta\varepsilon = -RT \ln$ (cmc). From the cmc value of $5.4 \cdot 10^{-6}$ M for the nonionic sample M—D′(R)—M with R = —$(CH_2)_3$—$(EO)_{16}$—O—CH_3 a $\Delta\varepsilon$-value of 12.1*RT* can thus be obtained. Because this surfactant contains seven CH_3— and three CH_2—groups, one can conclude a similar average value of 1.2*RT* for one methyl or methylene unit, as has been found for the nonionic hydrocarbon surfactants. Assuming this value to be correct, one can further conclude that the hydrophobic character of the siloxane groups is due to the methyl and methylene groups in the first approximation; in this case the —Si—O— groups would be neither hydrophilic nor hydrophobic. With these considerations it should be possible to estimate at least roughly the cmc values for nonionic and zwitterionic short chain siloxane surfactants. The cmc values in Tables 1–5 show that this is indeed correct in the first approximation. Ionic surfactants at low ionic strengths (i.e., without added excess electrolyte) have considerably higher cmc values because of the electrostatic repulsion between the charged groups; the same is true for hydrocarbon surfactants.

Table 5 shows furthermore, however, that this simple rule fails completely for long chain siloxane surfactants. Samples with around 20 and more siloxane units should have cmc values in the range of 10^{-20} M and lower; it can easily be seen that surfactants with such low cmc values would not be soluble in water and micellar equilibria would not be adjusted within months. The experiments show that this is not the case, and also long chain siloxane surfactants have cmc values many orders of magnitude higher than the expected values; the reasons for these deviations were explained above.

III. THE AGGREGATION BEHAVIOR OF SILOXANE SURFACTANTS

A. Principles of Surfactant Aggregation in Aqueous Solutions

From the theory of micelle formation [47] it has been derived that the form of surfactant aggregates is determined by the packing parameter $P = a\ (l/v)$, where a is the area of the hydrophilic head groups at the micellar surface, l is the length of the hydrophobic group, and v its volume. Values of v and l are determined by the constitution of the surfactant molecules, while a depends also on electrostatic and steric properties of the head group and is not necessarily equal to its cross section. From geometrical considerations it can be shown that for $P \geq 3$ spherical micelles are formed, while for $P \approx 2$ rodlike micelles, for $P \approx 1$ disklike micelles or lamellar structures, and for $P < 1$ reversed structures are formed. The radius of spherical micelles is given by the length of the hydrophobic groups and is independent of the surfactant concentration; for anisometric aggregates the short radii are also equal to the length of the hydrophobic groups, while the long diameters increase with the surfactant concentration and the number of aggregates remains constant. These general principles have been confirmed for many different types of surfactant.

B. The Aggregation Behavior of Siloxane Surfactants in Aqueous Solutions

Only a few investigations of the size and shape of micellar aggregates of siloxane surfactants in aqueous solutions are known. Such data, which were obtained by static and dynamic light scattering, small-angle neutron scattering (SANS), electric birefringence, and rheological measurements, and electron microscopy, are summarized for short and long chain surfactants in Table 6 [6,18].

The results show that as opposed to the case with hydrocarbon surfactants, there is no correlation between the length of the hydrophobic siloxane backbone and the radii of the micellar aggregates. Some surfactants form micelles with relatively small radii in the range between 20 and 60 Å and with aggregation numbers in the range of 10 up to a few 100 monomers per micelle, and other compounds form micelles with much larger aggregates and aggregation numbers in the range of several thousand monomers per micelle. In both, the apparent dimensions of the aggregates do not increase but either remain constant or decrease with increasing concentration. Table 6 shows furthermore that short chain trisiloxane surfactants and long chain graft copolymers can be present in the two groups; in practically all cases the measured radii are much larger than the length of the hydrophobic siloxane chains.

The constitution of the hydrophilic group does not seem to affect the aggregation behavior of the siloxane surfactants. Ionic, zwitterionic, and nonionic siloxane surfactants can form both smaller and larger aggregates. The only clear correlation between the structure of the samples and their aggregation number that is permitted by the available data seems to come from the chain length of the hydrophilic groups: surfactants with rather short hydrophilic chains prefer the formation of large aggregates while samples with long chains form preferentially the smaller aggregates. This correlation could however be an artifact of the few available data. That is, additional experiments in this field are necessary to confirm or to exclude these conclusions.

A detailed analysis of the data in Table 6 shows for the systems with small aggregates an excellent agreement between the results obtained by the different methods [6]. The data show furthermore that some of the aggregates show a typical behavior of spherical micelles (i.e., the radii are practically independent of the surfactant concentration). The apparent decrease of the aggregate size with increasing surfactant concentrations could be quantitatively explained by the increasing intermicellar repulsion, which is due to hard sphere interactions [48–50]. The most convincing support for this statement can be derived from SANS measurements, which are shown for two examples of nonionic siloxane surfactants in Fig. 10. Neglecting the range of small values of the scattering vector q, which is affected by the intermicellar repulsion, Fig. 10 shows that the curves are parallel for all investigated concentrations with one exception, and the distances between the curves are proportional to the surfactant concentrations. This shows unambiguously that the micelles do not change their size and shape in the studied concentration range between 1 and 20 wt%.

The statements that the aggregates behave like spherical micelles and that the measurements can be evaluated consistently with the assumption of spherical particles, do not necessarily permit conclusion that the surfactants really form spherical micelles. This can be seen from the particles radii, which are practically always much larger than the lengths of the hydrophobic groups; it would also be very difficult to imagine the packing of graft copolymers in micelles with the structure of spherical hydrocarbon micelles. It is thus possible that the micelles of the siloxane surfactants have an ellipsoidal shape with an axial ratio close to 1; the experimental accuracy is not sufficient to distinguish between such aggregates and exact spheres. Also, the arrangement of the monomers in the micelles cannot be concluded from the experimental data at present.

Hydrocarbon surfactants can either form rodlike micelles as first aggregates at the cmc or show a sphere-to-rod transition at a characteristic concentration c_t

TABLE 6 Values for Molar Weight M_W, Radius of Gyration R_G, Hydrodynamic Radius R_H, Radius R, and Aggregation Number n at Various Concentrations for the Aggregates of Different Siloxane Surfactants at 25°C, Determined by Static Light Scattering (LS), Electric Birefringence (EB), Rheological (η), and Small-Angle Neutron Scattering (SANS) Measurements

			Radii (Å)			
Siloxane surfactant	c (wt%)	M_W (g/mol)	R_G	R_H	R	n
M—D′(R)—M						
R = —$(CH_2)_3$—$(EO)_{16}$—O—CH_3						
(spherical micelles)	0[a]	7×10^4	—	—	29.7	70_{LS}
	0[a]	7×10^4	—	—	29.7	70_{SANS}
	1	1.3×10^5	20.2	—	33	128_{SANS}
	2	1.2×10^5	20.3	—	32.6	119_{SANS}
	4	1.2×10^5	20.2	—	32.5	121_{SANS}
	6	1.3×10^5	—	—	33.2	133_{SANS}
	8	1.2×10^5	—	—	31.7	120_{SANS}
(probably no spheres)	10	2.5×10^5	—	—	40.0	251_{SANS}
M—D′(R)—M						
R = —$(CH_2)_3$—O—CH_2—CHOH—CH_2—$N^+(CH_3)_3$ $CH_3CO_2^-$						
(spherical micelles)	10	3.5×10^4	—	—	79	23.8_{SANS}
M—D_5—D'_5(R)—M						
R = —$(CH_2)_3$—$(EO)_{12,5}$—OH						
(spherical micelles)	5	6.9×10^4	14.8	—	25.6	18_{SANS}
	10	6.2×10^4	—	—	26.4	17_{SANS}
	20	4.9×10^4	—	—	25.5	15_{SANS}
M—D_{13}—D'_5(R)—M						
R = —$(CH_2)_3$—$(EO)_{12}$—OH						
(spherical micelles)	0[a]	2.1×10^5	—	—	42.7	48_{LS}
M—D_{18}—D'_5(R)—M						
R = —$(CH_2)_3$—$(EO)_{12}$—OH						
(spherical micelles)	0[a]	2.1×10^5	—	—	40.6	38_{SANS}
	2	2.6×10^5	27.1	—	43.9	54_{SANS}
	4	2.5×10^5	27.1	—	43.7	53_{SANS}
	8	2.4×10^5	27.0	—	43.0	50_{SANS}
	12	2.2×10^5	—	—	40.9	47_{SANS}
	16	2.2×10^5	—	—	40.4	47_{SANS}
	20	2.1×10^5	—	—	39.7	46_{SANS}
M—D_{13}—D'_5(R)—M						
R = —$(CH_2)_3$—O—CH_2—CHOH—CH_2—$N^+(CH_3)_2$—CH_2—CO_2^-						
(vesicles)	0[a]	2.8×10^7	620	550	—	$11{,}200_{LS}$
	0.025	2.8×10^7	614	—	466, 620;[b]	$11{,}160_{LS}$
	0.05	2.3×10^7	575	—	607, 808[b]	$9{,}160_{LS}$
	0.075	2.1×10^7	579	—	554, 737[b]	$8{,}370_{LS}$
	0.1	1.9×10^7	568	—	612, 814[b]	$7{,}570_{LS}$
	0.15	1.9×10^7	575	—	561, 747[b]	$7{,}520_{LS}$

TABLE 6 Continued

Siloxane surfactant	c (wt%)	M_W (g/mol)	Radii (Å) R_G	R_H	R	n
	0.2	1.3×10^7	567	—	598, 796[b]	$5{,}180_{LS}$
	0.25	6×10^6	372	521_η	474, 631[b]	$2{,}390_{LS}$
	0.5	—	—	486_η	409, 644[b]	—
	0.75	—	—	446_η	502, 668[b]	—
	1	—	—	430_η	554, 737[b]	—
	2	—	—	422_η	609, 810[b]	—
	3	—	—	411_η	476, 634[b]	—
	4	—	—	444_η	592, 787[b]	—
	5	—	—	554_η	695, 924[b]	—
M—$D_{13}D'_5$(R)—M R = —$(CH_2)_3$—O—$SO_3^-H_3N^+$—$CH(CH_3)_2$ (vesicles)	0[a]	9.8×10^6	600	—	—	$4{,}100_{LS}$
	0.5	—	—	—	377, 501[b]	—
	0.75	—	—	—	354, 471[b]	—
	1	—	—	—	327, 436[b]	—
	2	—	—	—	388, 516[b]	—
	3	—	—	—	391, 521[b]	—
	5	—	—	—	443, 589[b]	—
M—$D_{13}D'_5$(R)—M R = —$(CH_2)_3$—O—$SO^-{}_3H_3N^+$—$CH(CH_3)_2$ + 50 mM NaCl (vesicles)	0[a]	9.0×10^6	410	—	—	$3{,}700_{LS}$
M—D_{11}—D'_5(R)—M R = —$(CH_2)_3$—O—CH_2—CHOH—CH_2—$N^+(CH_3)_2$—$CH(CH_3)_2CH_3CO_2^-$ (vesicles)	0[a]	3.0×10^6	550	—	—	$1{,}160_{LS}$
	0.35	—	—	—	131, 174[b]	—
	0.5	—	—	—	359, 478[b]	—
	0.75	—	—	—	383, 510[b]	—
	1	—	—	—	514, 683[b]	—
	2	—	—	—	540, 719[b]	—
	3	—	—	—	561, 746[b]	—
M—D_{11}—D'_5(R)—M + 50 mM NaCl R = —$(CH_2)_3$—O—CH_2—CHOH—CH_2—$N^+(CH_3)_2$—$CH(CH_3)_2CH_3CO_2^-$ (vesicles)	0[a]	1.5×10^7	400	—	—	$5{,}600_{LS}$

[a]Extrapolated to $c \to 0$.
[b]First value, vesicle radius; second, half-length of disks, measured by EB.

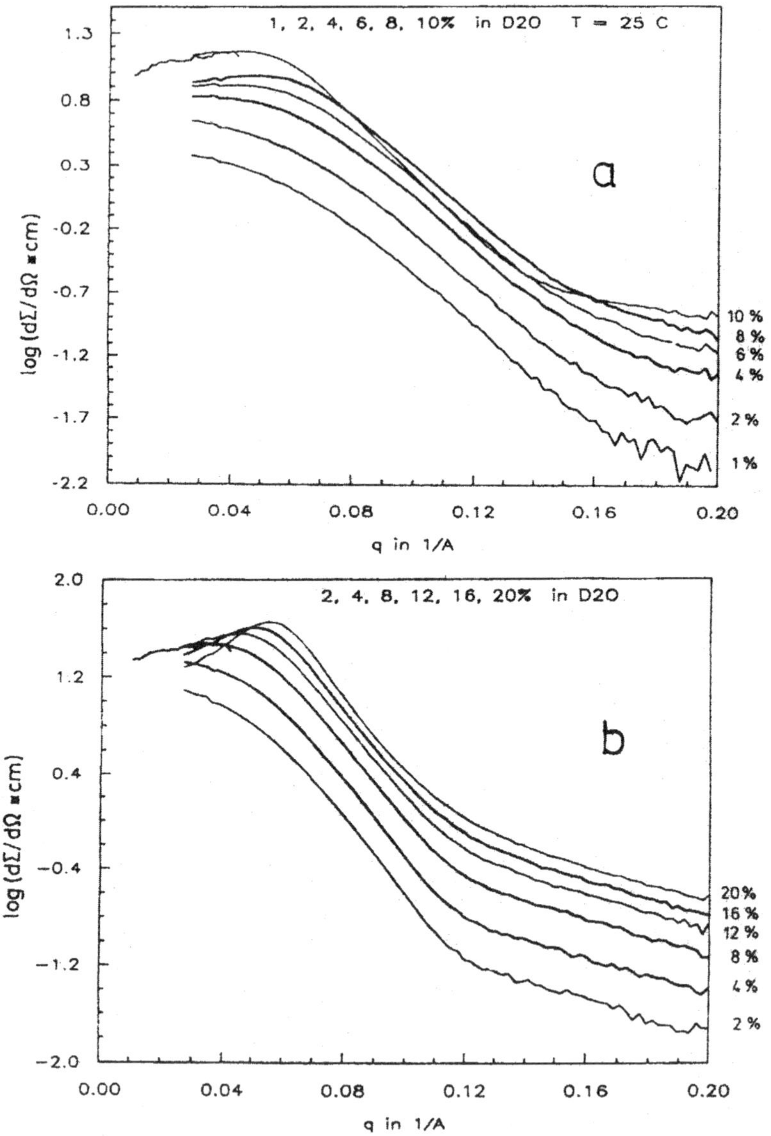

FIG. 10 The differential scattering cross section $d\Sigma/d\Omega$ (logarithmic) in SANS measurements at 25°C for aqueous solutions with various concentrations of (a) a nonionic trisiloxane surfactant (M—D′(R)—M with R = —$(CH_2)_3$—$(EO)_{16}$—O—CH_3) and (b) a nonionic graft polymer (M—$D_{18}D'_5$(R)—M with R = —$(CH_2)_3$—$(EO)_{12}$—OH) as a function of the scattering vector q.

[51,52]. A behavior in which the length of rodlike micelles increases with increasing surfactant concentrations while the concentration remains constant could not be detected for siloxane surfactants; also a transition concentration could not be found for such samples. There is, however, strong evidence for the existence of such aggregates, which must be formed above a characteristic concentration.

A solution of a nonionic trisiloxane surfactant with small aggregates does not show electric birefringence at concentrations below 5 wt% [6]. Above this concentration a small electric birefringence signal with rise and decay times below 1 μs can be detected; this could be due to the existence of small rods (lengths < 300 Å). Furthermore the SANS curve for the same sample (Fig. 10b) shows a significantly different shape at a concentration of 10 wt% in comparison to the parallel curves at lower concentrations. This could indicate a transition of the shape of the aggregates. The strongest support for the existence of rodlike aggregates comes from the result that many samples with small aggregates at low concentrations show a hexagonal phase at concentrations above about 50 wt% [6], while none of the studied samples shows a cubic phase at higher concentrations. Therefore it can be concluded that a sphere-to-rod transition must take place at least at the concentration at which the isotropic micellar solution borders the two-phase isotropic/hexagonal region. The high concentration and the resulting strong intermicellar interactions, however, make unambiguous detection of the shape of the micelles impossible.

Siloxane surfactants with shorter hydrophilic groups have been shown to possess a different behavior. It can be seen indeed in Figs. 3b and 11 that plots of the surface tension and the electric conductivity as a function of the concentration are similar to corresponding plots for samples with long hydrophilic groups. Static and dynamic light scattering measurements, electric birefringence measurements, and rheological investigations, however, indicate the presence of very large anisometric aggregates above the cmc [18].

Quantitative evaluation of the experiments shows that the giant aggregates are not long rods but polydisperse multilamellar vesicles. This is confirmed by transmission electron micrographs (e.g., in Fig. 12). It can be seen from Fig. 12 that many vesicles are not closed and that besides the polydisperse vesicles with average diameters around 200 nm, the solution contains fragments of flat bilayers and small micellar aggregates with diameters in the range of 20 nm. A similar observation has been made for solutions of polymeric nonionic siloxane surfactants [53]. It can be stated therefore that these siloxane surfactants start to form vesicles above the cmc. Also, systems have been found that enter a two-phase region with a lamellar phase at the apparent cmc [18]; this characteristic concentration is therefore often called critical aggregation concentration (cac) instead of critical micelle concentration (cmc).

Rheological measurements have shown that the solutions with the vesicles can be viscoelastic and can often have a yield stress value (Fig. 13). Such experiments

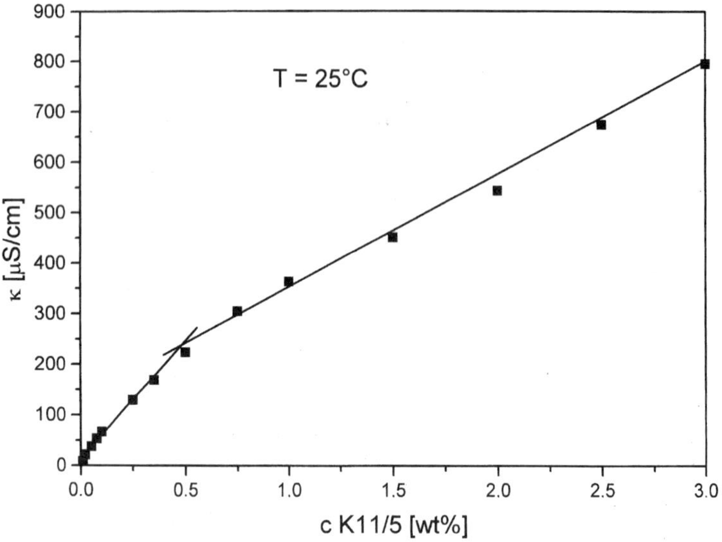

FIG. 11 The specific conductivity κ at 25°C for aqueous solutions of a cationic (M—D_{11}—D'_5(R)—M with R = —$(CH_2)_3$—O—CH_2—CHOH—CH_2—$N^+(CH_3)_2$—$CH(CH_3)_2$ $CH_3CO_2^-$) graft polymer as a function of the concentration *c*.

can show that the equilibration time in these solutions can be very long. A freshly prepared solution can show viscoelasticity but no yield stress value, and the yield value can develop very slowly if the samples are allowed to stand for 2 months. Slow kinetic processes in the siloxane surfactant solutions must therefore be taken into account in investigations.

C. Aggregation in Weakly Polar and Nonpolar Solvents

Siloxane surfactants can also be surface active in weakly polar solvents like polyether polyols; this is also true for siloxane surfactants with alkyl chains instead of the hydrophilic groups in nonpolar solvents like hydrocarbons. For example, Fig. 7 showed some surface tension behavior similar to that in water: the surface tension decreases with increasing concentration and remains constant at a characteristic concentration. This could be interpreted by the assumption that above this concentration a reversible aggregation of the surfactant monomers takes place.

Unfortunately a direct proof for the existence of aggregates in these solutions is lacking. It was shown in an academic experiment that the aggregation numbers of aggregates in a weakly polar solvent are small, if aggregates are present at all

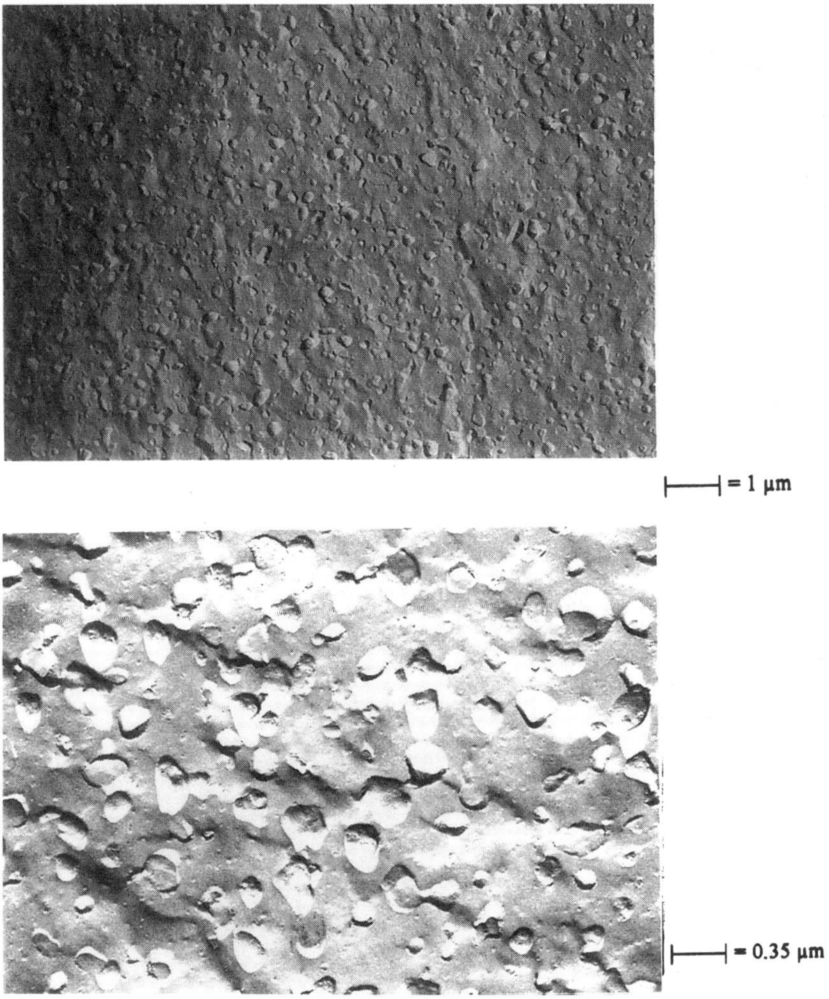

FIG. 12 Freeze-fracture TEM micrographs of the aggregates in an aqueous solution of 4 wt% of M—D_{13}—D'_5 (R)—M with R = —$(CH_2)_3$—O—CH_2—CHOH—CH_2—N^+ $(CH_3)_2$—CH_2—CO_2^- (zwitterionic graft polymer) with different magnifications.

[32]; thus detection of micellar aggregates in such solvents with rather high molar weights is very difficult. It is possible that instead of indicating an aggregation process, the break of the surface tension curves against the surfactant concentration is due only to saturation of the surface adsorption film with densely packed surfactant molecules, with the surfactant being dissolved in the monomeric state also above the break.

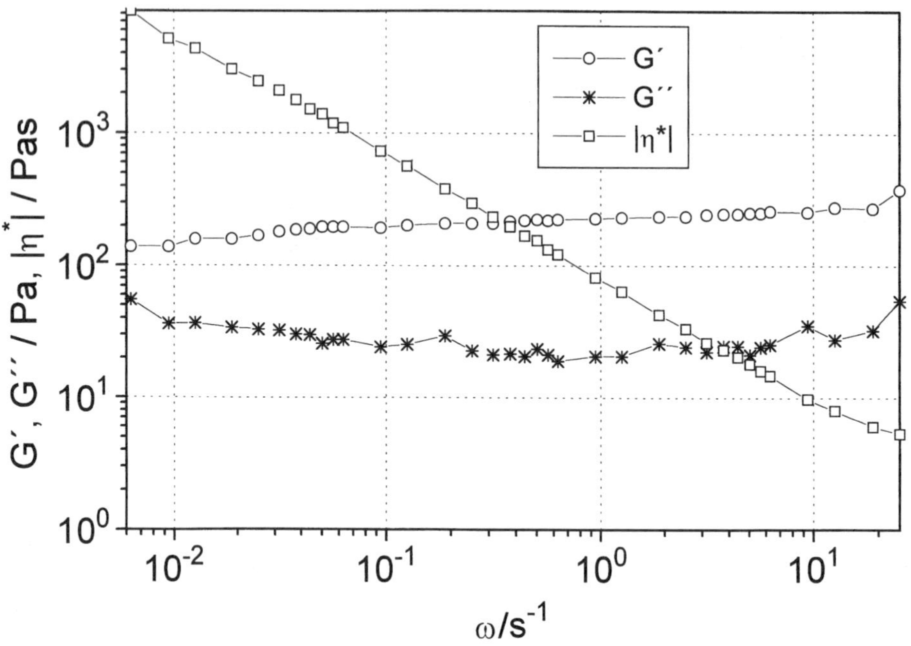

FIG. 13 The storage modulus G', the loss modulus G'', and the complex viscosity $|\eta^*|$ at 25°C of an aqueous solution of 70 wt% of a cationic graft polymer, (M—D_{11}—D'_5 (R)—M with R = —$(CH_2)_3$—O—CH_2—CHOH—CH_2—$N^+(CH_3)_2$—$CH(CH_3)_2$ $CH_3CO_2^-$), as a function of the oscillation frequency ω.

D. Phase Behavior of Binary Systems of Siloxane Surfactants and Water

The phase behavior of binary systems consisting of siloxane surfactants, and water, which is discussed in detail in a separate chapter, is briefly mentioned here. Phase diagrams of such systems are usually very simple. Surfactants that form small and spherical micelles above the cmc normally form a hexagonal phase at surfactant concentrations around or above 50 wt%, as shown, for example Fig. 14a [6]. This phase is separated from the isotropic solution by a two-phase region. On heating, the hexagonal phase melts and shows a transition to an isotropic one-phase region above a characteristic temperature. The surfactants with vesicles above the cmc form lamellar phases at higher surfactant concentrations, as can be seen in Fig. 14b [18,54]; for nonionic trisiloxane samples M—D′(R)—M, where R is —$(CH_2)_3$—$(EO)_x$—OH with short EO chains (x = 5–8), the formation of an L_3 or sponge phase is reported by He et al. [55]. The lamellar phases can consist

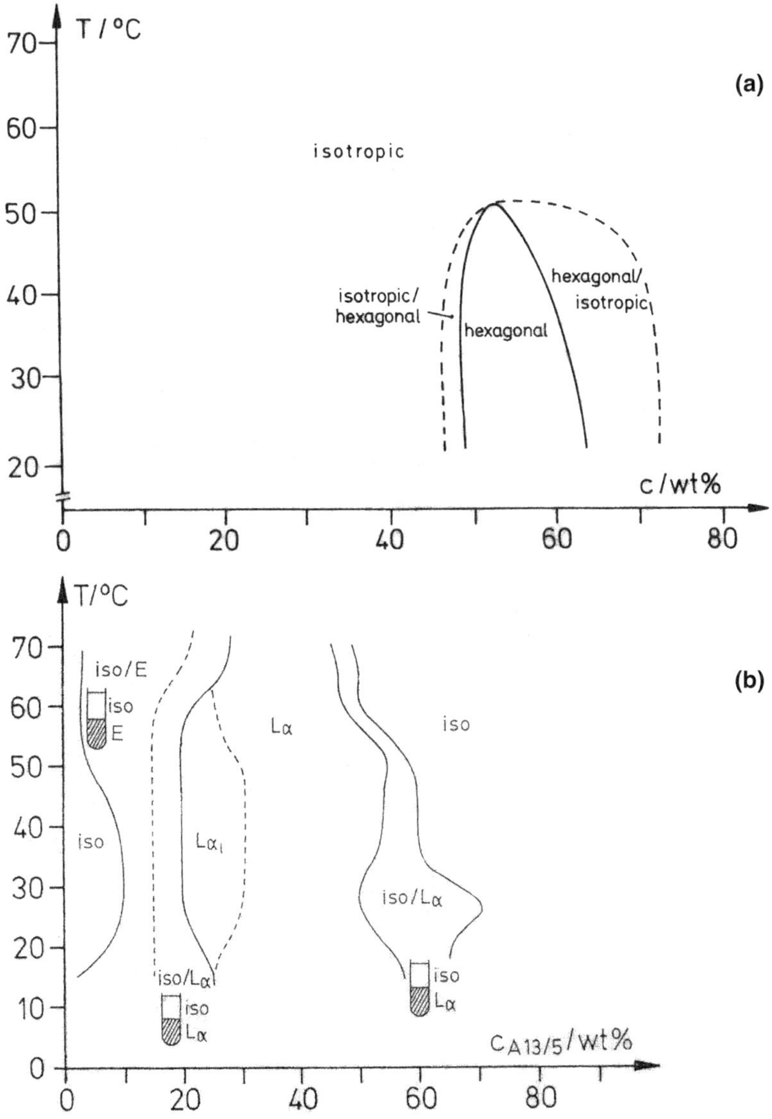

FIG. 14 Phase diagram of two binary systems (a) a nonionic trisiloxane surfactant, M—D′(R)—M with R = —(CH$_2$)$_3$—(EO)$_{16}$—O—CH$_3$) and water, and (b) an anionic graft polymer, M—D$_{13}$D′$_5$ (R)—M with R = —(CH$_2$)$_3$—O—SO$_3^-$ H$_3$N$^+$—CH(CH$_3$)$_2$, and water.

of vesicles ($L_{\alpha 1}$ phase) or at higher concentrations also of normal flat bilayers ($L_{\alpha h}$ phase). Normally there is no phase boundary between the two lamellar phases; they can be distinguished by their textures under the polarization microscope or by their rheological behavior, because the vesicle phases show often yield stress values contrary to the $L_{\alpha h}$ phases [56,57]. These phases are also separated from the isotropic solutions by two-phase regions and they show also a thermotropic behavior, but with much higher transition temperatures. More complicated phase diagrams with more than one mesophase are not very often observed in the binary systems; for example, Hill et al. [53] found a lamellar and an inverse hexagonal phase for the nonionic system M′(R)—D_{13}—M′(R), where R is —$(CH_2)_3$—$(EO)_8$—OH. In this system the inverse hexagonal phase disappears when the number of EO groups increases to 12.

Finally, we point out a characteristic feature of the phase diagrams. Most of the pure siloxane surfactants do not crystallize at room temperature but remain liquid. The liquid surfactants can be miscible with water in the whole range from 0 to 100 wt% above a certain temperature (Fig. 14a). The phase at the surfactant-rich side is called on L_2 phase in the literature [55], but the exact structure of this phase, which has no phase boundary with the micellar solution, is not unambiguously known. Electric conductivity measurements on a nonionic graft polymer showed a strong increase on addition of small amounts of an aqueous electrolyte solution [58]. This is the indication of a bicontinuous structure with water channels through the surfactant phase. On the other hand, He et al. [55] postulated a cubic phase with densely packed reverse micelles in the L_2 phases of nonionic trisiloxane surfactants M—D′(R)—M with R = —$(CH_2)_3$—$(EO)_x$—OH. In these systems the authors had also found a correlation between the chain length of the hydrophilic groups and the phase behavior; for $x = 5$–8, an L_3 and some lamellar phases were found, while for $x = 12$ hexagonal and lamellar phases are observed, and for $x = 18$ only a hexagonal phase.

IV. INTERACTIONS OF SILOXANE SURFACTANTS WITH ADDITIVES

A. Mixed Surfactant Systems in Solution

Several studies comparing mixed systems of siloxane and hydrocarbon surfactants have been reported. The hydrophobic moieties of the surfactants of both types, (i.e., silicones and hydrocarbons) are relatively miscible, in contrast to hydrocarbons and fluorocarbons. Most of the studies concentrate on the influence of the composition of the surfactant solutions on the measured cmc values, but the influence of the composition on the micellar aggregates and on phase behavior was investigated, as well.

Depending on the constitution of the two surfactants, ideal, antagonistic, or synergistic behavior was observed. Hill [59] studied the surface tensions of mixtures of nonionic siloxane surfactants [trisiloxane surfactants M—D′(R)—M; R = —(CH$_2$)$_3$—(EO)$_x$—OH with x = 7, 12, and a graft polymer M—D$_{100}$—D′$_{18}$(R)—M; R = —(CH$_2$)$_3$—(EO)$_{12.5}$—OH] with a cationic (C$_{12}$TMABr), an anionic (SDS), and a nonionic [C$_{12}$(EO)$_7$] surfactant. The surface tensions at the cmc for the mixtures are generally close to the low values for pure siloxane surfactants, and some of the mixtures show minima of the surface tension at the cmc. These results indicate a preferential adsorption of the siloxane surfactants at the surfaces.

Hill found that the nonionic–nonionic combinations show antagonistic (positive nonideal, higher cmc values than for ideal mixing), the nonionic–cationic combinations show ideal, and the nonionic–anionic combinations show synergistic (negative nonideal, lower cmc values than for ideal mixing) mixing behavior. Regular solution theory [60–62] failed to give a quantitative explanation of the measured cmc values. This means that the interaction between the surfactants depends on the mixing ratio of the two components. The predicted cmc curves do not coincide with the experimental results. The measured cmc values for the nonionic–nonionic combinations are too high to be accounted for by the model. The antagonistic mixing of the latter combinations could not be explained by a phobicity between the siloxane and the hydrocarbon groups, as in the case of mixtures of hydrocarbon and perfluoro surfactants. The results were tentatively explained by suggesting that the different molecular sizes and shapes of the molecules prevent an effective packing of the two surfactant molecules in the mixed micelles. This effect is compensated in the mixtures with the ionic surfactants by the decrease of the electrostatic repulsion between the ionic hydrophilic groups.

Ohno et al. studied mixed systems of a nonionic graft polymer M—D$_9$—D′$_4$(R)—M [R = —(CH$_2$)$_3$—(EO)$_{12}$—OH] and sodium dodecyl sulfate (SDS) or lithium perfluorooctanesulfonate (LiFOS) using surface tension, electric conductivity, and solubilization measurements [63]. They observed an attractive interaction between the siloxane and the hydrocarbon or perfluoro surfactant, respectively. They attribute the interaction to the hydrophilic groups and the possible incorporation of the two types of surfactant into the mixed micelles in spite of the immiscibility of silicone oils and fluoroalkanes. The authors did not determine the influence of the added surfactants on the sizes and shapes of the micelles, although they measured the radii (54 Å) and the aggregation numbers (30) of the micelles of the pure siloxane surfactant; these values are close to the ones for similar systems given in Table 6.

Stürmer [18] studied mixtures of the zwitterionic graft polymer

(M—D$_{13}$—D′$_5$(R)—M [R = —(CH$_2$)$_3$—O—CH$_2$—CHOH—
CH$_2$—N(CH$_3$)$_2$$^+$—CH$_2$—CO$_2$$^-$)]

with various other surfactants, including polymeric anionic or cationic siloxane surfactants

(M—D$_{13}$—D′$_5$(R)—M [R = —(CH$_2$)$_3$—O—SO$_3$$^-$ or H$_3$N$^+$—CH—(CH$_3$)$_2$])

and

$M—D_{11}—D'_5(R)—M$ [R = —$(CH_2)_3$—O—CH_2—CHOH—CH_2—
$N(CH_3)_2^+$—CH—$(CH_3)_2$ $CH_3CO_2^-$]

as well as nonionic trisiloxane surfactants (M—D′(R)—M [R = —$(CH_2)_3$—$(EO)_x$—OH, with x = 8, 12], a nonionic technical perfluoro surfactant ($C_{6-12}F_{13-25}C_2H_4$—$(EO)_{4-7}$—OH: Fluowet OTN), and a cationic (C_{14}TMABr), an anionic (SDS), a zwitterionic (C_{14}DMAO), and a branched nonionic (i-C_{12}—$EO)_{4,9}$—OH) hydrocarbon surfactant. He carried out surface and interfacial tension measurements, and he determined furthermore the micellar sizes and shapes at a constant concentration far above the cmc as a function of the mixing ratio. With respect to the surface and interfacial tension values for most of the mixtures, ideal mixing behavior was observed: that is, the corresponding value changed from the side of the zwitterionic graft polymer to the side of the additive according to a linear relation. Only with the perfluoro surfactant and with the branched nonionic surfactant a weak synergism could be detected, while with the anionic graft polymer a significant antagonism was observed, which was attributed to a partial incompatibility of the two head groups.

The change of the aggregates (vesicles) of the zwitterionic graft polymer with increasing amounts of additives depends on the micellar structures of the additives. Surfactants that form globular micelles [C_{14}TMABr, SDS, C_{14}DMAO, and also M—D′(R)—M; R = —$(CH_2)_3$—$(EO)_{12}$—OH] lead at first to an increase in the number density of the vesicles, and then to a transition of the vesicles to small spherical micelles with incorporated siloxane surfactant. This behavior is similar to that of phospholipid vesicles on addition of surfactants [64], except that no rodlike aggregates can be detected during the transition. With other additives that form vesicles themselves, ideal mixing behavior is observed: that is, the sizes of the vesicles vary linearly from the side of the zwitterionic siloxane surfactant to the side of the additive. From these results it can be concluded that in all studied mixtures, mixed aggregates are formed. From the measurements, however, no conclusion about the arrangement of the different molecules in the mixed aggregates can be drawn at present.

B. Siloxane Surfactants and Nonpolar Additives

The micelles of siloxane surfactants should solubilize nonpolar substrates like hydrocarbons into their hydrophobic interior, because hydrocarbons and silicone oils with low molar weights are miscible. It is also known that silicone oils can be solubilized in normal hydrocarbon micelles like other nonpolar solvents [65]. The solubilization of these silicone oils in surfactant solutions with rodlike micelles leads to a decrease of the lengths of the rods with increasing additive concentrations and finally to a transition to spherical micelles, which can swell to microemulsion droplets; the same behavior had been found for the solubilization of hydrocarbons in solutions with rods [38]. Silicone oils with large molecules are not soluble enough in the micellar solutions to enforce the rod-to-sphere transition.

Micellar siloxane surfactant solutions often have rather high interfacial tension values against hydrocarbons around or above 5 mN/m. According to earlier observations on normal hydrocarbon surfactants [36,37] siloxane surfactants should therefore solubilize only small amounts of hydrocarbons; this has been confirmed experimentally [5]. Only for the few systems with low interfacial tension values around 0.01 mN/m can a better solubilization capacity be expected.

In this context it is interesting to note that a hexagonal phase of a siloxane surfactant can solubilize small amounts of hydrocarbons and shows a transition to an isotropic gel phase that is very likely a cubic phase [6]. The rod-to-sphere transition occurs thus also in the siloxane surfactant solutions by the solubilization of hydrocarbons.

Investigations of Robinson et al. [66] on a ternary system of a trisiloxane surfactant [M—D′(R)—M; R = —$(CH_2)_3$—$(EO)_8$—OH], water, and cyclohexane have shown that this system has an L_2 phase in the cyclohexane corner, and this phase, which can incorporate 22 water molecules per surfactant molecule, is not a water-in-oil microemulsion. SANS measurement showed that the aggregates have a bilayer structure, and the thickness of the bilayers increases with increasing water content. The thickness of the bilayers at low water contents corresponds to the length of two extended hydrophilic EO groups, which are interdigitated or coiled. It is conceivable that the phase contains reverse vesicles.

C. Siloxane Surfactants and Cosurfactants

A few studies on ternary systems of siloxane surfactants, water, and cosurfactants (e.g., aliphatic alcohols) have been reported. Hoffmann et al. observed [6] that addition of *n*-decanol to a hexagonal phase in the graft polymer M—D_{18}—D'_5(R)—M with R = —$(CH_2)_3$—$(EO)_{12}$—OH leads to the appearance of a lamellar phase at higher concentrations, and in the transition region between the hexagonal phase and the lamellar phase a narrow region of a nematic phase, which is very likely an N_D phase, can be detected. Such nematic phases are frequently found in binary systems of hydrocarbon surfactants and water on addition of aliphatic alcohols [67–69]; their existence shows that a transition of rods to disks is enforced by the added cosurfactant in accordance with the theory of micelle formation.

Stürmer [18] studied the ternary system of the polymeric zwitterionic siloxane surfactant M—D_{13}—D'_5(R)—M [R = —$(CH_2)_3$—O—CH_2—CHOH—CH_2—$N(CH_3)_2^+$—CH_2—CO_2^-] with *n*-decanol and water. This surfactant, which shows a vesicle phase on its own in water, forms at low surfactant concentrations with *n*-decanol a large one-phase region of an L_α phase analogous to corresponding ternary systems with zwitterionic hydrocarbon surfactants [70] (Fig. 15). While some of the vesicles in the binary system are not closed and the phase contains lamellar fragments and small micelles, the lamellar phase in the ternary system contains closed multilamellar vesicles (Fig. 16).

The lamellar L_1^* and L_α phases in Fig. 15 seem not to be separated by a phase boundary. Both phases contain vesicles that are densely packed in the L_α phase, and the phases can be distinguished by their rheological behavior, as demonstrated in Fig. 17. The L_1^* phase is a Newtonian liquid without viscoelasticity, while the L_α phase shows viscoelasticity and also a yield stress value, which is due to the densely packed vesicles. The L_1^* phase is probably not a single phase but a two-phase region (L_1/L_α) that does not phase-separate. Contrary to the ternary system with a zwitterionic hydrocarbon surfactant, an L_3 phase was not found in the ternary system with the siloxane surfactant.

The rheological properties of the vesicle phases with the siloxane surfactant can be changed by charging the surfactant bilayers with small amounts of cationic or anionic surfactants [71]. Such modifications can be achieved by adding anionic or cationic hydrocarbon surfactants to the vesicle phase.

Anionic or cationic siloxane surfactants can also be used; it is more difficult, however, to obtain stable single-phase vesicle phases with the siloxane surfactants. The bilayers of the vesicles become stiffer with increased charge density, and this leads to significantly increased viscoelastic properties and yield stress values of the phases. These phases with charged vesicles need a certain amount of the cosurfactant to show viscoelasticity and yield stress values; if the cosurfactant content is too low, the systems show Newtonian flow behavior. It must be pointed out, however, that also in this case the kinetic processes in the phase are very slow; that is, it may be many weeks before the equilibrium state is reached. This can be seen from the rheological properties of systems that can change if allowed to stand for weeks.

V. SUMMARY

The hydrophobic group of siloxane surfactants consists of a siloxane backbone of variable chain length to which one or more anionic, cationic, zwitterionic, or nonionic hydrophilic groups are attached. These siloxane surfactants are strongly polar solvents, and are surface active in water where they are highly effective in most cases; minimum surface tensions around 20 mN/m can be obtained in aqueous solutions with many different siloxane surfactants. If the hydrophilic groups are replaced by hydrophobic ones like alkyl or perfluoroalkyl chains, the siloxane surfactants can also show surface activity in weakly polar and nonpolar solvents, in contrast to normal hydrocarbon surfactants.

Hydrophilic siloxane surfactants in aqueous solutions show aggregation behavior like that of hydrocarbon surfactants above a critical micelle concentration and

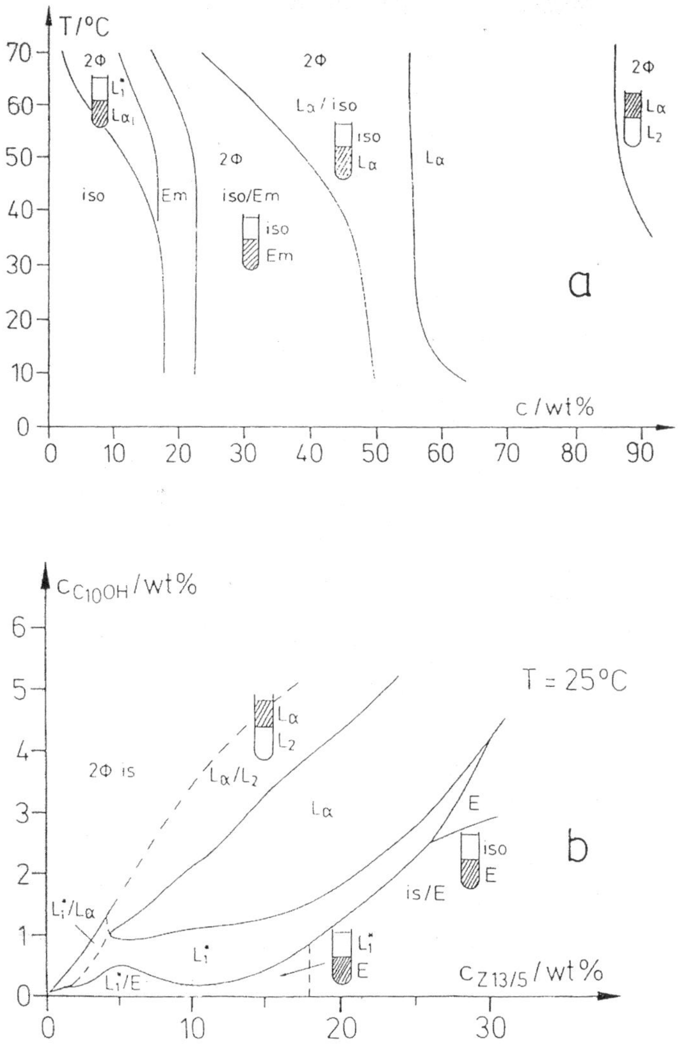

FIG. 15 Phase diagrams of (a) the binary systems of a zwitterionic graft polymer (M—D_{13}—D'_5(R)—M with R = —$(CH_2)_3$—O—CH_2—CHOH—CH_2—$N(CH_3)_2^+$—CH_2—CO_2^- and water and (b) the ternary system of the same graft polymer, *n*-decanol and water, at 25°C.

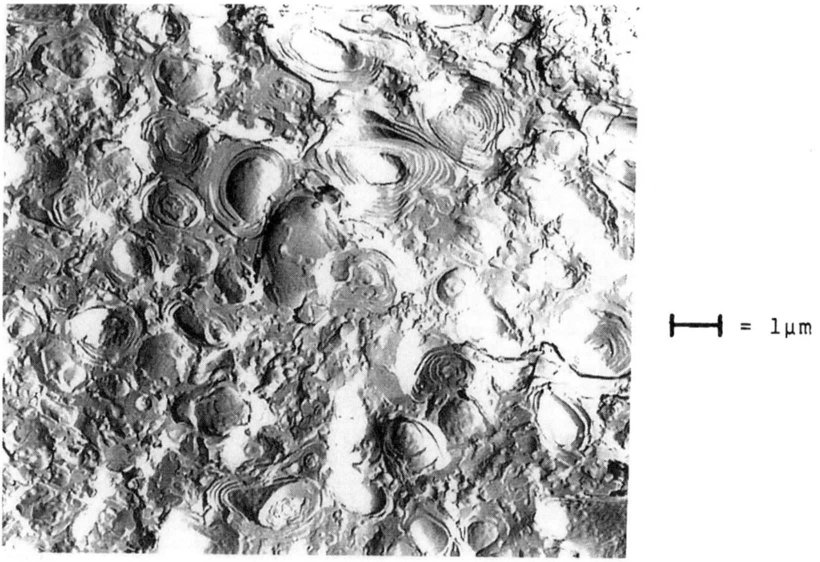

FIG. 16 Freeze-fracture TEM micrograph of the aggregates in an aqueous solution of 5 wt% of M—D_{13}—D'_5 (R)—M with R = —$(CH_2)_3$—O—CH_2—CHOH—CH_2—N^+ $(CH_3)_2$—CH_2—CO_2^- (zwitterionic graft polymer) and 1 wt% of *n*-decanol.

form lyotropic liquid crystalline mesophases at higher surfactants concentrations. The structure of the aggregates and the types of the mesophase appear to be correlated to the constitution of the siloxane surfactants. Samples with long hydrophilic chains form preferentially globular micelles above the cmc and a hexagonal phase as first mesophase, while samples with shorter hydrophilic groups show large vesicles above the cmc and a lamellar phase as first mesophase. The length of the hydrophobic siloxane chain does not seem to affect the aggregation and phase behavior of the siloxane surfactants.

Aggregation of hydrophobic siloxane surfactants in weakly polar and nonpolar solvents above a critical concentration (cac) is also very likely but has not yet been proven unambiguously.

Mixtures of siloxane surfactants with normal hydrocarbon surfactants can show ideal, antagonistic, or synergistic mixing behavior, which is determined mainly by the nature of the hydrophilic groups (because the hydrophobic siloxane and hydrocarbon groups do not show mutual phobicity). Hydrocarbons can be solubilized in micelles of siloxane surfactants, and the solubilization capacity is determined by the interfacial tension between the surfactant solution and the hydrocarbon. The same is true for the solubilization of silicone oil in hydrocarbon

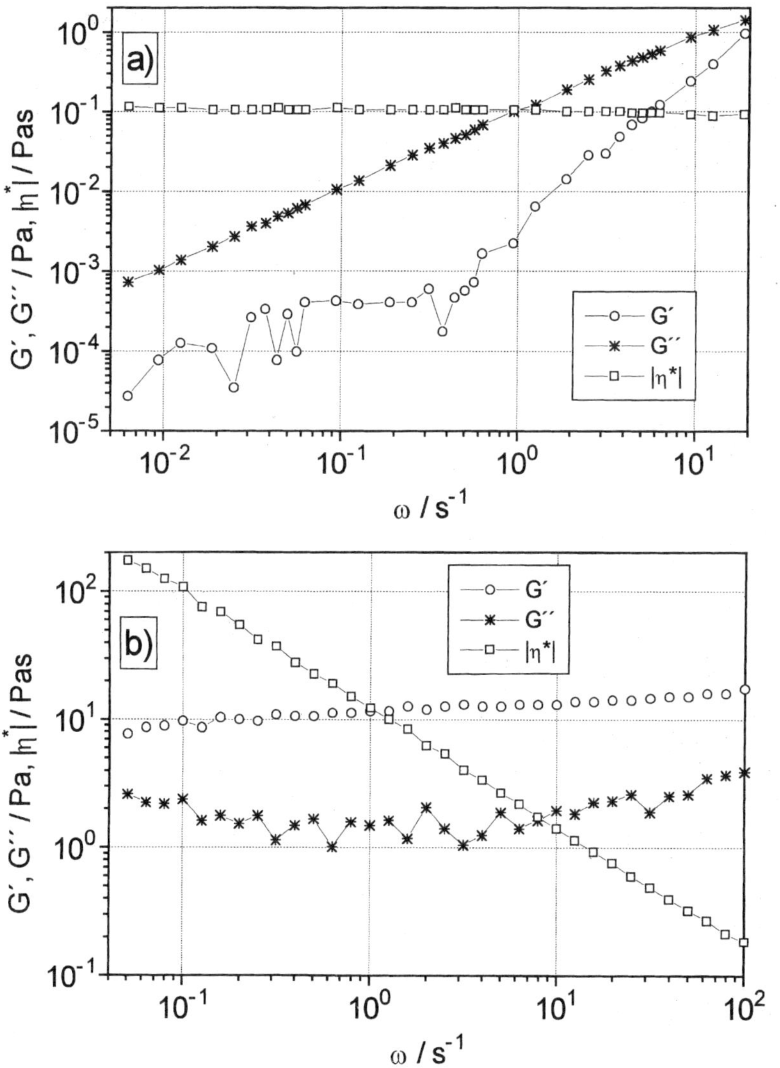

FIG. 17 The storage modulus G', the loss modulus G'', and the complex viscosity $|\eta^*|$ at 25°C of an aqueous solution of 10 wt% of a zwitterionic graft polymer ($M—D_{13}—D'_5(R)—M$ with $R = —(CH_2)_3—O—CH_2—CHOH—CH_2—N^+(CH_3)_2—CH_2—CO_2^-$) with (a) 1 wt% *n*-decanol or (b) 1.5 wt% *n*-decanol as a function of the oscillation frequency ω.

micelles. The solubilization can bring about a transition of rodlike micelles to spherical ones in both cases.

As in the case of hydrocarbon surfactants, the addition of cosurfactants like aliphatic alcohols to solutions of siloxane surfactants can favor the formation of vesicles and lamellar phases. Such systems are usually strongly viscoelastic and can show yield stress values, while solutions with globular micelles mostly are Newtonian liquids with low viscosity values.

REFERENCES

1. W. Noll, in *Chemie und Technologie der Silicone*, Verlag Chemie, Weinheim, 1968.
2. B. Grüning and G. Koerner, *Tenside Surf. Deterg. 26*, 312 (1989).
3. S. A. Snow, W. N. Fenton, and M. J. Owen, *Langmuir 6*, 385 (1990).
4. S. A. Snow, W. N. Fenton, and M. J. Owen, *Langmuir*, *7*, 868 (1991).
5. P. Robisch, Master thesis, University of Bayreuth, 1989.
6. M. Gradzielski, H. Hoffmann, P. Robisch, W. Ulbricht, and B. Grüning, *Tenside Surf. Deterg. 27*, 367 (1990).
7. W. Noll, C. Sucker, and A. De Montigny, *Kolloid Z. Z. Polym. 251*, 643 (1973).
8. H. W. Fox, E. M. Solomon, and W. A. Zisman, *J. Phys. Chem. 54*, 723 (1950).
9. A. Stürmer, C. Thunig, H. Hoffmann, and B. Grüning, *Tenside Surf. Deterg. 31*, 90 (1994).
10. C. Tanford, in *The Hydrophobic Effect*, John Wiley & Sons, New York, 1980.
11. J. W. Gibbs, *Trans. Connecticut Acad. 3*, 375 (1876).
12. M. J. Rosen, *J. Am. Oil Chem. Sci. 49*, 293 (1972).
13. M. J. Rosen, in *Surfactants and Interfacial Phenomena*, John Wiley & Sons, New York, 1978.
14. G. Koerner, G. Rossmy, and G. Sänger, *Goldschmidt Inform. 29*, 2 (1974).
15. F. Willis, *J. Colloid Interface Sci. 35*, 1 (1971).
16. N. L. Jarvis, *J. Colloid Interface Sci. 29*, 647 (1969).
17. K. P. Anathapadmanabhan, E. D. Goddard, and P. Chandar, *Colloids Surfaces 44*, 281 (1990).
18. A. Stürmer, Ph.D. thesis, University of Bayreuth, 1995.
19. G. Nilsson, *J. Phys. Chem. 61*, 1135 (1957).
20. E. G. Schwarz and W. G. Reid, *Ind. Eng. Chem. 56*, 26 (1964).
21. K. Shinoda, T. Nakagawa, B. Tamamushi, and T. Isemura, in *Colloidal Surfactants*, Academic Press, New York, 1963.
22. K. Shinoda, M. Hato, and T. Hayashi, *J. Phys. Chem. 76*, 909 (1972).
23. M. J. Owen, *Ind. Eng. Chem. Prod. Res. Dev. 19*, 97 (1980).
24. W. A. Zisman, in *Symposium on Adhesion and Cohesion Processes*, (P. Weiss, ed.), Elsevier, New York, 1962, p. 176.
25. T. E. Gentle and S. A. Snow, *Langmuir 11*, 2905 (1995).
26. I. Traube, *Annalen 265*, 27 (1891).
27. H. Kobayashi and M. J. Owen, *J. Colloid Interface Sci. 156*, 415 (1993).

28. L. Cantu, M. Corti, V. Degiorgio, H. Hoffmann, and W. Ulbricht, *J. Colloid Interface Sci. 116*, 384 (1987).
29. B. Schwandner, Ph.D. thesis, University of Bayreuth, 1986.
30. H. Hoffmann and A. Stürmer, unpublished results.
31. T. Svitova, H. Hoffmann, and R. M. Hill, *Langmuir 12*, 1712 (1996).
32. A. Stürmer, master's thesis, University of Bayreuth, 1991.
33. S. A. Snow, W. N. Fenton, and M. J. Owens, *J. Cell. Plast. 26*, 172 (1990).
34. N. Hasegawa, H. Morita, T. Doi, S. Hayashida, and Y. Yamaguchi, *Int. Prog. Urethanes 6*, 116 (1993).
35. D. Schaefer, *Tenside Surf. Deterg. 27*, 154 (1990).
36. G. Oetter and H. Hoffmann, *J. Dispersion Sci. Technol. 9*, 459 (1988–89).
37. H. Hoffmann, *Prog. Colloid Polym. Sci. 83*, 16 (1990).
38. H. Hoffmann and W. Ulbricht, *J. Colloid Interface Sci. 129*, 388 (1989).
39. J. Smid-Korbar, J. Kristl, and M. Stare, *Int. J. Cosmet. Sci. 12*, 135 (1990).
40. G. Schmauks, G. Sonnek, R. Wüstneck, M. Herbst, and M. Ramm, *Langmuir 8*, 1724 (1992).
41. R. Wagner, G. Sonnek, R. Wüstneck, A. Jänicke, M. Herbst, L. Richter, and L. Engelbrecht, *Tenside Surf. Deterg. 31*, 344 (1994).
42. F. Tiberg and A. M. Cazabat, *Europhys. Lett. 25*, 205 (1994).
43. E. D. Goddard, K. P. Ananthapadmanabhan, and P. Chandar, *Langmuir 11*, 1415 (1995).
44. P. Mukerjee and K. J. Mysels, in Critical Micelle Concentrations of Aqueous Surfactant Systems, NSRDS-NBS 36 (1971).
45. G. S. Hartley, in *Aqueous Solutions of Paraffin Chain Salts*, Hermann, Paris, 1936.
46. E. Hutchinson, A. Inabe, and L. G. Bailey, *Z. Phys. Chem. (Frankfurt) 5*, 344 (1955).
47. I. N. Israelachvili, D. J. Mitchell, and B. W. Ninham, *J. Chem. Soc. Faraday Trans. 2 72*, 1525 (1976).
48. N. W. Ashcroft and J. Lekner, *Phys. Rev. 145*, 83 (1966).
49. J. H. H. M. Moonen, Ph.D. thesis, University of Utrecht, 1987.
50. A. Vrij, *J. Chem. Phys. 69*, 1742 (1978).
51. C. Tanford, *J. Phys. Chem. 76*, 3020 (1972).
52. J. Kalus,H. Hoffmann, K. Reizlein, W. Ulbricht, and K. Ibel, *Ber. Bunsenges. Phys. Chem. 86*, 37 (1982).
53. R. M. Hill, M. He, Z. Lin, T. Davis, and L. E. Scriven, *Langmuir 9*, 2789 (1993).
54. M. He, Z. Lin, L. E. Scriven, H. T. Davis, and S. A. Snow, *J. Phys. Chem. 98*, 6148 (1994).
55. M. He, R. M. Hill, Z. Lin, L. E. Scriven, and H. T. Davis, *J. Phys. Chem. 97*, 8820 (1993).
56. D. Roux, F. Nallet, and O. Diat, *Europhys. Lett. 24*, 53 (1993).
57. C. Thunig, H. Hoffmann, U. Munkert, H. W. Meyer, and W. Richter, *Prog. Colloid Polym. Sci. 93*, 1 (1993).
58. H. Hoffmann and W. Ulbricht, unpublished results.
59. R. M. Hill, in *Procedings of the 65th Colloid and Surface Science Symposium* (P. M. Holland and D. N. Rubingh, eds.) ACS Symp. Ser. Vol. 501, American Chemical Society, Washington, DC, 1992, p. 278.

60. D. N. Rubingh, in *Solution Chemistry of Surfactants*, Vol. 1 (K. L. Mittal, ed.), Plenum Press, New York, 1979, p. 327.
61. J. H. Hildebrandt, J. M. Prausnitz, and R. L. Scott, in *Regular and Related Solutions*, Van Nostrand, New York, 1970.
62. J. H. Hildebrandt and R. L. Scott, in *Regular Solutions*, Prentice-Hall, Englewood Cliffs, NJ, 1962.
63. M. Ohno, K. Esumi, and K. Meguro, *J. Am. Oil Chem. Sci. 69*, 80 (1992).
64. E. Schönfelder and H. Hoffmann, *Ber. Bunsenges. Phys. Chem. 98*, 842 (1994).
65. H. Hoffmann and A. Stürmer, *Tenside Surf. Deterg. 30*, 335 (1993).
66. D. C. Steytler, D. L. Sargeant, and B. H. Robinson, *Langmuir 10*, 2213 (1994).
67. G. Hertel and H. Hoffmann, *Prog. Colloid Polym. Sci. 76*, 123 (1988).
68. G. Hertel and H. Hoffmann, *Liq. Cryst. 6*, 1883 (1989).
69. G. Hertel, Ph.D. thesis, University of Bayreuth, 1989.
70. G. Platz, C. Thunig, and H. Hoffmann, *Ber. Bunsenges. Phys. Chem. 96*, 667 (1992).
71. H. Hoffmann, U. Munkert, C. Thunig, and M. Valiente, *J. Colloid Interface Sci. 163*, 217 (1994).

5

The Science of Silicone Surfactant Application in the Formation of Polyurethane Foam

STEVEN A. SNOW Dow Corning Corporation, Midland, Michigan

ROBERT E. STEVENS Performance Chemicals Technology, Air Products and Chemicals, Inc., Allentown, Pennsylvania

I. INTRODUCTION AND HISTORICAL PERSPECTIVE

The largest commercial application of silicone surfactants is their use as additives for the production of polyurethane foam. Worldwide volume for silicone surfactants in polyurethane foam has been recently estimated at 30,000 metric tons/year [1]. The growth of the silicone surfactants market parallels the growth of the polyurethane foam market. The two are so intertwined that it is difficult to envision one without the other.

Polyurethane foam was first produced in the laboratories of the I. G. Farben Industries in Germany by Dr. Otto Bayer and his coworkers in 1941 and commercialized in 1954 [2]. These early foams were often stabilized by polydimethylsiloxane oils because of their high surface activity in this essentially "organic" medium [3,4]. The early foams, based on polyester polyols, were relatively high density materials that had outstanding properties. They were commercially suited for some applications but too expensive for many others [5]. Polyether polyols were introduced in 1957, but foams from these polyols originally had to be produced in two steps: formation of a prepolymer from the polyol and the isocyanate, and then foaming via the addition of water, catalyst, and stabilizers [6]. The prepolymer foams were also stabilized by silicone oils [3]. By 1958, new polymerization catalysts and new foam stabilizers allowed for the introduction of the "one-shot" process for preparing flexible polyurethane foams [7]. This process enabled the economical production of the wide variety of polyurethane foam that is common today.

In 1958 the preparation of organosilicone–polyether copolymers was patented by the Union Carbide Corporation [8]. While the announcement of one-shot foam processing does not mention silicone surfactants [7], Saunders explains that "The stabilizing effect of the alkylsilane–polyoxyalkylene copolymer (e.g., 'Silicone L-520,' Union Carbide) is truly remarkable, and must be given equal or greater credit than the tin catalysts for making possible the one-shot polyether foams" [4]. It should be noted that although it is possible to prepare flexible slabstock polyurethane foam without tin catalysts, such foams cannot be prepared without the use of silicone surfactants [9]. The patent literature indicates that Fritz Hostettler of Union Carbide originated the use of silicone polyether copolymers in polyurethane foam [10–12]. Although these patents did not issue until the 1960s, the application dates are in 1957 and 1958. Literature references to the use of silicone polyether copolymers in foam did not appear until 1960, long after the process had been established [9,13]. Since those exciting early years, every major development in polyurethane foam technology has been accompanied by developments in silicone polyether copolymer technology.

Polyurethane foam is produced for a broad range of applications and in many forms, from a rigid, pneumatic resin to a flexible, porous elastomer. Rigid polyurethane foam is primarily used as insulating material in construction (both wall and roofing insulation), appliances, piping, transportation, and packaging [14,15]. The major markets for flexible polyurethane foam are as cushioning materials in furniture, bedding, carpet underlay, automotive applications, and packaging [16]. The organosilicone–polyether copolymer surfactants used to stabilize these foams are currently available from a number of companies including Air Products and Chemicals, OSi Specialties (now part of the Witco Corporation), and Th. Goldschmidt AG [1].

Although reviews have been written concerning the use of silicone polyether copolymer surfactants in polyurethane foam, none has appeared in over 10 years

[17–22]. This chapter discusses the molecular structures of these surfactants as they are used in the various applications of polyurethane foam and reviews the scientific research that has been carried out to define the many complex functions these surfactants perform in the processing of polyurethane foam.

II. POLYURETHANE FOAM PROCESSES

A full discussion of polyurethane foam processing is beyond the scope of this chapter. Fortunately, a number of very good monographs have covered the subject over the past 40 years [2,4–6,14,15,23–27]. A brief overview is provided.

Bikerman [28] has provided some definitions that are helpful in framing this discussion:

Liquid foam—a gas dispersion in a liquid
Solid foam—a gas dispersion in a solid
Sponge—a bicontinuous mixture of a gas in a solid

As described below, processing of polyurethane foam involves all these compositions.

The preparation of a polyurethane foam involves the formation of gas bubbles in a liquid system (a combination of polyol, isocyanate, water, blowing agents, catalysts) and the growth and stabilization of those bubbles as the polymer forms and cures [3,4]. The initial liquid foam cures to a material ranging from a solid foam to a sponge. A rigid polyurethane foam is best described as a solid foam, while flexible polyurethane foams are best described as sponges. The process by which a foam is transformed into a sponge is defined as cell opening.

The general reaction chemistry is depicted in Fig. 1. The primary reactions involved in these systems are the formation of urethane (***1***) and urea (***2***) from the reactions of polyol and water with the isocyanate. These reactions are termed the "gelling" and "blowing" reactions, respectively. Possible secondary reactions involve further reaction of the urethane or the urea with isocyanates to form allophanates (***3***) or biurets (***4***). Another reaction that is very important in rigid foam is the trimerization of isocyanates to form isocyanurates (***5***). All these reactions are significantly exothermic. The relative rates of these reactions can be controlled through the judicious choice of catalysts [29–31]. Typical catalysts include amines and organotin compounds.

A "polyurethane" is more precisely a copolymer of "soft" polyol segments and "hard" urea segments. The urea moieties interact very strongly with each other via hydrogen bonding, forming a cross-linked network. This network can phase-separate from the mixture.

The viscoelasticity of the mixture during the foaming process is a complicated function of the rate of polymerization, the temperature, the extent of urea phase separation, and other factors. The viscosity profile of the foaming mixture was recently reported by McCluskey and coworkers [32]. A key milestone in the pro-

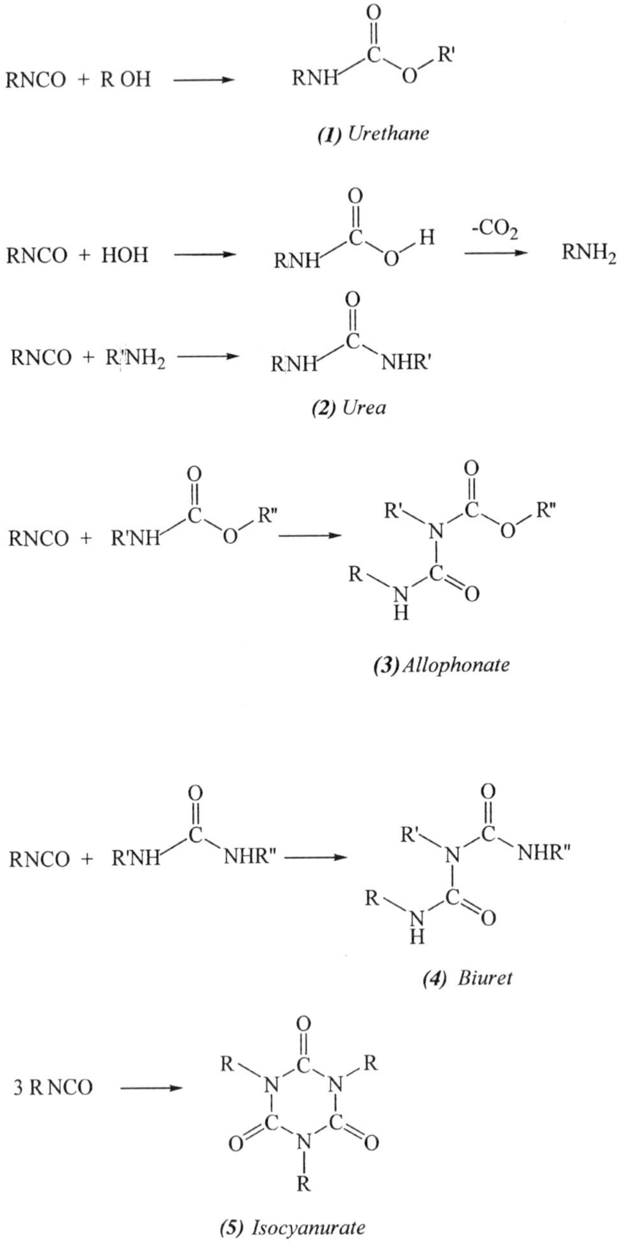

FIG. 1 Polyurethane reaction chemistry.

file is the rapid gelation that occurs during the phase separation of the polyurea segments.

The mechanical properties of the final cured foam are a function of both the soft (polyether) and the hard (urea) segments. For the soft segment, critical factors influencing the mechanical properties are the molecular weight and functionality of the polymer polyol. For the hard segment, the extent of phase separation is a critical factor. The final foam can range from a soft elastomer to a rigid resin.

III. MOLECULAR STRUCTURES OF SILICONE POLYETHER COPOLYMER SURFACTANTS USED FOR POLYURETHANE FOAM

A. General Structures

In practice, a wide range of silicone surfactant structures are necessary to ensure compatibility with the tremendous variety of raw materials used to produce polyurethane foams having a broad range of physical properties. This structural diversity also reflects the necessity for the surfactant to perform many different tasks in different foam formulations. For example, with flexible slabstock foam, the primary function of the surfactant is to stabilize the growing foam bubbles (cells). However, with close-celled, rigid foam, an important surfactant function is to emulsify the incompatible ingredients within the foam mixture.

Fortunately, the flexibility of siloxane chemistry allows for the synthesis of an infinitely wide range of surfactant structures. Just as important, within this wide range, surfactant structure can be tailored toward a specific formulation or application. The siloxane and polyether average degree of polymerization (or molecular weight—MW) can be precisely controlled. The ratio of polyether MW to siloxane MW allows for the proper balance of surfactant solubility and surface activity in a given formulation or application. This balance can be fine-tuned to an even greater degree of precision by altering the ratio of ethylene oxide and propylene oxide in the polyether. Ethylene oxide is more polar and "hydrophilic" than propylene oxide. Branching can be introduced in either the siloxane or the polyether to change the properties of the final surfactant. Finally, different end groups can be placed on the siloxane and/or the polyether to alter the performance of the surfactant. The synthesis of these materials has been dealt with in detail elsewhere [33].

The four most common structural types of silicone surfactant, shown in Figs. 2–5, are as follows:

$$Me_3SiO\text{-}(SiMe_2O)_x\text{-}SiMe_3$$

FIG. 2 Structure of silicone oil.

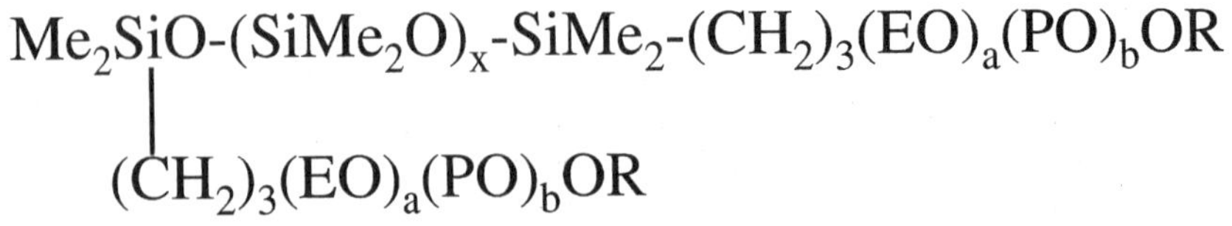

FIG. 3 ABA block copolymer structure.

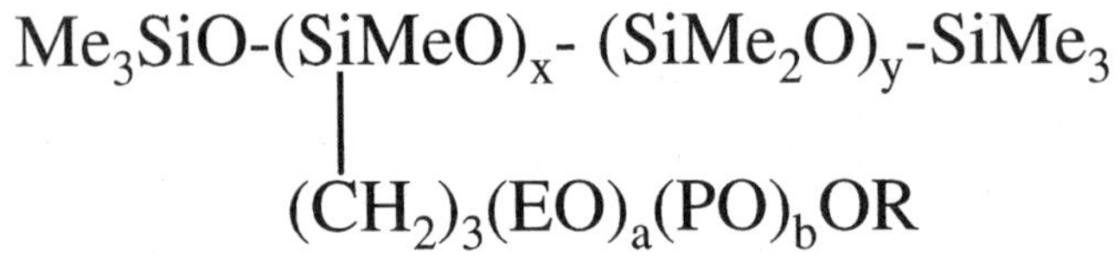

FIG. 4 Graft copolymer structure.

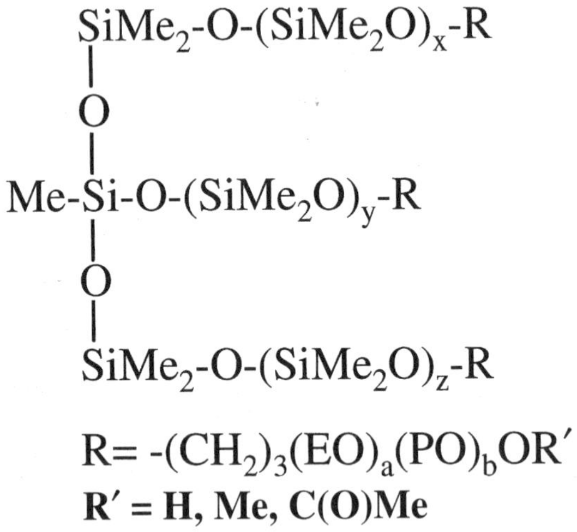

FIG. 5 Highly branched structure.

Silicone oil—or polydimethylsiloxane
ABA block copolymer structure
Graft copolymer structure
Branched structure

An overview of the structural ranges of the silicone surfactants used in the various types of polyurethane foam is presented in Table 1. General ranges only are presented. Given surfactants used in a particular application may fall outside these ranges. The materials used in each application are discussed in more detail below.

TABLE 1 Structural Ranges of Silicone Surfactants for Polyurethane Foam

Application	Surfactant molecular weight	Weight percent silicone	Polyether molecular weight	Weight percent EO in polyether
Flexible molded	300–1500	30–100	0–800	0–100
Rigid	1500–15,000	20–60	400–2000	60–100
Flexible slabstock	20,000–80,000	15–30	1000–4000	35–65

B. Flexible Foam Surfactants

Flexible polyurethane foams are sponges (i.e., they are open celled). There are two major types of flexible foam: slabstock and molded.

1. Flexible Slabstock Foam Surfactants

In general, the surfactants for flexible slabstock foam are higher in molecular weight than those used for other types of foam. Stability of the liquid foam as it rises is of primary concern in these systems. The surfactants that made one-shot polyether flexible foam processing possible were of the branched structural type [4]. Graft copolymer surfactants for flexible slabstock foam became available in the mid-1960s [34,35]. Surfactants of both structural types are used today.

The graft copolymer surfactants were considered to be higher in potency (less surfactant is necessary to stabilize the foam) but lower in processing latitude (a narrower range of formulations and catalyst levels over which the surfactant can be used) than the branched surfactants. Recent advances in the graft copolymer surfactants have produced materials that have both high potencies and broad processing latitudes [36,37].

The polyethers used in these surfactants have inert end groups to prevent reaction of the surfactant with isocyanate. Normal end groups include esters and ethers. The polyethers are of fairly high molecular weight (1000–4000 Da) and usually are close to equimolar in ethylene oxide and propylene oxide content. Surfactants containing all–ethylene oxide polyethers do not lower the surface tension of polyether polyols and as such are not useful in flexible slabstock systems [18]. Surfactants containing all–propylene oxide polyethers have also been shown to be ineffective in flexible slabstock systems [18].

2. Flexible Molded Foam Surfactants

In marked contrast to the flexible slabstock surfactants, the materials used in flexible molded foam are small molecules with molecular weights usually less than 1500 Da. In molded formulations, the liquid foam is stabilized chiefly by the high bulk viscosity of the system as provided by high molecular weight polyols and isocyanates. The role of the surfactant in these formulations is to provide fine uniform cells and liquid foam stability where the foam is stressed against the surfaces of the mold. Depending on the formulation, silicone fluids or oils can provide sufficient stabilization, and silicone polyether copolymers may not be necessary [38,39]. Because of the high viscosity of the liquid foam in flexible molded systems, cell opening is critical. Special cell-opening surfactants have been developed for this purpose [40,41].

C. Rigid Foam Surfactants

The surfactants used to process rigid polyurethane foam are intermediate in size between the materials used in flexible molded and flexible slabstock formulations. Graft copolymer surfactants dominate this application. As in flexible slabstock foam, the original materials for rigid foam were highly branched structures with polyethers bonded to the siloxane via Si—O—C linkages. However, stability problems associated with hydrolysis of the Si—O—C linkage led to the replacement of such materials with the more hydrolytically stable graft copolymer surfactants, which contain Si—C backbone–polyether linkages.

The major roles of silicone surfactants in rigid formulations are cell size control (providing a fine-celled structure with a narrow cell size distribution) and emulsification. Emulsification is extremely important in rigid applications because the raw materials for rigid foams are incompatible [42]. Owing to the importance of emulsification, the polyethers used in rigid foam surfactants contain higher levels of ethylene oxide than the polyethers used for flexible slabstock foam surfactants [43]. Generally the polyethers are hydroxyl-terminated, but if the formulator uses surfactant in the isocyanate side of the formulation, materials with ester or ether end caps are available [44].

There has been intense effort in last 10 years to develop surfactants that allow the use of non-CFC blowing agents. This effort has led to the introduction of a number of new surfactants [45–47].

IV. SURFACTANT BEHAVIOR IN POLYURETHANE FOAMS

A. Surfactant Behavior in Interfacial Processes

1. Overview of Interfacial Processes

Many complex processes occur during the formation of a polyurethane foam. These processes can be conveniently divided into interfacial and bulk processes. This section gives an overview of interfacial processes, with an emphasis on how the silicone surfactant affects them.

Figure 6 outlines a simplistic view of the interfacial processes during the life of a foam. As indicated in Fig. 6, foams are thermodynamically unstable systems. A significant amount of gravitational and surface energy is adsorbed during foam

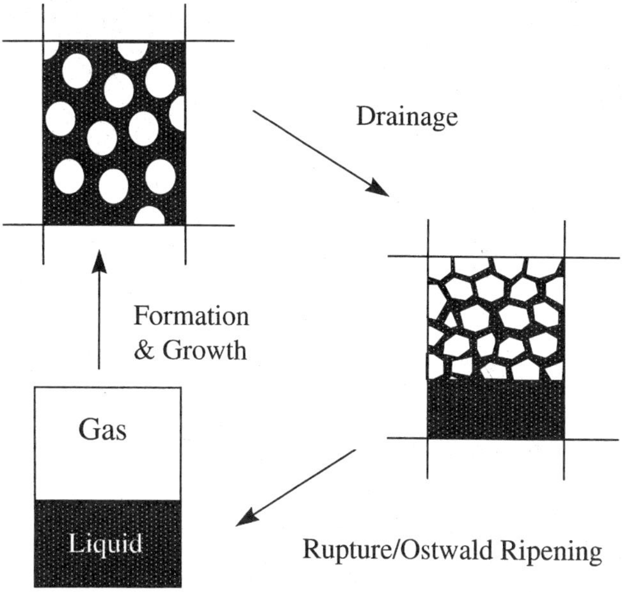

FIG. 6 Foam processes.

formation. In polyurethane foams this energy is provided by high shear mixing and the release of chemical energy during the formation of the polyurethane.

The earliest interfacial process is the initial formation of bubbles in the liquid. Kanner and coworkers [48] demonstrated that there is no spontaneous nucleation of the bubbles in polyurethane foams; the bubbles have to be stirred in. These initial bubbles are small, their diameters being on the micrometer scale.

Once bubbles have formed, they must remain stable during their growth phase. As CO_2 gas is formed in the blowing reaction, it expands these tiny bubbles. They can also expand when auxiliary blowing agents such as methylene chloride volatilize. The expansion of the bubbles increases the overall surface area and therefore surface energy of the foam. At a constant overall surface area, the total surface energy absorbed can be lessened by reducing the energy per unit area of the liquid (or equivalently, the surface tension). In polyurethane foams, the surface tension of the liquid is reduced by the addition of silicone surfactants.

Both molded (fixed-volume) and free-rise (volume not fixed) polyurethane foams are produced commercially. For a molded foam, at a fixed energy input, reduction of the surface tension of the liquid results in the formation of more and necessarily smaller bubbles as the available surface area is increased. For a free-rise foam, a reduction in surface tension can result in an increase in foam volume and/or a decrease in bubble size. It has been shown that addition of a silicone surfactant to a polyether polyol allows five to seven times more gas to be mixed with the polyol than when the surfactant is absent [48]. This increase in foam volume is consistent with the reduction of liquid surface tension by the surfactant.

The stability of a foam is inversely proportional to the rate at which surface and gravitational energy is released. Any process that reduces the surface area of a foam releases energy. As shown in Fig. 6, these processes include bubble coalescence and the diffusion of gas from smaller to larger bubbles ("Ostwald ripening"). Gravitational energy is released during the drainage of liquid down the foam.

Bubble and foam stabilization is an important function of the surfactant in polyurethane foams, particularly in flexible slabstock polyurethane foam. A surfactant can be perceived to play the role of a retarding agent for the processes of bubble coalescence, Ostwald ripening, and liquid drainage. Retarding bubble coalescence involves stabilizing the liquid film between the bubbles to both thinning processes and rupture. The film can thin in a small area (stretching) or across a large area (drainage). Thinning of the film can be counteracted by a high bulk viscosity, a sufficiently viscoelastic surface layer, or by surface transport of surfactant and bulk material to the thinned regions, essentially "healing" them. This transport proceeds along gradients from low to high surface tension and is commonly referred to as the Marangoni effect. Universal drainage of the film distorts the initial spherical bubble into complex polyhedral shapes containing flat walls (Fig. 6). Two common shapes are the pentagonal dodecahedron and the tetrakaidecahedron [49].

The rate of thinning of vertically supported, thin liquid films of polyether solutions of various silicone surfactants was measured [50]. It was found that the rate of thinning for polyol solution films of an "MQ resin," an extremely branched siloxane not containing polyethers, was much slower than the rate for the films of the same polyol containing commercial silicone polyether copolymer surfactants. The retardation of drainage rate was correlated to an increase in surface viscosity.

The thinning of liquid films between bubbles promotes film rupture. The rupture process possibly involves the formation of a hole in the liquid film followed by the growth of the hole. Both bulk and surface transport processes could be involved in film rupture. Therefore, the surfactant may play a role in film rupture. Literature data link the film rupture process to the phase separation of polyurea segments [51–54]. It is possible that the polyurea acts as a nucleating agent for the process of hole formation.

For a closed-cell, rigid polyurethane foam, it is necessary to suppress bubble coalescence (cell opening) throughout the foaming process. However, to produce an opened-cell flexible polyurethane foam, some degree of bubble coalescence must occur to allow air to flow through the foam.

A surfactant can also retard the destruction of the foam by retarding the rate of gas diffusion from smaller to larger bubbles. This diffusion is driven by a pressure gradient; the pressure within smaller bubbles is greater than that within larger bubbles. A surfactant can alter the rate of gas transport between bubbles by changing the mechanical properties of the thin liquid film between the bubbles. Measurements by Owen et al. [55] have shown that the rate of bubble–bubble gas diffusion in stable polyurethane foams is negligible.

2. Surfactant Adsorption at the Liquid–Air Interface

The most critical process surfactants undergo in foams is their adsorption at interfaces, particularly the liquid–air interface. In fact, the adsorption of a surfactant at the liquid–air interface is a prerequisite for a foam to form and have a finite lifetime. Surfactant adsorption at any interface, including the liquid–air interface, changes the mechanical properties of the interface. The key mechanical properties of an interface are tension (energy) and viscoelasticity. Changes in these mechanical properties strongly influence the interfacial processes that occur during foam formation, growth and destruction.

(a) Reduction of Surface Tension (Energy). When a surfactant spontaneously adsorbs at an interface, tension at the interface is eased, and energy is released. On a molecular level, when the surfactant adsorbs at the interface, it aggregates to form an coherent monolayer and conformationally orients, through bond rotation and bending, to allow for the lowest surface energy, or tension, possible.

For a surfactant to aid in the growth and stabilization of a polyurethane foam, it must reduce the surface tension of the foaming liquid, which is predominantly a polyether polyol. The surface tension of these polyols ranges from 33 to 40 mN/m. This value is so low that it cannot be further reduced by the adsorption of hydrocarbon-based surfactants. Essentially, they are not surface active in this medium. However, silicone surfactants can reduce the polyol surface tension to a much lower value of 21–25 mN/m [3,4,17–22,48,56–64].

The reduction of polyol surface tension is a function of specific structural features of the siloxane surfactant. First, the siloxane backbone of the surfactant is virtually insoluble in the liquid medium, and therefore is driven to adsorb at the liquid–air interface. Second, the siloxane backbone is "flexible," allowing for a low energy orientation of methyl groups on the air side of a liquid–air interface. This flexibility is the result of the strikingly low Si—O bond rotation and bending energies in the siloxane backbone [22]. Furthermore, the methyl group has intrinsically low surface energy, a result of weak van der Waals attractive forces [22]. Third, the minimal surface tension is a function of the both the length of the siloxane backbone and, specifically, the ratio of siloxane monomeric (Me_2SiO) units to polyether chains per molecule. It has been demonstrated that to reach the limiting value of 20.5–21 mN/m, there must be between 10 and 20 Me_2SiO units in the siloxane backbone per polyether chain [57]. This is the minimum chain length necessary for the siloxane backbone of the copolymer to have the flexibility of the backbone of unsubstituted polydimethylsiloxane, which has a surface tension of 20.8 mN/m [22]. For surfactants that do not meet that criterion, an increase in the ratio of Me_2SiO units to polyether chains per molecule reduces the surface tension [57,60,62]. Finally, the minimum surface tension has been shown to increase with the percent branching in the siloxane portion [60]. Branching in the siloxane portion probably restricts the flexibility of the siloxane backbone to some extent.

A more detailed analysis of surface free energy involves separate measurement of surface enthalpies and entropies. Dubjaga concluded that the reduction of surface free energy in polyether–silicone surfactant solutions is due primarily to increases in surface entropy [62]. During the process of adsorption of the silicone surfactant at the polyol–air interface, one can imagine the replacement of polyol polymers at the surface layer by polydimethylsiloxane (PDMS) polymers. The greater flexibility of PDMS, as mentioned earlier, would increase the number of energetically accessible "states" or entropy of the surface layer.

(b) Surface Molecular Configurations. The Gibbs equation, which may be used to analyze the quantitative connection of surface tension and surfactant adsorption [65], is as follows:

$$\frac{d\gamma}{d \log C} = -2.303\ RT\Gamma \tag{1}$$

where γ is surface tension, C is surfactant concentration, R is the gas constant, T is the temperature, and Γ is the surface adsorption of the surfactant.

The surface adsorption of the surfactant, Γ, which is expressed in surface concentration units (mol/cm^2), is inversely proportional to the area per surfactant molecule at the interface, as demonstrated by Eq. (2):

$$A = \frac{10^{16}}{N\Gamma} \tag{2}$$

where A is the area per molecule and N is Avogadro's number.

Studies of the area per molecule of different surfactants as a function of their molecular structures give insight on the configuration of the molecule at the surface. Owen et al. investigated the surface adsorption for a series of ABA (polyether–silicone–polyether) block copolymers at the liquid (water or tripropylene glycol or LG56 polyol)–air interface [57,58]. These copolymers varied only in the length of the "B" block, the siloxane portion. The area per molecule (A/M) was a function of the solvent type, with the values in water (60 Å^2) smaller than in the two glycols (240 Å^2). It was concluded that the A/M was a function of the nature of solvation of the polyether chain.

In water the polyether was assumed to be extensively hydrated, through strong hydrogen-bonding interactions, and therefore in an extended configuration. However, the A/M did not vary with the siloxane chain length. These results were consistent with a configuration in which the siloxane chain was "doubled" or "folded over." Molecular models predicted an A/M of 62 Å^2 for this configuration, in excellent agreement with the experimental data. In the glycols, since the A/M is four times as large, and presuming that the siloxane configuration is the same as it is in water studies, the A/M must reflect structural changes in the polyether por-

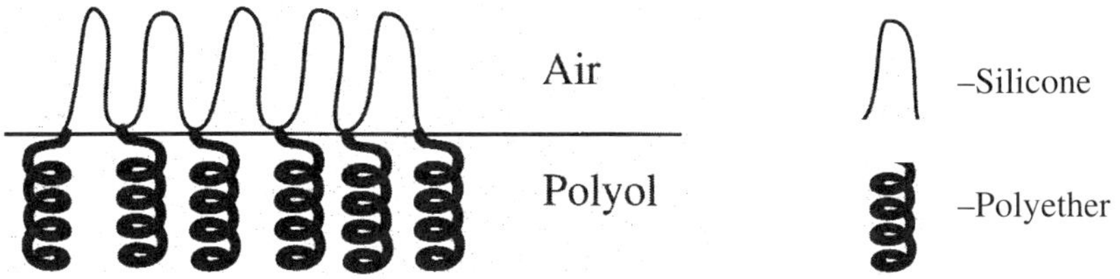

FIG. 7 Orientation of a silicone polyether copolymer at the air–polyol interface.

tion of the copolymers. The diminishing of solvent–polyether hydrogen bonding when water is replaced by the glycols could cause a coiling of the polyether chains, and a significant increase in the A/M. This configuration is shown in Figure 7.

Dubjaga [62] investigated the effect on A/M at the polyether–air surface of increasing the siloxane length in siloxane–polyether graft copolymers. The area per molecule decreased from approximately 220 Å^2 to 110 Å^2 when the length of the siloxane chain was increased from 40 units to 53 units. This halving of the area might be due to a folding up of the siloxane chain, since the longer chain would have a greater freedom of movement than the shorter chain. Values for the A/M for graft copolymers in polyethers of approximately 100–110 Å^2 were also measured by Dubyaga et al. [63]. They proposed that the siloxane chain was in a "spiral" conformation in these graft copolymers.

The A/M of silicone surfactants at the air–liquid interface, if the surfactant is insoluble in the liquid, can be measured using spread monolayer film techniques. These monolayers, once formed, can be subjected to highly controlled compression/expansion cycles, with the result being the generation of surface area versus pressure plots. Kanner et al. investigated the monolayer properties of a number of silicone surfactants on water [21]. Although these surfactants were not insoluble in water, the authors rationalized that the rates of surface adsorption/desorption were so slow that they could be ignored. In this study, good polyurethane foam stabilizers yielded spread monolayers on water with a well-defined "transition zone." The zone was interpreted as representing the transition between two surface states of the monolayer. A dispersed state could transform into a condensed state as the molecules are pushed closer together. Alternatively, the transformation could be due to a concerted change in the configuration of the molecules such as the folding up of an extended siloxane chain to a doubled-over configuration.

(c) Dynamic Surface Tension Measurements. One might challenge the relevance of the measurements of equilibrium surface tension toward the actual foam process. During foam formation bubbles are growing, continuously increasing the overall surface area. It is uncertain, and probably doubtful, whether at any point in time an equilibrium exists between surfactant dissolved in the bulk liquid and that

which is adsorbed at the surface. Therefore, one should also be concerned with the rate of attainment of air–liquid interfacial equilibrium. This quantity is more commonly expressed as a "dynamic" surface tension (the faster the rate of attainment of equilibrium, the lower the dynamic surface tension).

Dynamic surface tension plays a role in foam stabilization. The Marangoni effect on bubble stabilization was mentioned in Sec. IV, A.1. Maximum foam stability is obtained when surface tension gradients are maintained long enough to allow for surface transport to a thin spot. Surface tension gradients can also be relieved by the diffusion of surfactant from the bulk of the liquid to the surface. The rate of this process is inversely proportional to the dynamic surface tension. Therefore, there is an optimal rate of surfactant diffusion/surface adsorption for maximum foam stability or, correspondingly, an optimally low dynamic surface tension.

The rate of attainment of equilibrium surface tension in surfactant solutions of polyols has been shown to be slow [3,57]. This finding suggests that the rate of surfactant diffusion is reduced both by large surfactant molecular volume and the high bulk viscosities of the polyols. The dynamic surface tension of various linear and branched siloxane surfactants in polyol was investigated by Owen and coworkers [66]. The dynamic surface tension was inversely proportional to the surfactant concentration in the range of 0.2–2 wt% surfactant. These results were confirmed by two other studies [63,64]. A strong correlation of the dynamic surface tension and the percentage of siloxane in the copolymer was demonstrated [63,64,66]. This correlation was interpreted as indicating that the siloxane moiety, being quite immiscible with the solvent polyol, is energetically driven to adsorb at the surface. Increasing the size of the siloxane increases this driving force.

(d) Surface Viscoelasticity. The adsorption of the surfactant at the bubble surface can also result in an increase in surface viscoelasticity. A higher surface viscoelasticity increases the energy necessary to stretch or otherwise disturb the film. An increase in surface viscosity could also retard the rate of drainage of liquid from the films.

Many of the investigators in this field have commented on the viscoelastic nature of polyurethane foam cell membranes (or cell windows, or lamella, or liquid films) and how silicone surfactants affect this viscoelasticity [2–4,16,18–23, 32,48,50–53,55,59,61–64,66–69].

A number of techniques have been employed to quantify the surface viscoelasticity of surfactant–polyol solutions. One technique is to measure the rupture length of a liquid film vertically withdrawn from a polyol–surfactant mixture [61–63]. This rupture length was considered to be proportional to the surface viscoelasticity of the film. It was shown that the addition of silicone surfactants to pure polyols resulted in a significant increase in the rupture length.

It has been demonstrated that the surface viscosity of unsubstituted polydimethylsiloxane is barely above zero [22]. This suggests that the polyether portion of the surfactant is the source of the surface viscosity [22,55,69]. These findings suggest that the ratio of siloxane to polyether in the surfactant will result in a specific balance, a compromise, of surface tension reduction and surface viscosity. Increasing the proportion of silicone to polyether decreases the surface tension and the surface viscosity, and vice versa. It is speculated that the surface viscosity contribution from the polyether portion of the surfactant is due to the self-association of the polyether chains, and/or their interaction with the solvent, perhaps through hydrogen bonding [55].

Surface viscosity can be broken down into more fundamental quantities (e.g., surface shear viscosity and surface dilatational viscosity). Investigators have attempted to measure surface shear viscosities of polyol solutions of silicone surfactants [55,64,69]. Surface shear viscosity was found to increase with surfactant concentration and to be higher for graft copolymer structures relative to ABA block copolymer structures. The increase in viscosity with increases in surfactant concentration is consistent with the formation of a more tightly packed surface monolayer with more attractive interactions throughout the layer. Therefore more energy is required to move the molecules relative to each other, the essence of the viscosity measurement. Regarding the higher surface viscosity of graft copolymers versus the block copolymers, it has been shown that at low molecular weights branched siloxane polymers are more viscous than linear siloxane polymers [70]. At constant chain length, relative to neighboring linear structures, neighboring branched structures have more points of contact or "topological constraints" [70]. This increase in topological constraint increases the energy of relative molecular movement and therefore the viscosity.

From their spread film studies, Kanner et al. [21] speculated that the presence of a well-defined "transition zone" correlated to the ability of the surfactant layer to withstand compression and/or expansion, which could aid in foam stabilization. This speculation was validated by their observation that silicone surfactants could withstand much higher surface pressures than unsubstituted PDMS before collapse of the monolayer.

3. Surfactant Adsorption at Liquid–Liquid and Polymer–Polymer Interfaces

(a) Adsorption at Liquid–Liquid Interfaces. The activity of silicone surfactants within the polyurethane foam at interfaces other than the polyol–air interface has also been investigated. Interfaces of interest include liquid–liquid interfaces, which are present in the early stages of foam formation because water is immiscible in isocyanates and partially miscibile in (primarily) polyoxypropylene-containing polyether polyols. Silicone surfactants can potentially adsorb at these interfaces, reducing interfacial tension, decreasing the particle size of the dispersed phase liquid, and increasing interfacial viscoelasticity. Another intriguing possibility is that the surfactant could potentially mediate the interfacial reactions occurring during foam formation (e.g., the isocyanate–water reaction).

Hager and coworkers investigated the role of silicone surfactants in the interfacial mixing of these various liquids [71]. They found that the surfactants promoted the interfacial mixing of water and polyol. Fortuitously, the ability of the surfactant to aid in the mixing of polyol and water correlated to its foam stabilizing capacity. Kopusov and coworkers demonstrated that silicone surfactants alter the polyurethane polymerization, probably through activity at liquid–liquid interfaces [67].

(b) Adsorption at Polymer–Polymer (Urea–Polyol) Interfaces. An important feature of the formation of certain polyurethane foams, in particular open-celled flexible polyurethane foams, is the aggregation and ultimate phase separation of urea molecular segments [2,16,23,32,51–53,56,71–73]. The aggregation process is driven by strong hydrogen bonding between the segments.

This aggregation process has significant implications for the application properties of flexible polyurethane foam. The load-bearing properties of the foam are a strong function of the size and concentration of these "hard" polymer segments, and the phase separation of urea appears to occur just prior to gas escape ("sigh-back" or "blow-off") in these foams [51–54].

Assuming that gas escape corresponds to a significant degree of cell opening (film or bubble rupture), it seems likely that urea precipitation is linked to cell opening. One possibility is that the urea precipitate acts as a "nucleating agent" for the formation of a hole or fracture tip in the bubble film.

Research on the effect of the silicone surfactant on the phase separation of urea segments has been published by Rossmy and coworkers [51–53] and Neff [54]. The key discovery was that when an effective silicone surfactant is present, the urea phase separation does not appear to cause foam collapse, as it does when the surfactant is not present. In essence, the silicone surfactant "stabilizes" the precipitating urea, and the degree of stability roughly correlates to the percentage of ethylene oxide in the surfactant. Presumably, the ethylene oxide in the surfactant can interact strongly with urea via the formation of hydrogen bonds.

Snow and coworkers studied the potential effect of a wide range of silicone and organic surfactants on urea in flexible polyurethane foams [73]. Some of the silicone surfactants studied were specifically designed to interact strongly with urea through hydrogen bonds and other strong polar bonds. However, specific surfactant–urea effects were not observed.

B. Surfactant Behavior in Bulk Processes

1. Surfactant Interactions with Other Foam Ingredients in the Bulk Liquid Phase

Although silicone surfactants are primarily applied in polyurethane foams because of their significant interfacial activity, they also can also potentially undergo specific interactions with other molecules in the bulk liquid phase. These interactions can lead to changes in bulk processes during foam formation such as polymerization and viscoelasticity buildup.

For example, in rigid and molded polyurethane foams, silicone surfactants are sometimes added to modify the bulk rheological properties of the foam. This includes their use as plasticizers, flow modifiers, and thickening agents. The physical mechanisms underlying this behavior have not been extensively discussed in the literature.

In one study Lipatova and coworkers [56] demonstrated that the presence of silicone surfactants decreased the viscosity of a polyurethane elastomer (not a foam) in early stages of formation, shifting the beginning of gelation to a higher degree of isocyanate group reaction. The time elapsed before gelation was shown to be a function of the surfactant concentration. The polymerization rate was shown to be a complex function of surfactant concentration.

However, the authors pointed out that these phenomena might be due to interfacial effects, since some of the reactants, particularly water, and some of the products (precipitated urea) were not miscible in the system. The surfactant could have been acting as an emulsifier, as a phase transfer catalyst, or as a particle stabilizer.

Acceleration of polymerization rate by specific silicone surfactants was measured and reported by Kopusov and Zharkov [67]. In this case, acceleration could possibly be attributed to the ability of the surfactant to efficiently emulsify water–isocyanate mixtures.

Surfactant–tin catalyst interactions have been suggested by a number of authors (see, e.g., Ref. 18), although this phenomenon has not been investigated in great detail.

2. Surfactant Aggregation in the Bulk Liquid Phase

In bulk liquid phases, primarily water, surfactants aggregate to form micelles and more complex structures, including liquid crystalline phases [28,74–76]. In aqueous foams, the presence of micelles [77–79] and liquid crystalline phases [78,79] influences the stability of thin liquid films and foams. In studies related to polyurethane foam, it has been speculated that silicone polyether surfactants micellize in polyether—polyol solutions [57–60]. This speculation is based on the observation that plots of surface tension versus surfactant concentration for these solutions often show breakpoints similar to those seen at the critical micelle concentration (cmc) for surfactants in aqueous solutions.

Assuming that micellization does occur in polyethers, it was speculated that a micellar network is formed during the expansion of a polyurethane foam [59]. It was hypothesized that this network could allow for the transmission of "power" from the curved liquid–air interfaces to the bulk phase. This transmission process was suggested as a major stabilizing process within the foam.

However, supporting experimental evidence for the presence of micelles, such as light scattering studies, has not been presented to date. Furthermore, the presence of surfactant liquid crystalline phases in these foams has not been observed. The existence of a break in a plot of γ versus log C might signify only a solubility boundary. Therefore, the evidence for aggregation of these surfactants within the bulk liquid phase of these foams is inconclusive.

V. SUMMARY

The formation of polyurethane foam requires the application of silicone-based surfactants, specifically silicone polyether copolymers. These copolymers are active at the liquid–air, liquid–liquid, and polymer–polymer interfaces within these foams. A requirement for these surfactants, allowing them to stabilize the foam, is to reduce the surface tension of the liquid polyether polyols by 8–12 mN/m. This reduction cannot be achieved by hydrocarbon-based surfactants. The adsorption of these surfactants at the polyol–air interface appears to yield a molecular configuration of the surfactant where the siloxane portion is folded over itself. This adsorption also increases the surface viscoelasticity, which aids in stabilizing the foam. These surfactants also appear to be active at the water–polyol and urea–polyol interfaces. This activity increases the miscibility of water in polyol and prevents a catastrophic collapse of the foam after the onset of urea phase separation.

Overall, these surfactants have many complex functions within a polyurethane foam.

REFERENCES

1. Reed, D. Essential ancillaries: A billion-dollar business facing a diversifying future, *Urethanes Technol. 1995*, January/February, pp. 22–23.
2. Herrington, R.; Hock, K., eds., *Flexible polyurethane foams*, Dow Plastics, Midland, MI, 1991.
3. Frensdorff, H. K. Polyurethane foams: Stability, collapse, shrinkage, *Rubber Age 1958*, 812–818.
4. Saunders, J. H. The formation of urethan foams, *Rubber Chem. Technol. 1960*, 33, 1293–1322.
5. Saunders, J. H.; Frisch, K. C. *Polyurethanes Chemistry and Technology*, Part I, *Chemistry*, Wiley-Interscience, New York, 1962.
6. Saunders, J. H.; Frisch, K. C. *Polyurethanes Chemistry and Technology*, Part II, *Technology*, Wiley-Interscience, New York, 1964.
7. Anonymous. One-shot way to urethane foam, *Chem. Eng. News*, Dec. 1, 1958, pp. 48–49.
8. Bailey, D. L.; O'Connor, F. M. Siloxane–oxyalkylene block copolymers, U.S. Patent 2,834,748 (1958).
9. Erner, W. E.; Farkas, A.; Hill, P. W. One-shot urethane foam, *Mod. Plast.* February 1960, pp. 107–111.
10. Hostettler, F. Verfahren zur Herstellung von Polyurethanschaumstoffen, German Patent 1,091,324 (1960).

11. Hostettler, F. Curing composition comprising an organotin compound and a siloxane–oxyalkylene copolymer, U.S. Patent 3,194,770 (1965).
12. Hostettler, F. Process of making polyurethane foams, U.S. Patent 3,194,773 (1965).
13. Sandridge, R. L.; Morecroft, A. S.; Hardy, E. E.; Saunders, J. H. Properties of a semiflexible urethane foam system, *J. Chem. Eng. Data 1960*, 5, 495–498.
14. Oertel, G., ed. *Polyurethane Handbook*, 2nd ed., Carl Hanser, Munich, 1994.
15. Woods, G. *The ICI Polyurethanes Book*, 2nd ed., John Wiley & Sons, Chicester, 1990.
16. Herrington, R. M.; Turner, R. B. The formation, cell-opening and resultant morphology of flexible polyurethane foams, *Adv. Urethane Sci. Technol. 1992*, 11, 1–67.
17. Kanner, B.; Reid, W. G.; Petersen, I. H. Synthesis and properties of siloxane–polyether copolymer surfactants, *Ind. Eng. Chem. Prod. Res. Dev. 1967*, 6(2), 88–92(8).
18. Boudreau, R. J. How silicone surfactants affect polyurethane foams, *Mod. Plast.* January 1967, pp. 133–147, 234–240.
19. Schwarz, E. G. Silicone surfactants for urethane foams: Mechanism, performance, and applications, in *Applied Polymer Symposia #14: Silicone Technology*, John Wiley & Sons, New York, 1970, pp. 71–93.
20. Kanner, B.; Prokai, B. Silicone surfactants for urethane foam, *Adv. Urethane Sci. Technol. 1973*, 2, 221–239.
21. Kanner, B.; Goddard, E. D,; Kulkarni, R. D. Surface chemical aspects of polyurethane foaming, *Proceedings 5th International SPI Conference 1980*, pp. 647–655.
22. Owen, M. J. The surface activity of silicones: A short review, *Ind. Eng. Chem. Pro. Res. Dev. 1980*, 19, 67–103.
23. Macosko, C. W.; Artavia, L. Polyurethane foam formation, in *Physics of Low Density Cellular Plastics* (Hilyard, N. C.; Cunningham, A.; eds.), In Press.
24. Woods, G. *Flexible Polyurethane Foams*, Applied Science Publishers, NJ, 1982.
25. Bruins, P. F., ed. *Polyurethane Technology*, Wiley-Interscience, New York, 1969.
26. Buist, J. M.; Gudgeon, H., eds. *Advances in Polyurethane Technology*, Wiley-Interscience, New York, 1968.
27. Frisch, K. C.; Saunders, J. H., eds. *Plastic Foams*, Parts I and II, Marcel Dekker, New York, 1972.
28. Bikerman, J. J. *Foams*, Springer-Verlag, New York, 1973.
29. Listemann, M. L.; Savoca, A. C.; Wressel, A. L. Amine catalyst characterization by a foam model reaction, *J. Cell. Plast. 1992*, 360–398.
30. Listemann, M. L.; Wressel, A. L.; Lassila, K. R.; Klotz, H. C.; Johnson, G. L.; Savoca, A. C. The influence of tertiary amine structure on blow-to-gel selectivity, in *Polyurethanes World Congress 1993, Vancouver, Canada*, Society of the Plastics Industry, New York, 1994, pp. 595–608.
31. Illger, H.-W.; Dorner, K.-H.; Hettel, H. Reaction kinetics study of high resilient polyurethane foams, *Proceedings of the FSK/SPI 1987, Aachen, Germany*, Society of the Plastics Industry, New York, 1988, pp. 305–310.
32. McCluskey, J. V.; O'Neill, R. E.; Priester, R. D.; Ramsey, W. A. vibrating rod viscometer: A valuable probe into polyurethane chemistry, *J. Cell. Plast. 1994*, 30, 224–241.
33. LeGrow, G.; Petroff, L. J. This volume, Chapter 2.
34. Haluska, L. A. Branched siloxane–Alkyleneoxide copolymers, U.S. Patent 3,271,331 (1966).

35. Haluska, L. A. Polyurethane foam preparation using silioxane glycol branch copolymers, U.S. Patent 3,398,104 (1968).
36. Nicholson, W. R.; Plevyak, J. E. A versatile silicone surfactant for use in major segments of flexible polyurethane foam—A new concept and product, *Proceedings of the SPI 27th Annual Technical/Marketing Conference 1982, Bal Harbour, Florida*, Society of the Plastics Industry, New York, 1983, pp. 303–307.
37. Budnik, R. A.; Cobb, R. L.; Mehta, K. R.; Farris, D. D.; Brasington, R. D. New conventional slabstock surfactant, *Proceedings of the SPI 33rd Annual Technical/Marketing Conference 1990, Orlando, Florida*, Society of the Plastics Industry, New York, 1991, pp. 326–330.
38. Pavlenyi, J.; Baskent, F. O. High comfort foams: A new approach to performance cushioning, *Proceedings of the SPI 27th Annual Technical/Marketing Conference 1982, Bal Harbour, Florida*, Society of the Plastics Industry, New York, 1983, pp. 290–296.
39. Brune-Fischer, A.; Burkhart, G.; Zellmer, V. New concepts in designing silicone surfactants for HR-molded foams, *Proceedings of the SPI 35th Annual Technical/Marketing Conference 1994, Boston*, Society of the Plastics Industry, New York, 1995, pp. 267–273.
40. Battice, D. R.; Lopes, W. J. New cell opening surfactants for molded high resiliency polyurethane foam, *Proceedings of the SPI 30th Annual Technical/Marketing Conference 1986, Toronto, Canada*, Society of the Plastics Industry, New York, 1987, pp. 145–148.
41. Harakal, M. E.; Ernst, B. H.; Womack, F. D.; Battice, D. R. Novel surfactants for melamine filled flexible polyurethane foams, *Proceedings of the SPI 32nd Annual Technical/Marketing Conference 1989, San Francisco*, Society of the Plastics Industry, New York, 1990, pp. 515–519.
42. Kollmeier, H. J.; Schator, H.; Zaeske, P. Importance of silicone surfactants for the formation of rigid polyurethane foams, *Proceedings of the SPI 26th Annual Technical/Marketing Conference 1981, San Francisco*, Society of the Plastics Industry, New York, 1982, pp. 219–220.
43. Kollmeier, H. J.; Schator, H. New developments in silicone surfactants for rigid polyurethane foams *J. Cell. Plast.* July–August 1985, pp. 239–242.
44. Kollmeier, H. J.; Schator, H.; Zaeske, P. Correlation between surfactant performance and raw materials in the formation of rigid polyurethane foams, *J. Cell. Plast.* July–August 1983, pp. 255–258.
45. Bodnar, T. W.; Thornsberry, J. D. New surfactants for reduced and Non-CFC blown rigid systems, *Proceedings of the SPI 33rd Annual Technical/Marketing Conference 1990, Orlando, Florida*, Society of the Plastics Industry, New York, 1991, pp. 52–57.
46. Lunney, S. R.; Szabat, J. F.; Landon, S. J.; Lombardo, J. L. New developments in reduced CFC-11 rigid foam appliance systems, *Proceedings of the SPI 33rd Annual Technical/Marketing Conference 1990, Orlando, Florida*, Society of the Plastics Industry, New York, 1991, pp. 571–579.
47. Burkhart, G.; Klincke, M. Innovative silicone surfactant technology for HFC-134a and cyclopentane blown rigid polyurethane foams, *Polyurethanes World Congress 1993, Vancouver, Canada*, Society of the Plastics Industry, New York, 1994, pp. 361–367.

48. Kanner, B.; Decker, T. G. Urethane foam formation—Role of the silicone surfactant, *J. Cell. Plast. 1969*, 5(1), 32–39.
49. Jones, R. E.; Fesman, G. Air flow measurement and its relations to cell structure, physical properties, and processibility for flexible urethane foam, *J. Cell. Plast.* January 1965, pp. 200–216.
50. Kanner, B.; Prokai, B.; Eschbach, C. S.; Murphy, G. J. New aspects of the stabilization of flexible polyether urethane foam by silicone surfactants, *J. Cell. Plast. 1979*, 15, 315–320.
51. Rossmy, G. R.; Kollmeier, H. J.; Lidy, W.; Schator, H.; Wiemann, M. Cell-opening in one-shot flexible polyether based polyurethane foams. The role of the silicone surfactant and its foundation in the chemistry of foam formation, *J. Cell. Plast. 1977*, 13(1), 26–35.
52. Rossmy, G.; Kollmeier, H. J.; Lidy, W.; Schator, H.; Wiemann, M. Mechanism of stabilization of flexible polyether polyurethane foams by silicone-based surfactants, *J. Cell. Plast. 1981*, 17(6), 319–327.
53. Rossmy, G.; Kollmeier, H. J.; Lidy, W.; Schator, H.; Wiemann, M. Flexible polyether polyurethane foams by silicone-based surfactants, *J. Cell. Plast. 1981*, 17, 28–37.
54. Neff, R. Reactive processing of flexible polyurethane foam. Ph.D thesis, University of Minnesota, 1995.
55. Owen, M. J.; Kendrick, T. C. Surface chemistry of polyurethane foam formation. III. Effect of gas diffusion between bubbles and surface viscosity on bubble stability, *J. Colloid Interface Sci. 1968*, 27(1), 46–52.
56. Lipatova, T. E.; Vengerovskaya, SH. G.; Feinerman, A. E.; Sheinina, L. S. Adsorption layers of surface active substances and their influences on polyurethane network formation *J. Polym. Sci.: Polym. Chem. Educ. 1983*, 21, 2085–2094.
57. Kendrick, T. C,; Kingston, B. M.; Lloyd, N. C.; Owen, M. J. The surface chemistry of polyurethane foam formation. I. Equilibrium surface tensions of polysiloxane–polyether block copolymer solutions, *J. Colloid Interface Sci. 1967*, 24, 135–140.
58. Kendrick, T. C.; Owen, M. J. Surface tensions of polysiloxane–polyether block copolymer solutions, *Chim. Phys. Appl. Prat's. Ag. Surf. C. R. Congr. Int. Deterg. 1969*, 2,1, 571–580.
59. Rossmy, G.; Sanger, G.; Seyffert, H. Foam formation, *VDI-Be. 1972*, 182, 173–176.
60. Hamann, H.; Ritter, J. Contribution to the knowledge of polyorganosiloxane surfactants, *Plaste Kautsch. 1983*, 30(7), 364–366.
61. Dahm, M. The role of surfactants during polyurethane foam formation, Publication 1462, National Academy of Sciences–National Research Council, Washington, DC, 1966, pp. 52–63.
62. Dubjaga, J. G.; et al. Investigation of polyurethane foam formation in the presence of organosilicon surfactants with differing molecular ratios between siloxane and the oxyalkylene portion, *Plaste Kautsch. 1979*, 26(11), 616–619.
63. Dubyaga, E. G.; Komarova, A. B.; Tarakanov, O. G. Surface properties of solutions of oxyalkylenedimethylsiloxane block copolymers in simple oligoethers, in relation to composition of oxyalkylene blocks, *Colloid J. USSR 1986*, 47(6), 881–886.
64. Dubyaga, E. G.; Konoplev, A. V.; Zakharova, T. A. Colloidal properties of oxyalkylenedimethylsiloxane graft copolymers in an oligoether, *Polym. Sci. 1992*, 34(8), 705–710.

65. Rosen, M. J. *Surfactants and interfacial phenomena*, Wiley-Interscience; New York, 1978, Chapters 2 and 5.
66. Owen, M. J.; Kendrick, T. C.; Kingston, B. M.; Lloyd, N. C. The surface chemistry of polyurethane foam formation. II. The role of surface elasticity, *J. Colloid Interface Sci. 1967*, 24, 141–150.
67. Kopusov, L. I.; Zharkov, V. V. Influence of foam stabilisers on reactions in the manufacture of flexible polyurethane foams, *Int. Polym. Sci. Technol. 1981*, 8(3), T34–T35.
68. Hamann, H.; Tschernko, G. The theory of cell opening in the formation of elastic polyurethane foams, *Plaste Kautsch. 1979*, 26(11), 619–624.
69. Ritter, J.; et al. Surface viscosity and maximum shearing stress of polyether solutions of organosilicon surfactants, *Plaste Kautsch. 1979*, 26, 624–625.
70. Gordon, G. V. Unreported results.
71. Hager, S. L.; Craig, T. A.; Jorgenson, M. W.; Artavia, L. D.; Macosko, C. W. Interfacial mixing of urethane foam chemicals, *J. Cell. Plast. 1994*, 30, 44–58.
72. Milliren, C. M. A new non-CFC MDI-based flexible foam technology, *Plast. Eng.* January 1991, pp. 23–25.
73. Snow, S. A.; Fenton, W. N.; Owen, M. J. The addition of polyoxyethylene/polyoxypropylene block copolymers to silicone surfactant systems to improve the porosity of flexible polyurethane foam, *J. Cell. Plast. 1990*, 26, 172–182.
74. Isenberg, C. *The science of Soap Films and Soap Bubbles*, General Publishing Company, Toronto, Ont., Canada, 1992.
75. Myers, D. *Surfactant Science and Technology*, VCH Publishers, New York, 1988.
76. Tadros, T. F., ed. *Surfactants*, Academic Press/Harcourt Brace Jovanovich, New York, 1984. See especially Chapter 8 (Emulsions and Foams), by Brian Vincent.
77. Baets, P. J. M.; Stein, H. N. Influence of surfactant type and concentration on the drainage of liquid films, *Langmuir 1992*, 8, 3099–3101.
78. Siegel, D. P. Thin film rupture via inverted micellar intermediates, *J. Colloid Interface Sci. 1984*, 99(1), 201–207.
79. Chu, X. L.; Nikolov, A. D.; Wasan, D. T. Monte Carlo simulation of inlayer structure formation in thin liquid films, *Langmuir 1994*, 10, 4403–4408.

6

Silicone Polymers for Foam Control and Demulsification

RANDAL M. HILL Central Research and Development, Dow Corning Corporation, Midland, Michigan

KENNETH C. FEY Advanced Materials Business Development Group, Dow Corning Corporation, Midland, Michigan

I. INTRODUCTION

The formation of foams and emulsions is common in nature, in domestic life, and in industry. For example, milk is a naturally occurring emulsion of lipid in water, stabilized by surface active proteins. Hand dishwashing detergents are designed to foam well in addition to emulsifying grease. Laundry detergents are expected to foam, but not too much. Many kitchen recipes generate foams or emulsions, a notable example being hollandaise sauce. The manufacture of stable plastic foam using siloxane surfactants is the topic of a chapter in this volume. However, in many industrial processes, formation of stable foams and emulsions must be prevented. Undesirable foam can be a costly problem that reduces plant capacity, causes environmental problems and safety hazards, and results in loss of valuable products and raw materials. Examples range from preventing the foaming of frying oils in potato chip production and controlling foam during fermentation processes to breaking water-in-crude-oil emulsions. To deal with these problems, a variety of foam control and demulsification technologies have been developed [1–7]. A wide variety of industries make use of cost-effective foam control as a process aid and to provide a vital function in their products. Foam control process aids are the largest single category of process aids used in the chemical industry [8,9].

Polydimethylsiloxane (PDMS, or silicone oil) is surface active in aqueous as well as nonaqueous media; it has been used both to stabilize and destabilize foam as well as to break emulsions [10]. Many silicone-containing copolymers, including the siloxane polyethers, which are the topic of much of the rest of this volume, have also been used as foam control agents and as demulsifiers. This chapter briefly reviews the nature of foams and emulsions, and the mechanisms by which antifoams and demulsifiers are believed to function in both aqueous and nonaqueous applications. Selected foam and emulsion control problems that involve the use of silicone polymers and copolymers are discussed.

II. FOAM STABILITY

Foam is a dispersion of a gas in a liquid. The factors that influence the stability of foams have been extensively discussed in the literature [11–13]. Stable foams are always associated with the presence of some surface active agent (surfactant). Transient foams can also occur in distillation processes near miscibility boundaries, and in viscous liquids. Foam differs from other colloidal dispersions in that the particle size (bubble size) can be orders of magnitude larger. However, the stability of foam, like that of an emulsion or dispersion of solid particles, depends on the adsorbed film at the particle surface. Investigation of foaming and foam control problems usually focuses on the thin film (bubble wall) rather than on the bubble as a particle. The stability of this thin film, and therefore of the foam, depends on surface elasticity and surface viscosity, and on electrostatic and steric interactions [11]. Gelatinous surface films formed by protein–polysaccaride complexes can form exceedingly stable foams. The rate of drainage of liquid out of the foam (due to the density difference between the liquid and air) also influences foam stability—thinner films tend to be more fragile. Bulk viscosity slows drainage and therefore stabilizes foam. Evaporation of the liquid, or diffusion of gas in and out of the bubbles, also tends to decrease foam stability.

Foam may be divided into two types: *kugelschaum*, a dispersion of small spherical bubbles separated by thick films, and *polyhederschaum*, closely packed polyhedral bubbles separated by thin films. This distinction is useful because the processes that cause instability of the foam are somewhat different for the two types.

III. ANTIFOAMING AND DEFOAMING

An agent that acts to inhibit the formation of foam is called an antifoam. Use of such an agent inhibits the initial formation of foam, but any foam that does form may be relatively stable. An antifoam acts on the kugelschaum type of foam. An agent that ruptures foam that is already formed and has begun to drain is called a defoamer. The action of a defoamer is easily seen in the rapid decay of foam when agitation is stopped. Defoamers act on the polyhederschaum type of foam. It is emphasized, however, that these two terms are not always used this way in either the trade or the literature. In fact, they are often used interchangably along with the term "foam control." A defoamer should exhibit rapid knockdown of a foam, while persistence of action is more characteristic of antifoaming. In practice, the same type of chemical additives tend to be used for both defoaming and antifoaming. We will use the term "foam control" to refer generically to all types and degrees of foam control exhibited by such process aids.

IV. EMULSION STABILITY

An emulsion is a dispersion of two immiscible liquids, usually oil in water or water in oil. The preparation of stable emulsions is an extremely important industrial problem and has been extensively discussed in numerous reviews [14–16]. Like foams, emulsions are kinetically stabilized systems. Once an emulsion has formed, several processes begin to occur. The particles rise (or sink) moving closer together, and Brownian motion causes particles to collide. Particles may stick together (aggregation or flocculation) and remain stable, or proceed via film thinning to coalescence. The stability of emulsions can be understood in terms of (1)

Stokes' law, (2) the role of surfactants, and (3) thin-film drainage and stability. Stokes' law defines the rate at which a particle in a liquid medium rises (or sinks):

$$v = \frac{2}{9} \frac{R^2 \Delta\rho g}{\ }$$

In this equation, v is the sedimentation velocity, R is the particle radius, $\Delta\rho$ is the difference in density between the particle and the continuous phase, g is acceleration due to gravity, and η is the viscosity of the continuous phase. The equation shows that reducing the particle size, matching density, and increasing viscosity lead to slower sedimentation and therefore increased emulsion stability. Surfactants decrease the interfacial energy, which decreases the energy cost to achieve smaller particle size. Surfactants also influence thin-film properties as noted earlier, generally slowing the rate of thinning and providing a repulsive barrier to coalescence. There is no simple relationship between a surfactant's ability to lower interfacial tension and its contribution to the repulsive barrier.

V. DEMULSIFICATION

An agent that destabilizes an emulsion, causing it to break, is called a demulsifier. Most ways of breaking an emulsion attack the surfactant film. For example, addition of salt screens electrostatic repulsions, allowing particles to coalesce. Various mechanical devices that impose centrifugal force and accelerate sedimentation can also be used to break emulsions. A variety of chemical additives are available that function as demulsifiers. An agent that breaks a water-in-oil emulsion, allowing the water to be removed, can also be called a dewatering agent.

VI. SILICONE FOAM CONTROL AGENTS

Silicone foam control agents are based on polydimethylsiloxane (PDMS) as shown below:

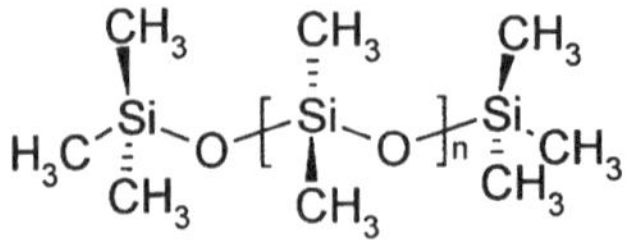

where n is generally greater than 100 and may range above 1000. PDMS possesses a number of unusual properties related to the flexibility of the Si—O—Si backbone and the very low cohesive energy of the methyl groups, which make it useful as a foam control agent [4,10]. PDMS has a low surface tension (20.5 dyn/cm).

It is nonvolatile and remains fluid at high molecular weights. It is immiscible with many organic liquids, and not easily emulsified in water. Although it is surface active, it also has an extremely small value of surface viscosity [17]. All these properties contribute to its usefulness for controlling foams and breaking emulsions.

For aqueous applications PDMS is used in combination with a hydrophobic particulate such as hydrophobized silica (mostly used in the 0.2–30 μm particle size range, with particulate contents of 1–20%). Such a blend of oil with particulate is called an antifoam "compound." Antifoam compounds are generally more effective than the individual components used alone. In fact, the oil or the particulate used alone may have only minimal foam-breaking activity [18]. Some degree of synergy is observed for many different combinations of oils and hydrophobic particulates and for foaming media of various types [18–22]. Although the synergy is still not well understood, very effective foam control agents continue to be developed using such combinations for many different types of foam problems. Hydrophobic particles (used alone) in a foaming system aggregate quickly to form "clumps," whereas compounds remain effectively dispersed. This suggests that one important function of the oil is to promote effective dispersion of the particles.

VII. SILICONE COPOLYMERS

A number of silicone-containing copolymers are also used as foam control agents and demulsifiers. For example, silicone polyethers (SPEs) are organopolysiloxane–polyoxyalkylene copolymers [23] [the molecular structure of a rake-type (or graft) siloxane polyoxyethylene copolymer is shown below]:

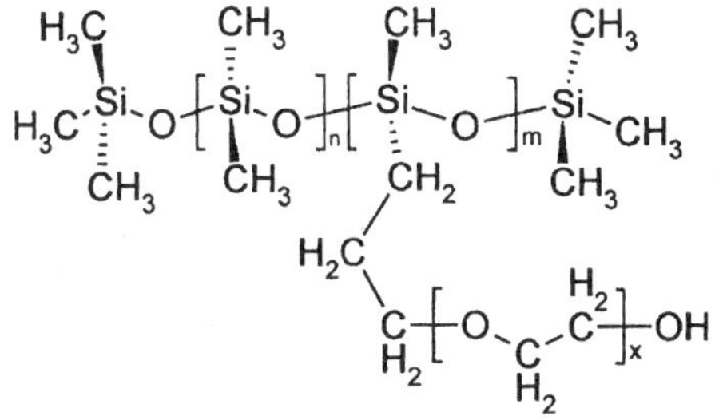

These materials are formed via grafting reactions between a silicone backbone and glycol groups through either Si—C or Si—O—C bonding schemes [24]. Like organic polyoxyalkylene copolymers such as the Pluronics, SPEs that have cloud points are effective foam control agents at temperatures near and above their cloud points. SPEs find application as foam control agents in coatings, paints, and inks

[25]. Some of these applications depend on cloud point defoaming and some do not. SPEs also find utility in nonaqueous applications such as crude oil production, in lubricants, and in jet and diesel fuels [26–28]. Specific applications are discussed later.

Fluorosilicones are copolymers in which some fraction of the methyl groups has been replaced by a fluorocarbon group, usually trifluoropropyl [the molecular structure of polytrifluoropropylmethylsiloxane (PTFPMS) is shown below]:

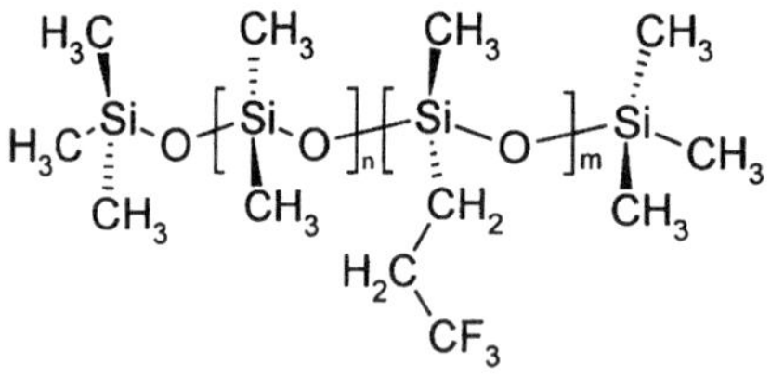

Fluorosilicones have been used in numerous nonaqueous foam control and demulsification applications [27]. They usually have somewhat higher surface tensions than PDMS [29], but their very low solubility in hydrocarbon liquids has significant advantages for applications such as crude oil–gas separation [30] distillation of middle cut fuels [31], and with organic or halogenated organic solvents [32]. There is a correlation between decreased foamability of crude oil and reduced dilatational surface rheology due to the presence of fluorosilicone antifoams [33].

VIII. DELIVERY

An antifoam product must often be prepared for optimal use through dilution or emulsification. This is why foam control agents are often supplied as emulsions for aqueous uses, and dispersed in an organic solvent for nonaqueous uses. Delivery of the optimal particle size distribution for a given foaming problem can be achieved by the formulator by means of controlled plant processing and criteria based on particle size and size distributions. This is typically achieved through predispersing the antifoam compound in an immiscible fluid that can remain liquid (i.e., water or polypropylene glycol [34]) or by predispersing the antifoam compound in a silicone polyether, which acts both as a dispersing fluid and as a surfactant when the mixture is finally dispersed into the foaming medium. Dispersion of silicone antifoam compounds in SPEs has been described in patents [35].

IX. FOAM RUPTURE MECHANISMS

The control of foam, and the optimal formulation of a foam control agent for a particular foaming problem, is a complex problem that often seems needlessly obscure to technologists unfamiliar with the field. A clear understanding of the surface chemical and kinetic factors that contribute to performance should help, and for this reason we include a brief, but hopefully comprehensive review of the mechanisms involved. The basic conditions for a material to function as an antifoam, or defoamer, are as follows:

1. It must be insoluble in the foaming medium.
2. It must be readily dispersible in the foaming medium.
3. It must have a lower surface energy than the foaming medium.

(The precise surface energy requirement is defined by the entering coefficient.)

There are four basic processes by which an antifoam agent ruptures aqueous foam: entering, bridging, dewetting, and rupture. These are depicted schematically in Fig. 1. Rupture of nonaqueous foam involves entering and spreading.

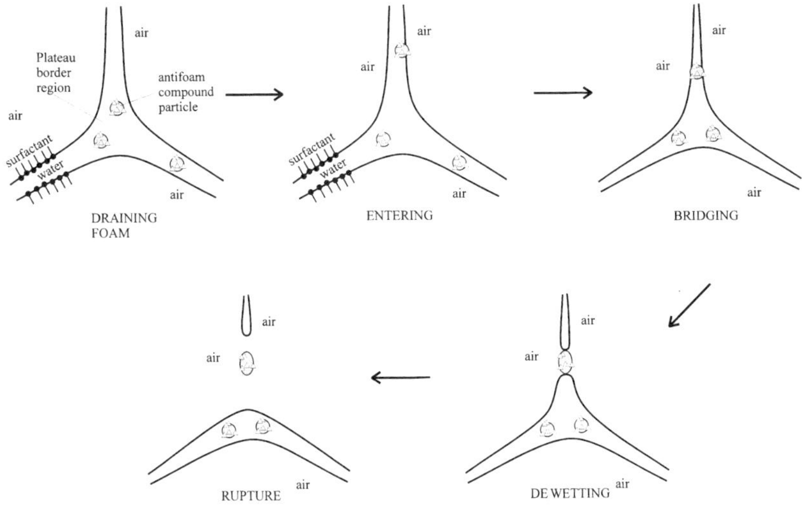

FIG. 1 Schematic of the processes by which an antifoam particle ruptures aqueous foam: a portion of a foam Plateau border region, with an antifoam particle entering, bridging, dewetting, and rupturing.

A. Entering and Spreading

To rupture a foam film, an oil droplet or a hydrophobic particle must first emerge from the liquid phase into the air–liquid interface. This process is called entering.

If the liquid is water, the tendency of an oil drop to enter and spread at the air–water interface is quantified by the entering coefficient:

$$E = \gamma_{w/a} + \gamma_{w/o} - \gamma_{o/a}$$

and the spreading coefficient:

$$S = \gamma_{w/a} - \gamma_{w/o} - \gamma_{o/a}$$

Where $\gamma_{w/a}$ and $\gamma_{o/a}$ are the surface tensions of the aqueous phase and the oil phase, respectively, and $\gamma_{w/}$o is the interfacial tension between the aqueous phase and the oil. Both quantities must be positive for the corresponding process to be energetically favorable.

Similar equations can be written for any two liquid phases. While these quantities are helpful for eliminating materials that cannot enter a given interface, and therefore cannot destabilize foam, real systems are almost always so far from equilibrium that E and S are not very predictive of actual performance.

Entering is obviously essential to foam rupture, and it is generally agreed that for a particle or droplet to cause rupture of a foam film, it is necessary (but not sufficient) for the entering coefficient to be positive. There has been much discussion (partly motivated by the need to explain the synergy observed for mixtures of oil and hydrophobic particles, which seems to require an active role for the oil) over whether a positive S is also necessary [1,36].

Although spreading was much discussed in earlier work on foam rupture, in all cases in which spreading coefficients have been measured, no correlation has been found between the magnitude of the spreading coefficient and the ability to rupture foam [18]. Recent experiments indicate that spreading may occur, but it is not essential to rupture aqueous foams [1,20]. However, the ability of an insoluble liquid phase to spread at the foam film–air interface has been shown to be a critical aspect of controlling nonaqueous foams [26]. The mechanisms involved in rupture of nonaqueous foam are somewhat different. Each of these processes is discussed in more detail in the sections that follow.

B. The Kinetics of Film Rupture

Although we use E and S to represent the thermodynamic gain associated with entering and spreading, actual foam is an extremely dynamic system. With new bubbles constantly being formed, the actual efficacy of a foam control agent will depend on the *rate* at which it is able to rupture foam films [37]. This in turn depends on the number of antifoam particles present, and on the kinetics of the entering and dewetting processes. As the antifoam particle approaches the

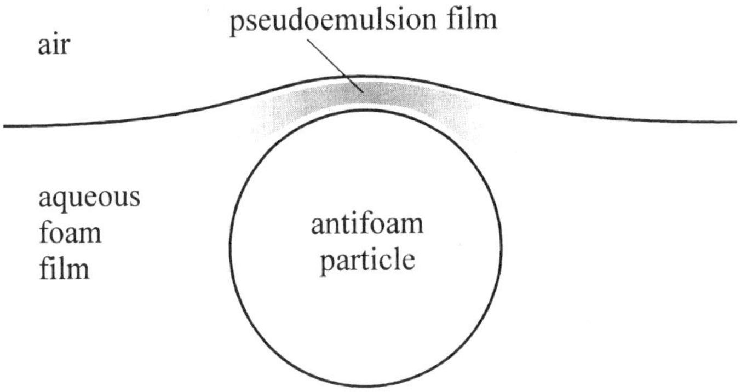

FIG. 2 An asymmetric thin film of water between oil and air called a pseudoemulsion film.

air–water interface, an unsymmetrical three-phase thin film (called a pseudoemulsion film) forms, consisting of antifoaming oil, foaming medium, and air (Fig. 2).

The rate at which entering takes place depends on the drainage rate of the thin film, and on the potential energy barrier the particle feels as it approaches the air–water interface [38]. Recent work has shown that the film can be stable for long periods of time even when the entering coefficient is large and positive [20]. If the film is stable, no entering and no foam rupture takes place. The geometry of the particle strongly influences the kinetics of entering—an irregular shape, or the presence of asperities on the particle surface, leads to much more rapid entering [1,21].

C. Bridging and Dewetting

Once a particle has entered one side of a foam lamella, it may within some time interval bridge the film. The rate of bridging depends on the relative size of the particle and the foam lamella, the contact angle θ (measured through the aqueous phase), and the rate of drainage of the liquid phase out of the foam. If θ is greater than or equal to 90°, an unstable configuration results for a spherical particle, leading to dewetting and film rupture. For nonspherical particles dewetting may occur even for $\theta < 90°$ [1].

When solid particles are used to control foam, whether by themselves or in combination with an oil, hydrophobicity is a necessary condition for antifoaming activity—Garrett et al. [18] found that only the most hydrophobic silicas were effective antifoams when used alone. Liquid oils used alone are often ineffective. However, mixtures of the two were effective even when the less hydrophobic silicas were used. Hydrophobic particles, including polytetrafluoroethylene particles and hydrophobized silicas, are known to decrease foamability (antifoaming) but to have little effect on foam stability after agitation has ceased (defoaming) [18]. Mixtures of solid particles and liquid oils reduced the amount of foam produced by shaking as well as causing the foam produced to break much faster—hence such mixtures function as both antifoams and defoamers. If particles must bridge foam films to rupture them, we would expect an effective particle to be at least as large as the thickness of the foam film. Typical foam film thicknesses are 0.01–1 μm. However, the most effective particle sizes are larger than this, in the range of 5–40 μm [20]. This discrepancy probably exists because the particles actually reside, and are mostly active in, the Plateau border (PB) regions of the foam (which are thicker) rather than in the flat film regions [20]. Koczo et al. [39] showed that the tendency of particles to become trapped in the PB regions strongly increases with increasing particle size.

D. Particle Size Versus Number of Particles

It is often observed that the performance of an antifoam depends on its particle size. However, changing the particle size also changes the number of particles and therefore affects the statistics of the foam rupture process [1,40]. Larger particles bridge and rupture lamellae faster. Reducing the particle size by half slows down bridging, but also increases the number of particles by a factor of 8. Thus, there is a trade-off between particle size and number of particles—having more particles leads to more entering and bridging events, but the time lapse between entering and rupture is longer.

E. Time Decay of Antifoam Efficacy

It is well known that the efficacy of antifoams decreases with time [1]. This is attributed to gradual breakdown of the particles into smaller and smaller particles, which eventually become too small to be effective. Increasing the viscosity of the oil, incorporating fumed silica (which builds the viscosity of the mixture), and cross-linking the silicone oil [41–44] are all documented ways of improving the resistance of the foam control agent to this process, and therefore improving its persistence. Incorporation of resins, and isocyanate condensation [45], lead to similar effects. Incorporating fumed silica into silicone oils influences the viscoelastic properties of the mixture, which in turn affects the initial dispersed particle size in the foaming medium (if the antifoam is added to the foaming medium neat), and the resistance of the particles to further breakdown in size and loss of efficacy.

F. Particle–Oil Synergy

For the systems they investigated, Garrett et al. [18] found that the antifoam efficacy of a mixture of liquid paraffin and a hydrophobic silica increased up to about

5% added silica. Further increasing the amount of silica had little effect. Garrett [46] and Frye and Berg [47] explain the synergistic performance of mixtures of particles and oil using the entering–bridging model described earlier. Garrett [1] argued that the behavior of the mixed oil–particle system can best be understood by realizing that the particles will likely reside in the oil droplet–aqueous solution interface. Being situated at the surface causes the particles to act like asperities, thus facilitating penetration through the energy barrier represented by the pseudoemulsion film. Another factor that contributes to the synergy between particles and oil is the way the conditions for dewetting change when an oil–particle combination is present—a smaller contact angle is required for dewetting [1,37]. If the particles are situated in the oil droplet's surface, then a (water–air–solid) contact angle as small as 40° can lead to dewetting.

X. SOLUBLE FOAM CONTROL AGENTS

There is a whole class of "soluble" foam control agents. These are block copolymers of polyoxyethylene (EO) polyoxypropylene (PO), or some other hydrophobic polymer such as PDMS. They function by phase-separating above their cloud point [48], forming insoluble droplets of a hydrophobic liquid [49]. The best performance is found for copolymers with a cloud temperature below the application temperature of the foaming medium. The cloud temperature of an EO-based surfactant depends on the pH and ionic strength of the aqueous phase and is extremely sensitive to the presence of other (especially ionic) surfactants [50–53]. Surfactants with molecular weights above a certain value seem to work the best; the MW that works the best depends on the molecular structure [54]. This subject has been reviewed by Blease et al. [54], along with discussion of two applications of soluble foam control agents, automatic dishwashing, and sugar beet processing. Bonfillon-Colin and Langevin have demonstrated that the entering–bridging–dewetting mechanism described earlier accounts for cloud point defoaming [49]. SEPs also exhibit a cloud point and consequent defoaming [55]. This class of soluble foam control agents has found increasing use in recent years. Some soluble polymeric surfactants are believed to destabilize foam by entering the surface as a soluble surface active species and altering the surface elasticity and viscosity required for stable foam. The film penetration and spreading processes discussed here do not explain the performance of such materials. Short chain alcohols are also sometimes used as soluble foam control agents, primarily for their defoaming ability [8].

XI. MECHANISMS IN NONAQUEOUS DEFOAMING

In aqueous foams, long-range repulsive interactions and kinetic barriers to entering are common, leading to the situation already described in which the kinetics of the entering and bridging process dominates efficacy. In nonaqueous foams, the complex thin-film stabilization phenomena responsible for aqueous foam stability are mostly absent. Foam stability in nonaqueous foams is due to surface dilational viscosity and elasticity. Film thinning is resisted only by Gibbs elasticity and bulk viscosity, and therefore proceeds fairly quickly to an unstable state. Callaghan has shown that interfacial rheological properties can be simply correlated with foam control for hydrocarbon liquids [56]. Nonaqueous foam control agents must fine-tune a balance between immiscibility, surface activity, and dispersibility. The agent must be insoluble in the foaming liquid and able to spread at the interface between the foaming liquid and air. However, it must not be so incompatible that it is not readily dispersed into the foaming medium. This is especially important in applications such as diesel fuel, in which the foam control agent must remain stably dispersed for long periods of storage. Specific details for crude oil and diesel fuel are discussed shortly.

A. Demulsification Mechanisms

The mechanisms of demulsification have been far less studied than for antifoams, but they are thought to involve the following processes. Many problem emulsions are stabilized by high molecular weight species such as proteins or asphaltenes (crude oil), or even solid particles. A highly surface active additive, which is able to displace the stabilizer from the surface, can destabilize the emulsion if the additive itself is a poor emulsifier. Comparing two similar polymers, Zaki and Al-Sabagh [57] found that the one that gave lower interfacial tensions was the more effective demulsifier. These authors also found that an additive that solubilized the stabilizing substance away from the surface was able to destabilize the emulsion. They related their finding that copolymers with a particular range of hydrophile–lipophile balance (HLB) values were the most effective to the ability to promote emulsion inversion. High temperatures and lower viscosity both promote demulsification because of Stokes' law [57].

XII. APPLICATIONS

Many foam control agents are used primarily as processing aids in a wide variety of manufacturing processes. Foam control is used in applications such as polymer processing, gas–oil separation, diesel fuel, pulp and paper manufacture, powdered and liquid detergents, textile manufacture, paint and ink manufacturing and use, phosphoric acid manufacture, metalworking, lubricants, sugar processing, food manufacture (frying), wastewater treatment, fermentation (pharmaceuticals, food, and industrial chemicals), agrochemicals, leather manufacture, chemical processing, and distillation. As this list shows, the applications include a wide variety of aqueous and nonaqueous environments. Several examples from each of these general classes are given, to illustrate the principles just discussed.

A. Pulp and Paper Production

The conversion of wood to paper products involves two basic processing stages: first the digestion of the wood itself to "market pulp," which is then manufactured into paper products ranging from cardboard and newsprint to white paper and fine products. The overall process of paper manufacturing with an emphasis on foam control was reviewed by Allen et al. [58].

The pulping process involves the chemical degradation of lignins to saponified by-products, which are washed from the resulting fibers. Most pulping is done by the Kraft process, which involves aqueous alkaline hydrolysis of the lignin under high temperature and pressure. The lignin by-products, act as the foaming surfactants in this high alkalinity environment. Rotary washing and the agitation that accompanies the washing process provide a severe test for foam control agents. Controlling the foam is critical, since the amounts of lignin by-products as well as other residues carried with the pulp to downstream processes increase as foam control is lost. This results in poorer quality pulp and requires more severe bleaching, should the ultimate product be writing paper or some other fine grade of paper. Moreover, excessive foam buildup at this stage of the paper milling process can flood the brownstock washer's vacuum system, resulting in a reduction of the pulp production rate.

The paper making process, once bleaching has been completed, includes the use of various agents to size the fibers, to retain clay fillers on the fibers as the paper sheet is formed, and to impart strength and tear resistance as a function of the wood that has been used as well as the intended use for the grade of paper being produced. Antifoams are used in this portion of the process to minimize foaming in the vacuum collection tanks as the paper mat is laid down and partially dried on a moving web or screen.

Several factors influence foam control agent selection. Obviously, cost efficiency of the foam control system is paramount. However, the specific needs of the paper mill may emphasize different aspects of foam control. For example, defoam, or "knockdown," is needed to eliminate a head of foam that builds up because the level of foam control agent being fed is overwhelmed by the changes in the pulp stream. Other important factors include the ease of using the foam control agent, and the applicability of specific preexisting facilities for pumping, handling, and spraying of foam control agents.

Antifoam selection may also be influenced by the need to control residue left on the pulp after processing. Pitch, for example, is the accumulation of waxlike deposits on processing equipment and surfaces. It can in many cases result from the use of organic oils with stearamide-based components. And the concern over the generation of chlorinated dioxin and furan species during chlorine bleaching, which caused some level of public outrage in the 1980s, was due to the significant levels of precursor impurities in the mineral oils used at that time. Though oil-based products using "clean" oils retain approximately 80% of the brownstock defoamer market, defoamers based on silicone oils do not lead to pitch formation (since they do not require the use of stearamides) or to the generation of chlorinated dioxins. In the case of paper products that are intended for use in the processing or packaging of food, the U.S. Food and Drug Administration specifies the types of component that are to be used to ensure compliance with federal regulations.

Most silicone-based paper defoamers are silicone fluids mixed with treated silica. Specialty chemical formulators are fond of using this type of silicone component to improve the overall performance of their systems for difficult-to-control foaming problems. In addition, they can purchase antifoam compounds that are reported to have a higher level of knockdown performance and longevity because of their higher molecular weight and higher degree of cross-linking. Silicone fluids and compounds may be used as oil-based formulations or may be formulated as water based emulsions.

Some defoamers used in paper making operate on the basis of cloud point phenomena. As such, any material that exhibits a cloud point slightly below the operating temperature of the paper making process' is a candidate for testing. Fatty alcohols are widely used, and silicone polyethers, provided they exhibit the correct cloud point character, also can be used.

B. Paints and Coatings

Foam problems occur both during the manufacture of paints and coatings and in their application. This application was reviewed in a book that includes pictures of the various defect types [59]. Pigment dispersion is often accomplished by devices that also entrain air. Formulations always contain relatively high levels of efficient surfactants, both to wet out and stabilize pigment dispersions and to assist in wetting of the substrate during application. Application techniques include roller, brush, dip, and spray methods—the most effective defoamer depends on the application method and the specific formulation of the coating. Dye applications require special foam control agents matched to a specific chemical environment [60,61]. Silicone polyethers formulations have been patented for use in water-based printing inks [62] subjected to high temperatures and high shear.

Paint and coating applications are dominated by the use of organic defoamers. This is due in part to lower cost and in part to the lower potential for creating surface defects. However, silicone-based defoamers are a key segment in these markets because they work when organics do not. As such, they command higher value. Silicone fluids and antifoam compounds are used only in formulated versions to disperse them to a very high degree, reducing their potential to create surface defects. Because silicone compounds are easier to formulate with surfactants, silicone fluids are not widely used.

Silicone polyethers are generally more compatible than compounds with paint and coating formulations, and as such find use in a number of applications. And because they possess surfactant-like characteristics, the proper selection of the exact structure can diminish the potential for surface defects. To be effective, a defoamer has to be surface active and relatively insoluble in the medium in which it is used. This requirement calls for a particularly fine balance between controlling foam and not causing surface defects. Too much compatibility will reduce surface defect formation, but will also reduce defoaming potential. In many cases, a specific defoamer structure is required for a specific paint or coating formulation. Thus, the optimal defoamer helps prevent the formation of defects such as bubbles and voids, and therefore leads to a better quality coating. In addition to improved coating quality, the rate of application of the coating can often be increased by the use of an effective foam control agent. The defoamer must not cause fish-eyes, cratering, pinholes, or orange-peel appearance, and must not detract from final color appearance, gloss, and adhesion of the coating [63,64].

Two types of foam can be observed in these systems. *Macrofoam* refers to large bubble structures that are prone to break by conventional theory and means. These tend to occur during the manufacture of the paint, ink, or coating. The time required for macrofoam that forms during manufacture to be completely eliminated is not critical in many cases. *Microfoam*, however, is characterized by very stable bubbles that are very small (less than a few micrometers). This type of foam can easily lead to pinholing or other similar form of surface defect. Microfoam is generally encountered during application of coatings, paints, and inks. A good defoaming agent must assist the microfoam bubble to migrate to the surface before the coating begins to set, or it must aid in the coalescence of several bubbles, since a mass of larger size and surface area will migrate to the surface faster than individual bubbles [65,66]. In many cases, microfoam's presence is not recognized until one views the finished coating and observes the surface defects it has caused.

C. Nonaqueous Applications

Polydimethlysiloxane, some silicone polyoxyalkylene copolymers, and fluorosilicone polymers have lower surface tensions than hydrocarbon liquids and are surface active at the hydrocarbon–air interface. Polymers that are marginally soluble in the continuous phase can be profoamers below their solubility limit. This disadvantage can be dealt with either by increasing the molecular weight of the polymer or by changing its chemical functionality to eliminate the solubility. Silicone antifoams have been shown to reduce both the surface dilatational viscosity and the surface elasticity of crude oils [67]. For these reasons, in nonaqueous systems such as petroleum processing, silicone defoamers are often the only effective agents. Antifoam compounds incorporating solid particles are not necessary to facilitate entering in nonaqueous foaming problems. Mannheimer [68] has shown that the molecular weight distribution of PDMS leads to the presence of silicone both as dissolved polymer and as dispersed droplets in certain hydrocarbon oils. He attributed defoaming to the insoluble droplets and antifoaming to the soluble species.

1. Chemical Processing and Distillation

Mixing, distillation, and pressure differences occur frequently in chemical processing and may each give rise to entrained air and stable foam [69]. This field was reviewed by McGee [70], who listed some of the problems that may arise from undesirable foam in chemical processing: increased costs from preventing potential safety hazards, interference with process instruments, ineffective pumping, increased energy costs for drying, decreased tank capacity, incompletely filled product containers, and perceived negative environmental impact from foaming discharges. Distillation problems often use high molecular weight PDMS for chemical stability at high temperature.

2. Petroleum and Fuels

The causes of foaming and the use of foam control agents in the petroleum industry were reviewed by Callaghan [26]. Several aspects of petroleum production and processing (e.g., the separation of natural gas from crude oil) require the use of foam control agents. Under the high pressure conditions of the oil field, crude oil contains dissolved gases that must be separated from the crude oil at the surface. The gases consist of a mixture of volatile hydrocarbons including methane, ethane, propane, and butane. Different fields and different depths produce at different pressures and temperatures, and with different mixtures of gases. As the crude oil is depressurized, evolving gas generates foam, leading to oil losses via the gas stream and downstream equipment damage [26]. Foam is stabilized by a family of large molecules called asphaltenes, by dispersed particles, by metal ions, and by certain chemical forms of sulfur. The same substances can also stabilize droplets of water or brine, requiring the use of a demulsifier.

The foaming problem is partly dealt with by mechanical design of the gas–oil separator and partly by the addition of silicone defoamers. Because of the very wide diversity of crude oil composition, foam control agents often must be adapted for each well. High molecular weight PDMS is dissolved in a suitable solvent and injected into the crude oil stream before the oil is allowed to depressurize and cool. The use of low HLB siloxane polyoxyalkylene copolymers to facilitate separation of gas from oil was disclosed by Callaghan et al. [71]. Many patents describe silicone polymers and copolymers for demulsification of oil [5,6,72,73]. Since even high molecular weight PDMS is somewhat soluble in crude oil at high temperatures and pressures, fluorosilicones are finding niche applications with particularly demanding foam control problems. Foaming is also a problem in petroleum distillation, cracking, coking, and asphalt processing. Callaghan [26] reviews the test methods commonly used to evaluate foam control agents in the petroleum industry.

3. Diesel Fuels

Formulated petroleum products, such as transmission fluids, lubricating oils [74,75], and diesel and jet fuel all contain an extensive package of performance-improving additives, some of which cause foaming problems. Diesel fuels usually contain a defoamer to permit tanks to be filled more rapidly. Stability of the dispersion is critical because the fuel may be stored in large tanks for extended periods of time. Low molecular weight PDMS is soluble in diesel fuel. High molecular weight achieves insolubility, but rapidly separates. Polyoxyalkylenes are insoluble. By combining the two chemical functionalities, both foam control and stable dispersion can be achieved [8]. Diesel fuel quickly absorbs water (up to 100–1000 ppm may be present), which imposes special requirements on the antifoam used. An optimal foam control agent for diesel fuel must also function and remain stably dispersed with water in the fuel, which can be difficult with polyoxyalkylene copolymers [76]. A variety of different siloxane polyoxyalkylene copolymer structures have been disclosed for controlling foaming in fuels [77–79]. The use of siloxane polyoxyalkylene copolymers to remove water haze from fuels (demulsification) has also been claimed [80,81].

Formation of silica upon combustion of silicone antifoams in fuel is not generally viewed as a problem. It is sometimes asserted that high levels of silicones in fuels can lead to abrasion problems (especially in engines such as diesels, which are designed to run up to several hundred thousand miles before overhaul), and poisoning of catalytic converters. However, there are few, if any, substantiated reports of actual problems. Nevertheless, since the potential probably exists, the use of the most efficient technology and the lowest use levels is obviously desirable.

XIII. TEST METHODS

To evaluate a potential foam control agent for a particular problem, it is desirable to use laboratory test methods to screen candidates. A comprehensive classification of test methods for evaluating the foaming characteristics of surfactants is given by Domingo et al. [82], along with recommendations for a set of methods that are simple and easy to build and use and to standardize. The selection was based on their suitability for practical study of foams in the more important industrial and household applications. To understand the behavior of different foaming media, several foam models of varying degrees of complexity were developed by Szekrenyesy et al. [83,84]. This is a complex problem; the interested reader is referred to the references already mentioned and to Bikerman [85] for further discussion.

Although foaming models and laboratory test methods can never completely substitute for pilot-plant and full-scale evaluation, if carefully designed and used they can identify the best candidates for further testing. There are almost as many test methods as there are foaming problems, but they all share certain characteristics. The critical variables are:

1. *The means by which gas is dispersed into the liquid.* Foam can be generated either by sparging a gas into the bottom of a liquid containing vessel or by mechanical agitation of the liquid.
2. *The shear involved in the air entrainment process.* The very widely used

shake test involves a very low shear version of foam generation by mechanical agitation.

3. *The degree of agitation of the liquid and of the foam head during the process.* Automatic dishwashers subject both the liquid and the foam to a high shear recirculation.
4. *The foaming medium.* The type and concentration of surfactants as well as the presence of dissolved salts, acid or alkali, suspended solids or fibers, and so on, will impact the foam making and foam stabilizing processes.
5. *Temperature and pressure.* Temperature impacts the bulk and interfacial phase behavior and dynamic surface tension characteristics of surfactants. Pressure release, which occurs in processes of some types, such as some methods of textile dyeing or in paper pulp manufacture, can rapidly generate foam or cause rapid expansion of preexisting foam.

The expansion ratio (volume fraction air) and the bubble size tend to be controlled by variables 1 and 2, while the rate of drainage of liquid out of the foam is determined by variable 3. Both the foaming ability of the surface active species present and the ability of the foam control agent to antifoam or defoam are sensitive to all these parameters. Therefore, meaningful laboratory evaluation of foam control must seek to match these parameters as closely as possible to the actual foaming problem. It is extremely difficult in practice to match the dynamics of foam generation in a test method. It is also important to avoid contamination of the foaming medium.

The American Society for Testing and Materials publishes several methods for screening or testing antifoams that address different aspects of these variables: gas dispersion (ASTM D892-74), use of a falling liquid stream to generate foam (ASTM D1173-53), use of a blender (ASTM D3519-76), and use of a shaking bottle (ASTM D3601-77).

XIV. SUMMARY

Silicone polymers and copolymers are surface active materials that are useful for controlling foam and for demulsification. Silicone polymers combine unique surface activity properties with an almost endless variety of chemical and physical property modifications to optimize performance for particular applications. The formation and stability of foams and emulsions involves surfactants and surface active polymers and particles, drainage and sedimentation processes, and thin-film drainage and stability. Insoluble silicone foam control agents for aqueous foaming problems usually consist of mixtures of silicone oils with hydrophobized silica. These foam control compounds function by a complex process of entering, bridging, and dewetting to rupture foam lamellae. Spreading, though often discussed, plays an uncertain role in aqueous foam control. Demulsification usually involves a more direct attack on the stabilizing film using agents that are surface active but nonstabilizing. Soluble foam control agents such as the silicone polyether surfactants function by forming droplets of an insoluble phase above their cloud point. In nonaqueous foaming problems, an insoluble surface active polymer that is able to spread on the liquid but gives weak surface viscosity and elasticity properties seems to work the best. Delivery of the foam control agent is critical to performance and utility of the product in the customer's hands. Technologies for granulated forms, emulsions, and self-emulsifiable antifoam blends are well documented and described in the patent literature.

Effective application of silicone foam control agents in aqueous foaming problems such as pulp and paper production and in paints and coatings involves varying the structure of the silicone polymer, incorporation of the appropriate hydrophobic particulate, and delivery of the optimal compound particle size to the point in the process at which the problem may best be attacked.

Use levels, knockdown ability, and persistence are crucial features of performance. In nonaqueous foaming and demulsification, problems such as those arising in crude oil production, and with lubricants and fuels, obtaining optimal performance requires the achievement of a critical balance of insolubility (or incompatibility) between the silicone polymer and the hydrocarbon liquid, and dispersibility. In fuels, long-term storage stability in the presence of moisture is a critical feature. One of the key problems for the formulator is the need for laboratory evaluation methods that meaningfully predict the performance of different foam control agents under actual use conditions.

REFERENCES

1. P. R. Garrett, in *Defoaming: Theory and Industrial Applications* (P. R. Garrett, ed.), Vol. 45, Surfactant Science Series Marcel Dekker, New York, 1993, p. 1.
2. R. M. Hill and S. P. Christiano, in *Polymeric Materials Encyclopedia* (J. C. Salamone, ed.), CRC Press, Boca Raton, FL, 1996, p. 285.
3. D. T. Wasan and S. P. Christiano, in *Handbook of Surface and Colloid Chemistry* (K. S. Birdi, ed.), CRC Press, Boca Raton, FL, 1997, p. 179.
4. M. J. Owen, in *Encyclopedia of Polymer Science and Engineering*, Vol. 2, 2nd ed., John Wiley & Sons, New York, 1985, p. 59.
5. T. Easton and B. Thomas, U.S. Patent 4,888,107, to Dow Corning Ltd. (1989).
6. A. Sivakumar and M. Ramesh, U.S. Patent 5,560,832, to Nalco Chemical Company (1996).
7. W. Knauf, K. Oppenlander, and W. Slotman, U.S. Patent 5,759,409, to BASF (1998).
8. K. J. Byron, *Crit. Rep. Appl. Chem. 30*, 163 (1990).
9. A. P. Kouloheris, *Chem. Eng. 10*(26), 88 (1987).
10. M. J. Owen, *Ind. Eng. Chem. Prod. Res. Dev. 19*, 97 (1980).
11. R. K. Prud'homme and S. A. Khan, eds., *Foams: Theory, Measurements, and Applications* Vol. 57, *Surfactant Science Series*, Marcel Dekker, New York, 1996.
12. J. Lucassen, in *Anionic Surfactants—Physical Chemistry of Surfactant Action* (E. H. Lucassen-Reynders, ed.), Marcel Dekker, New York, 1981, p. 217.

13. A. Colin, J. Giermanska-Kahn, and D, Langevin, *Langmuir 13*, 2953 (1997).
14. P. Becher, ed., *Encyclopedia of Emulsion Technology*, Marcel Dekker, Vol. 1, *Basic Theory*; Vol. 2, *Applications*; Vol. 3, *Basic Theory/Measurements/Applications*; Vol. 4, Marcel Dekker, New York, 1983, 1985, 1987, 1996.
15. J. Sjöblom, ed., *Emulsions and Emulsion Stability*, Vol. 61, *Surfactant Science Series*, Marcel Dekker, New York, 1996.
16. D. F. Evans and H. Wennerstrom, *The Colloidal Domain*, VCH Publishers, New York, 1994.
17. N. L. Jarvis, *J. Phys. Chem. 70*, 3027 (1966).
18. P. R. Garrett, J. Davis, and H. M. Rendall, *Colloids Surf. A 85*, 159 (1994).
19. P. R. Garrett, *J. Colloid Interface Sci. 76*, 587 (1980).
20. K. Koczo, J. K. Koczone, and D. T. Wasan, *J. Colloid Interface Sci. 166*, 225 (1994).
21. M. P. Aronson, *Langmuir 2*, 653 (1986).
22. R. D. Kulkarni, E. D. Goddard, and B. Kanner, *Ind. Eng. Chem. Fundam. 16*, 472 (1977).
23. R. M. Hill, M. He, Z. Lin, H. T. Davis, and L. E. Scriven, *Langmuir 9*, 2789 (1993).
24. S. Vick, *Soap/Cosmet./Chem. Spec.* May 1984, p. 36.
25. H. F. Fink, *Tenside Surf. Deterg. 28*, 306 (1991).
26. I. C. Callaghan, in *Defoaming: Theory and Industrial Applications* (P. R. Garrett, ed.), Vol. 45, *Surfactant Science Series,* Marcel Dekker, New York, 1993, p. 119.
27. G. C. Sawicki and J. W. White, *Chemspec. Europe 89 BACS Symp.*, 1989.
28. G. Adams and M. Jones, Belgian Patent BE0904498, to Dow Corning Corp. (July 16, 1986).
29. M. J. Owen, in *Siloxane Polymers* (S. J. Clarson and J. A. Semlyen, eds.), Prentice Hall, Englewood Cliffs, NJ, 1993, p. 309.
30. I. C. Callaghan, S. A. Hickman, F. T. Lawrence, and P. M. Melton, *Spec. Publ. R. Soc. Chem. 59*, 48 (1987).
31. *Res. Disclosures 322*, 126 (1991).
32. N. Terae, T. Mutoh, and A. Yoshida, Japanese Patent J01022310, to Shin-Etsu Chemical Industries (Jan. 25, 1989).
33. I. C. Callaghan, C. M. Gould, R. J. Hamilton, and E. L. Neustadter, *Colloids Surf. 8*, 17 (1983).
34. R. Sullivan, U.S. Patent 3,304,266, to Dow Corning Corp. (1967).
35. Japanese Patent J 53034854, to Toray Silicone (March 31, 1978).
36. S. Ross, *J. Phys. Colloid Chem. 54*, 429 (1950), and references therein.
37. R. Aveyard, B. P. Binks, P. D. I. Fletcher, T. G. Peck, and C. E. Rutherford, *Adv. Colloid Interface Sci. 48*, 93 (1994).
38. R. D. Kulkarni and E. D. Goddard, *Croat. Chem. Acta 50*, 163 (1977).
39. K. Koczo, L. A. Lobo, and D. T. Wasan, *J. Colloid Interface Sci. 150*, 492 (1992).
40. G. C. Frye and J. C. Berg, *J. Colloid Interface Sci. 127*, 222 (1989).
41. K. Aizawa, S. Sewa, and N. Hideki, U.S. Patent 4,749,740 to Dow Corning Corp. (1988).
42. V. B. John, G. C. Sawicki, R. Pope, and R. J. Scampton, European Patent EP 0217501, to Dow Corning Corp. (April 8, 1986).
43. K. Aizawa, S. Sewa, and H. Nakahara, U.S. Patent 4,639,489, to Dow Corning Corp. (1987).

44. R. M. Hill, M. S. Starch, and M. S. Gaul, U.S. Patent 5,262,088, to Dow Corning Corp. (1993).
45. E. Pirson and J. Schmidlkofer, Belgian Patent BE0866716, to Wacker Chemie (Nov. 16, 1978).
46. P. R. Garrett, *J. Colloid Interface Sci. 69*, 107 (1979).
47. G. C. Frye and J. C. Berg, *J. Colloid Interface Sci. 127*, 222 (1989).
48. T. Nakagawa, in *Nonionic Surfactants* (M. J. Schick, ed.), Marcel Dekker, New York, 1966, p. 572.
49. A. Bonfillon-Colin and D. Langevin, *Langmuir 13*, 599 (1997).
50. F. E. Bailey and J. V. Koleske, in *Nonionic Surfactants* (M. J. Schick, ed.), Marcel Dekker, New York, 1966, p. 794.
51. M. J. Rosen, *Surfactants and Interfacial Phenomena*, 2nd ed., John Wiley & Sons, New York, 1989, pp. 191–195.
52. F. Schambil and M. J. Schwuger, in *Surfactants in Consumer Products* (J. Falbe, ed.), Springer-Verlag, New York, 1987, p. 137.
53. D. Coons, M. Dankowski, M. Diehl, G. Jokobi, P. Kuzel, E. Sung, and U. Trabitzsch, in *Surfactants in Consumer Products* (J. Falbe, ed.), Springer-Verlag, New York, 1987, p. 324.
54. T. G. Blease, J. G. Evans, L. Hughes, and P. Loll, in *Defoaming: Theory and Industrial Applications* (P. R. Garrett, ed.), Vol. 45, *Surfactant Science Series*, Marcel Dekker, New York, 1993, p. 325.
55. R. M. Hill, in *Specialist Surfactants*, I. D. Robb, ed.), Blackie Academic & Professional, Glasgow, 1997, p. 143.
56. I. C. Callaghan, in *Defoaming: Theory and Industrial Applications* (P. R. Garrett, ed.), Vol. 45, *Surfactant Science Series*, Marcel Dekker, New York, 1993, p. 119.
57. N. Zaki and A. Al-Sabagh, *Tenside Surf. Deterg. 34*, 12 (1997).
58. S. L. Allen, L. H. Allen, and T. H. Flaherty, in *Defoaming: Theory and Industrial Applications* (P. R. Garrett, ed.), Vol. 45, *Surfactant Science Series*, Marcel Dekker, New York, 1993, p. 151.
59. J. W. Simmons, R. M. Thornton, and R. J. Wachala, in *Handbook of Coatings Additives*, L. J. Calbo, ed.), Marcel Dekker, New York, 1987, p. 147.
60. T. Mutoh, N. Tarae, and M. Tanaka, U.S. Patent 5,106,535, to Shin-Etsu Chemical Co. Ltd. (1992).
61. Y. Tsuda and M. Komatsu, U.S. Patent 5,431,853, to Nikko Chemical Institute (1995).
62. A. Itagaki, S. Azechi, S. Kuwata, Y. Tsutsumi, and M. Kotaiky, U.S. Patent 5,486,549, to Shin-Etsu Chemical Co. Ltd. (1996).
63. H. U. Hempel, M. Grunert, H. Tesmann, and H. Muller, U.S. Patent 4,504,410, to Henkel (1985).
64. K. W. Householder and V. I. Doesburg, U. S. Patent 3,846,329, to Dow Corning Corp. (1974).
65. G. R. Larson, C. A. Puschak, and K. A. Wood, U.S. Patent 5,486,576, to Rohm and Haas Co. (1996).
66. J. C. Wuhrmann, H. Mueller, K.-D, Brands, A. Asbeck, and J. Heidrich, U.S. Patent 4,655,960, to Henkel (1987).
67. M. J. Owen, in *Encyclopedia of Chemical Technology*, 4th ed., Vol. 7, John Wiley & Sons, New York, 1993, p. 928.

68. R. J. Mannheimer, *Chem. Eng. Commun. 113*, (1992).
69. S. Ross, *Chem. Eng. Prog. 63*, 41 (1967).
70. J. McGee, *Chem. Eng. 96*, 131 (1989).
71. I. C. Callaghan, C. M. Gould, and W. Grabowski, U.S. Patent 4,711,714, to British Petroleum (1987).
72. G. Koerner and D. Schaefer, U.S. Patent 5,004,559, to Th. Goldschmidt AG (1991).
73. D. E. Graham, W. A. Lidy, P. C. McGrath, and D. G. Thompson, U.S. Patent 4,596,653, to British Petroleum (1986).
74. Japanese Patent J 57159892, to Matsushita Reiki, (Oct. 2, 1982).
75. F. Jaffe and A. Papay, Canadian Patent CA1022531 to Stauffer (Dec. 13, 1977).
76. I. Schlachter and G. Feldmann-Krane, in *Novel Surfactants* (K. Holmberg, ed.), Vol. 74, *Surfactant Science Series*, Marcel Dekker, New York, 1998, p. 201.
77. G. Adams and M. A. Jones, U.S. Patent 4,690,688, to Dow Corning Corp. (1987).
78. D. R. Battice, K. C. Fey, L. J. Petroff, and M. A. Stanga, U.S. Patent 5,767,192, to Dow Corning Corp. (1998).
79. R. Spiegler, M. Keup, K. Kugel, P. Lersch, and S. Silber, U.S. Patent 5,613,988, to Th. Goldschmidt AG (1997).
80. T. Easton and B. Thomas, U.S. Patent 4,854,938, to Dow Corning Corp. (1989).
81. D. H. Rehrer, U.S. Patent 4460380, to Exxon Research & Engineering Co. (1984).
82. X. Domingo, L. Fiquet, and H. Meijer, *Tenside Surf. Deterg. 29*, 16 (1992).
83. T. Szekrenyesy, K. Liktor, and N. Sandor, *Colloids Surf. 68*, 267 (1992).
84. T. Szekrenyesy, K. Liktor, and N. Sandor, *Colloids Surf. 68*, 275 (1992).
85. J. J. Bikerman, *Foams*, Springer-Verlag, New York, 1973.

7

Silicone Surfactants: Applications in the Personal Care Industry

DAVID T. FLOYD Surfactant Division, Goldschmidt Chemical Corporation, Hopewell, Virginia

I. SUMMARY

Silicone polymers provide unique combinations of properties desired by the personal care industry. The industry's acceptance of these raw materials is well documented: sales increased from only a few tons in the early 1970s to tens of thousands of tons in the 1990s, and the market is expanding further. We now see these polymers in almost every segment of the personal care industry, where their uses range from minor additives to major components of primary ingredients in formulations. This general and rapidly growing use has been attributed to the ability of the silicone polymers to improve functional characteristics of the end products and to contribute greatly to the aesthetics of the products they are compounded into. Silicone polymers are available in a wide variety of molecular configurations with varying properties. Their safety and performance in the industry are well documented. Silicone polymers can also contribute to the mitigation of the negative impact on the skin of other organic ingredients by modifying surface spread and the localized concentration of these organics.

These characteristics suggest that with respect to personal care products, silicone polymers are not only among the safest ingredients, they are also a unique category of chemicals. They are environmentally friendly and improve the application, functionality, and aesthetics of personal care formulations. Thus, they should not be viewed as belonging to the same class as other synthetic polymers or organic chemicals, but as an independent category.

II. INTRODUCTION

The widespread use of silicone polymers was driven primarily by concerns over atmospheric depletion of ozone by chlorofluorocarbons in the mid- to late 1970s [1]. Low molecular weight methyl siloxanes or cyclic siloxanes were used to reduce the dependence on these chlorofluorocarbons. In the decade of the 1980s, polymers were being promoted by the silicone industry. Intensive research was initiated to develop more polymers based on silicone chemistry. Applications research showed these chemicals to be promising, positive discoveries and reinforced their uniqueness in the personal care industry [2–5]. This chapter reviews the use of silicone polymers in the personal care industry.

III. BACKGROUND

To recognize the importance of these findings, one must understand how widely silicones have been used in personal care formulations. Prior to the 1970s, moderate viscosity linear methylsiloxane or dimethicone had been used at low additive levels, mostly to prevent foaming in soap-based skin lotions, with an annual use of a few tons. Then in the mid-1970s, environmental concern over depletion of ozone in the upper atmosphere led formulators to search for product forms to replace aerosols.

U.S. Patent 4,126,269, assigned to Armour Dial Corporation in 1976, describes the use of low viscosity cyclomethicones as volatile carrier fluids for antiperspirant sticks. The patent describes improvements in tactility, staining reduction, compatibility, and variable volatility. In the decades that followed, this unique combination of properties—controllable volatility, functionality, sensory perceptibility, and low toxicity—was responsible for introducing silicones into virtually every product segment of the industry. The patents in Table 1 highlight the beginning of this market penetration. In addition to these primarily antiperspirant, hair care, and skin care segments, applications have been documented in sun care, nail enamels, fragrances, cleansers, and deodorants.

Table 2 summarizes the typical use levels of dimethicone and cyclomethicone in various personal care products within specific market segments. It should be emphasized that many of these applications involve methyl silicones and that a large majority are the volatile polymers.

TABLE 1 Patent Activity for Methyl Silicones in the Personal Care Industry

Patent	Issued to	Year	Product
G.B. 201-8590	Gillette	1980	Roll-on antiperspirant
U.K. 2,102,288	Helene Curtis Ind.	1981	Volatile polydimethylsiloxane additive for hair conditioner
U.S. 4,529,586	Lever Research	1982	Polydimethylsiloxane shampoo conditioning additive
U.S. 4,054,670	Buhler	1977	Polydimethylsiloxane skin protecting lotion additive
E.P. 0103910	Procter & Gamble	—	Polydimethylsiloxane moisturizing lotion additive
U.S. 4,515,784	Richardson-Vicks	1985	Polydimethylsiloxane oily skin treatment additive
U.S. 4,337,859	Kolmar	1982	Volatile polydimethylsiloxane carrier for makeup

Silicones are among the high performance ingredient successes of the 1990s, with product labels telling the story. In recent years, silicone use has expanded to virtually all personal care product segments, while the number of new products containing these materials is significant and growing rapidly—from approximately 28% in 1985 to almost twice that amount by 1992 [1]. This expanding use is related to a unique combination of attributes. First, silicones serve a variety of functions, acting, for example, as emollients, water barriers, or emulsifiers. They also provide specific sensory characteristics, notably the smooth, silky, non greasy feel desired by consumers. Finally, formulators have come to trust the safety of silicones based on the evidence of a wide range of studies that document a low order of toxicity [6].

As noted, there is a trend toward the use of silicone polymers in personal care products. Silicone polymers are being used not only as additives to formulations but as replacements for hydrocarbon-based chemicals [7]. Organic ingredients such as esters, emulsifiers, alcohols, and fatty compounds have been affected by silicone polymer technology, with varying levels of replacement occurring in many product segments (Table 3).

Product labels reveal a direct correlation between market trends and the use of silicones, an important relationship when one considers that product attributes and performance claims are a critical means of communicating with consumers. Table 4 provides a sampling of some typical commercial product claims and their relationship to the use of silicones. Claims of this type respond directly to market trends and a growing list of consumer demands. They follow trends toward clear, no color or low color products that are durable or offer environmental protection,

TABLE 2 Typical Use Levels of Methyl Silicone in Personal Care Products

	Silicone polymer molecular weight[a]		
	Low	Moderate	High
Skin care			
Hand and body lotion	2–6	1	NA
Facial treatment	2–6	NA	NA
Aftershave	10–20	NA	NA
Sun protection	5–30	5	2
Hair care			
Shampoo	NA	1–4	1–4
Conditioner	2–4	2	1–2
Leave-in conditioner	85	NA	15
Fixative, spray	1–5	1	NA
Fixative, gel	0.5	0.5	NA
Mousse	1	1	NA
Color cosmetics			
Foundation	10	1–4	1–3
Mascara/liner	10	1–4	1–3
Powder, pressed	10	1–4	1.5
Powder, loose	NA	5	NA
Antiperspirant/deodorant			
AP spray	10	NA	0–1
AP stick	50	1–2	1–3
AP roll-on	60	1–2	1–3
Deodorant spray	30	1–2	1–3

[a]NA, not applicable.

TABLE 3 Silicone Polymer Segment Replacement

Product segment	Material replaced by silicone polymers
Antiperspirant	Ester
	Emulsifier
Deodorant	Ethanol
	Emulsifier
Sunscreen	Mineral oil
	Emulsifier
Fragrance	Ethanol
Skin care	Ester
	Emulsifier
	Mineral oil
	Fatty compound

TABLE 4 Label Claims and Their Relationships to Ingredients

Commercial product	Claim	Silicone ingredient(s)
Skin treatment lotion	Oil free, softer feeling skin, silky liquid	Cyclomethicone Dimethicone Dimethicone copolyol Alkyl dimethicone
Moisturizing lotion	Light in texture, smoother skin, gentleness, oil-free, noncomedogenic, soft skin, greaseless	Cyclomethicone Dimethicone Dimethiconol Alkyl dimethicone
Shampoo/conditioners	Two-in-one, glossing, easy combing	Silicone betaine Amodimethicone Dimethicone Dimethiconol
Hair spray	Adds luster, fast drying, repairs, softens and smoothes, less tacky	Cyclomethicone Phenyl trimethicone Dimethicone copolyol
Clear antiperspirants/deodorants	Smooth, dries quickly, less tack, long lasting, non whitening	Cyclomethicone Dimethicone Dimethicone copolyol

are easy to apply and have a light application, and are oil free, nongreasy, and noncomedogenic.

Given these and other consumer requirements, the ultimate objective of the cosmetic chemist is to formulate products for which the desired claims may be made. Ideally, the chemist should understand how chemical composition and structure influence the fundamental properties of an ingredient and how these basic properties can translate to product benefits. Benefits support product claims, and are substantiated by scientific testing, making them acceptable for advertising. Although this ideal level of understanding has not yet been attained for many raw materials for silicone technology, it is sufficiently established and understood.

The unique structure of the basic polydimethylsiloxane polymer—which is actually a combination of antithetical components—accounts for the fundamental properties of silicones and their resulting benefits to personal care products.

The personal care industry uses labeling authorized by the Food and Drug Administration to describe ingredients for package labels. This system, which is referred to as INCI (CTFA) Nomenclature, is used in this chapter because it is eas-

TABLE 5 Nomenclature Reference

IUPAC nomenclature	INCI (CTFA) nomenclature[a]
Polydimethylsiloxane	Dimethicone
Cyclic polydimethylsiloxane	Cyclomethicone
α, ω-Hydroxypolydimethylsiloxane	Dimethiconol
Polydimethylsiloxane polyoxyethylene copolymer	Dimethicone copolyol
Polyalkylmethylsiloxane	Alkyl dimethicone
Amino functional siloxane	Amodimethicone
Phenyl polydimethylsiloxane	Phenyl trimethicone

[a]International Cosmetic Ingredient (Cosmetic, Toiletry and Fragrance Association).

ier to reference on commercial personal care products. Table 5 details the corresponding IUPAC nomenclature for reference.

Silicone is a generic name for many classes of organic silicone polymers with repeating siloxane monomers. Methylsiloxanes represent a prevalent class used in personal care products. The Si—O—Si backbone of these polymers is an inorganic-like component that is reactive and hydrophilic. This siloxane chain (Fig. 1) is polar and has strong intramolecular forces, as evidenced by its high bond dissociation energy compared with C—C or C—O bonds. Bond strength is responsible for much of the chemical and thermal stability of the polymer. The Si—O bond length is significantly longer, and the bond angle flatter, than comparable C—O and C—C bonds. Thus, the obstacle to rotation is very low and the polymer chain is very flexible. This property makes different orientations possible and provides the ability to accommodate substituents of various sizes and/or to allow for easy diffusion of other molecules.

The primary pendants on the silicon atoms in silicone, the CH_3 or methyl groups, can be characterized as organic and hydrocarbon-based. These groups are inert and hydrophobic, with low intermolecular forces and surface energies. They also are highly surface active, a characteristic that is maximized by virtue of flexibility of the Si—O—Si backbone. The result of this flexibility is a methyl

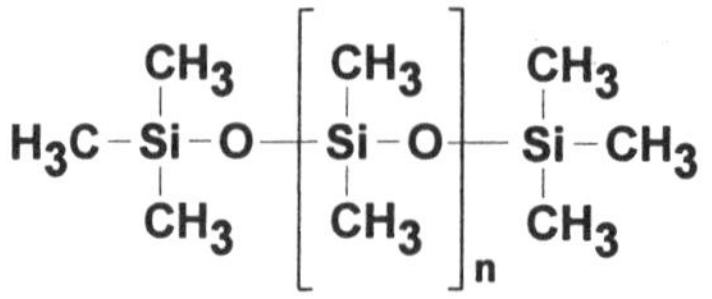

FIG. 1 Base silicone polymer.

"cloud," which facilitates the effective orientation of methyl groups to maximize their surface activity at interfaces.

The flexibility and reactivity of the Si—O—Si backbone also make possible a number of structural and compositional variations, which result in many families of silicones, including linear and cyclic varieties with variable degrees of cross-linking. The introduction of reactive groups directly onto silicon atoms or on alkyl or aryl groups attached to silicon allows synthesis of a wide variety of useful compounds. Since silicone polymers can exist over a wide range of sizes or molecular weights, an almost limitless number of compounds is possible. Each modification may influence different aspects of the final polymer, such as its rheology, mechanical properties, surface properties, and compatibilities.

Open space in the polymer allows gaseous molecules to easily diffuse through it, a property that is useful for forming "breathable" films. The flexibility of the polymer backbone allows methyl groups to maximize their hydrophobicity, insolubility, and surface activity. The solubility parameters of silicones are significantly lower than those of water and many organic materials. As a result, silicone polymers are highly water resistant and, except for certain low molecular weight species, incompatible with organic ingredients. These two properties contribute to barrier effects and substantivity. The same characteristics also inhibit silicone polymers from adhering to other materials, making them effective anti-adhesive and detackifying agents.

The orientation of methyl groups at the interface results in other important characteristics. Because intermolecular forces are very low, silicone polymers have a very low activation energy of viscous flow; they tend to spread easily on other surfaces and to remain fluid even at high molecular weights. The surface tension of the methyl group is one of the lowest of the organic constituents, giving silicone polymers a very low liquid surface tension. The critical surface tension for polydimethylsiloxane is actually higher than its liquid surface tension (24 vs. 20 dynes/cm), so the polymer will spread over its own absorbed film. These properties translate to advantages in antifoaming and pigment surface treatments, as well as easy wetting of surfaces such as skin and hair.

IV. APPLICATIONS OF SILICONE POLYMERS

Silicone polymers derived from a polysiloxane backbone were shown in Fig. 1. Reactive sites can be attached to this backbone. Typically the active site is a hydrogen group as in Fig. 2. These reaction sites can be pendant as shown in Fig. 2, a configuration often referred to as the "comb" structure. Terminal or α,ω substitutions can also be made. The reactive sites are typically reacted to organic moieties forming organomodified siloxane polymers [1,8–13]. The organic modifications can be nonionic or ionic (Fig. 3).

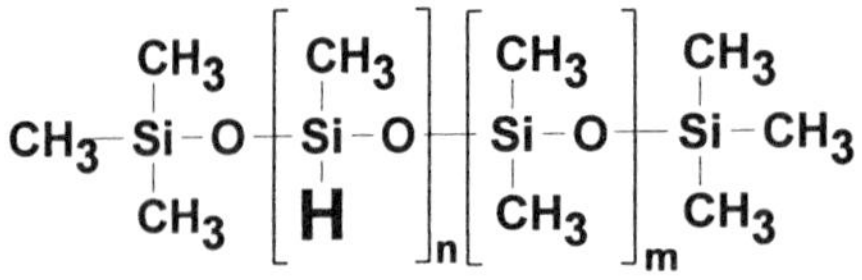

FIG. 2 Typical reaction site.

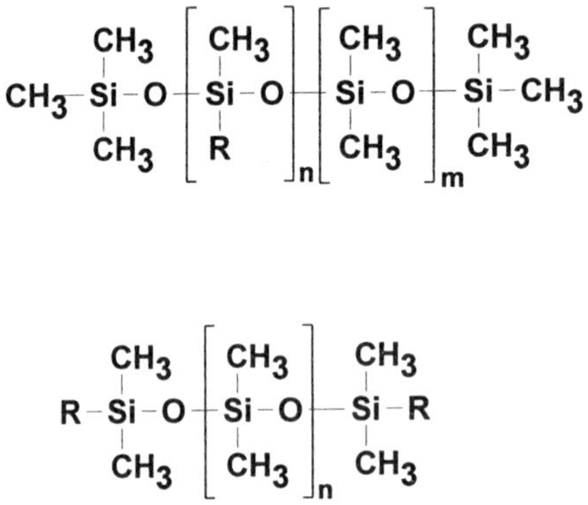

FIG. 3 Organomodified polysiloxanes: R = nonionic or ionic organic group.

Reacting organic moieties to siloxanes produce new molecules that incorporate the benefits of silicone polymers. These include surface conformation, liquidity, flexibility, surface tension effects, and safety with the activity, reactivity, and substantiation of the organic modification. The base siloxane polymer is limited by its solubility in various media such as water, alcohols, esters, and hydrocarbon oils. A well-chosen organic moiety can overcome this deficiency.

Organomodified silicone copolymers are of increasing interest in the sciences as well as for personal care applications. Their unique properties, which are quite different from those of conventional hydrocarbon polymers, have led to their service in widespread applications in the cosmetic and pharmaceutical industries. These different properties are due mainly to the nature of the base dimethylsiloxane structure [11–16]. This structure gives the polymers flexibility and good flow properties even at low temperatures. They have good thermal and oxidative resistance, low surface tensions, either hydrophobicity or lipophobicity depending on their modification, and a high permeability to many gases and biological inertness. With organomodified silicones, we can maintain the desired properties of dimethicone and reduce the negative aspects.

These organic modifications of the polydimethylsiloxanes are made by replacing one or more of the methyl groups with lipophilic or hydrophilic moieties to

TABLE 6 Comparison of Organic and Silicone Polymer Families

Category	Hydrocarbon-based type	Silicone polymer	Ref.
Oil phase	Mineral oil	Dimethicone, cyclomethicone	
	Esters	Alkyl siloxanes	
	Emulsifiers	Alkylether siloxanes	
		Dimethicone copolyols	
Anionics	Phosphate esters	Silicone phosphate esters	17
	Sulfates	Silicone sulfates	18
	Carboxylates	Silicone carboxylates	19
	Sulfosuccinates	Silicone sulfosuccinates	20
	Sulfonates	Silicone sulfonates	21
	Thiosulfates	Silicone thiosulfates	21,22
Amphoterics	Amino propionates	Silicone amphoterics	23
	Betaines	Silicone betaines	24,25
	Phosphobetaines	Silicone phosphobetaines	26,27
Cationics	Alkyl quaternary compounds	Silicone alkyl quaternary compounds	28
	Amido quaternary compounds	Silicone amido quaternary compounds	29
	Acetate quaternary compounds	Silicone quaternary compounds	30,31
	Imidazoline quaternary compounds	Silicone imidazoline quaternary compounds	32
	Carboxy quaternary compounds	Silicone carboxy quaternary compounds	33,34
Nonionics	Alcohol alkoxylates	Dimethicone copolyol	
	Alkanolamides	Silicone alkanolamides	35
	Esters	Silicone esters	36

form new polymers [2,9,11,15]. Some examples of the substitution of silicone polymers for organics are detailed in Table 6 [17–36].

As Table 6 indicates, many substitutions and variations can be made to form silicone polymers. One interesting aspect of silicone polymers is that they can be classified as surfactants because of their good surface tension modification effects. Obviously a wide range of silicone surfactants can be synthesized. Table 7 summarizes some of the possibilities for modification.

The applications for silicone polymers in personal care products are wide ranging and varied. Selective applications have been chosen as examples of how to use these polymers. The rest is left up to the creative imagination of the formulating chemist. The phenyl-modified siloxanes comprise a class of silicone polymers that has been widely used. The accepted labeling name for this class of polymers is

TABLE 7 Possible Silicone Surfactant Variations

Portion	Variations possible
Silicone	Molecular weight of the silicone
	Number of functional groups on the silicone molecule
	Type of functional groups on the silicone molecule
	Terminal or comb structure
Alkoxylate	Ethylene oxide content
	Propylene oxide content
	Location of propylene and ethylene oxide in the molecule
Fatty	Alkyl or alkyl amido
	Hydrophobicity (number of C atoms)
Ionic	Nonionic
	Cationic
	Amphoteric
	Anionic

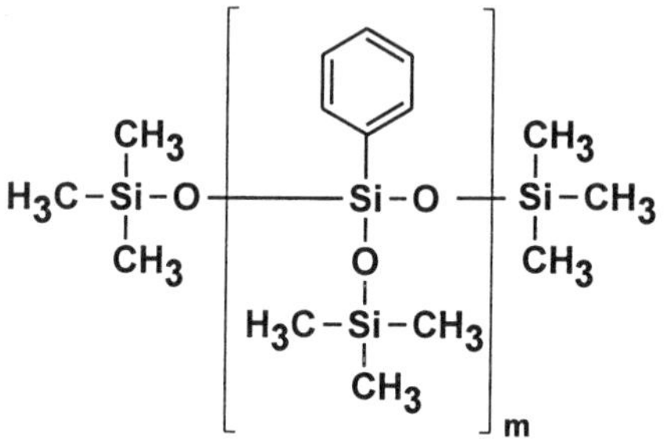

FIG. 4 Phenyl trimethicone.

phenyl trimethicone (Fig. 4). Its representatives are good glossing agents for hair and contribute a soft emollient, conditioning effect to skin. In response to the concern for volatile organic carbon emissions, manufacturers of hair sprays have reformulated their products as resin emulsions in hydroalcoholic systems. This formulation often causes a foam to develop in the spray nozzle of the aerosol dispenser. Phenyl trimethicone is efficient at low dosage levels of 0.1–0.2% in the formula to function as a defoamer, improving the aerosolization of the resin system for a more even deposition.

Hydroxy-terminated polysiloxanes have been used in the personal care industry for a number of years (Fig. 5). These tend to be high molecular weight polymers with viscosities of 2.0×10^5 to 5.0×10^6 mm^2/s. They are typically supplied as dispersions in cyclic polysiloxanes. These polymers, referred to in the personal

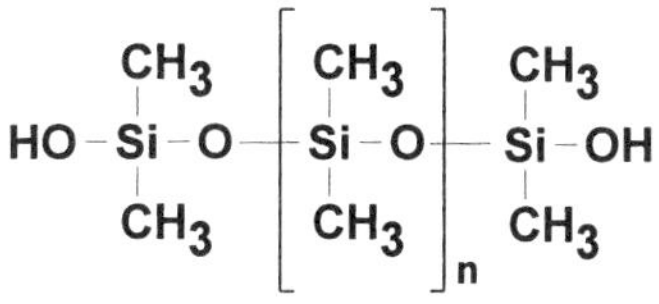

FIG. 5 Dimethiconol.

care industry as dimethiconols, are used on hair as superconditioners. They contribute to emolliency, gloss, or shine, and they promote good combing characteristics. They also give a hydrophobic protective coating to the hair shaft. They are often combined with the phenyl trimethicones in hair care applications.

Dimethiconols are also used in skin care emulsions, where they improve wash resistance and contribute to the barrier properties of the emulsions.

Low molecular weight hydroxy-terminated polysiloxanes have also been used to coat pigment particles to improve their dispersibility in lipid media. This application finds use in the area of pigmented cosmetics such as lipsticks and mascaras.

The polyether group in dimethicone copolyols can consist of homopolymeric chains of either ethylene or propylene oxide or can contain varying proportions of both ethylene and propylene oxide, as a copolymer of ethylene–propylene oxide (EO/PO). Through the chemical variations that can be realized in the composition of dimethicone copolyols, these products can provide unique surfactant properties. They can offer significant advantages as surface tension depressants, wetting agents, conditioners, glossers, emulsifiers, and foam builders in hair treatment preparations [37–42]. (Fig. 6).

Of particular interest, in terms of compatibility, is the orientation of polyether polysiloxanes in aqueous solutions. The dimethylsiloxane moiety is hydrophobic and aligns itself at the surface of the aqueous medium. The ethylene oxide moiety is hydrophilic and attaches itself into the aqueous medium (Fig. 7). The physical

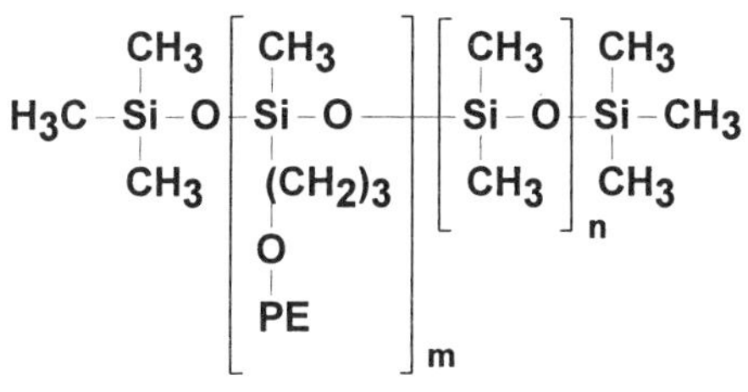

FIG. 6 Polyether polysiloxane copolymer [INCI (CTFA): dimethicone copolyol]: PE = $(—C_2H_4O)_x\,(—C_3H_6OH)_y$ —H.

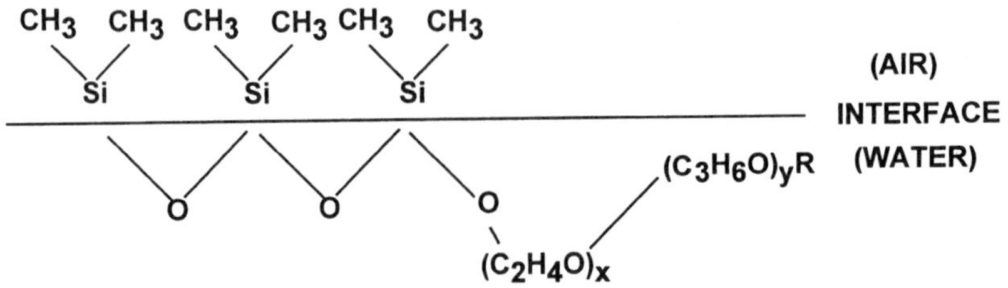

FIG. 7 Orientation of polyether polysiloxane copolymer in an aqueous solution.

properties pertinent to aqueous systems include surface tension, cloud point, foam height, wetting, solubility, and pH.

The dimethicone copolyols were evaluated for wet comb using a method similar to that described by Garcia and Diaz [43]. Relative combing force was used to compare the different copolyols. The combing work data were obtained by integration of the average curves for each of the five measurements, divided by the weight of the hair tress. The values reported were an average of the different tresses studied for each copolyol.

Combability measurements have become an important tool for an objective evaluation of the performance of hair conditioners and hair conditioning formulations [43–49]. The most important parameter is wet combability. The consumer normally combs wet hair before or during the drying procedure. Dry hair combability is influenced by several different parameters, such as temperature, time of drying, humidity of air, and the previously applied work of combing [46]. Therefore, we concentrated on the determination of wet combing behavior.

The determined relative work of combing in relation to the cloud point, the propylene oxide content, and the EO/PO ratio of the dimethicone copolyols is shown in Table 8. Figure 8 plots work of combing versus cloud point. The results

TABLE 8 Combing Force Relationships

Copolyol	Cloud point (°C)	PO in molecule (%)	EO/PO ratio	Average Si—O polymeric units	Relative combing work
B	90	0	100/0	20	621
G	71	11	77/23	80	558
C	64	17	73/26	25	561
E	42	38	48/52	50	543
F	31	48	35/65	25	546
D	10	53	20/80	25	491

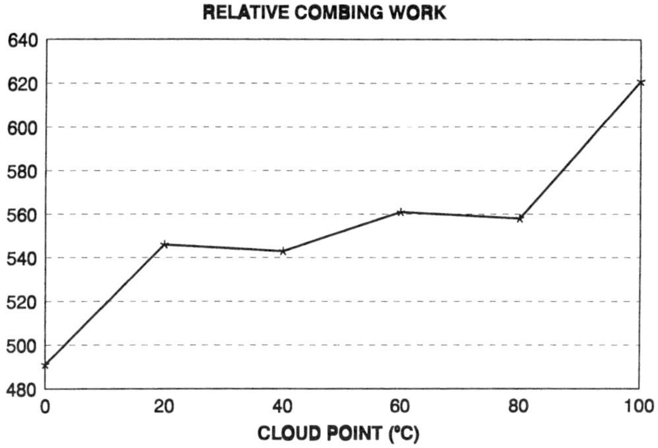

FIG. 8 Relation of combing force to cloud point.

show that conditioning effect or combability is related to the cloud point and the propylene oxide content of the dimethicone copolyol. It is known from earlier investigations [9], that the cloud points of dimethicone copolyol drop when the PO content is increased.

The silicone content was not significantly varied among the investigated dimethicone copolyol types. The strong influence of the great variations of EO/PO ratio overruled the small differences in the silicone content.

An understanding of surface activities and physical–chemical properties will allow the formulator to make choices that will produce the desired effect in a shampoo system. For example, combining a high molecular weight silicone polymer with a midrange polymer containing a high proportion of propylene oxide will provide for optimization of gloss, conditioning, and wet and dry combing [50].

The majority of silicone polymers in use in industry today are the nonionic type. It is also possible to bond ionic side chains to polydimethylsiloxane, either additionally or exclusively. This provides a completely new range of property combinations that is still not fully realized.

One method of synthesis that can be used to produce ionic silicone polymers is via epoxy siloxanes by reaction with special nucleophiles. The attractiveness of copolymers having ionic appendant groups lies in their ability to form far more intense physical–chemical interactions with liquid or solid substrate phases compared to nonionic copolymers.

Cationic and amphoteric groups lead to an increased substantivity on polar substrates, providing a very durable covering of surfaces. Therefore, these products have a potential not only to form protective films but also to act as emulsifiers for finely dispersed polar particles in organic media.

Additionally, polymethylsiloxanes modified with cationic or amphoteric groups combine a high gliding ability with antistatic properties, which are particularly useful in personal care and textile applications. Polyether side chains can also be included in addition to the ionic groups to achieve a solubility behavior for specific requirements.

The reaction of secondary amines with epoxy siloxanes results in tertiary amino siloxanes. When reacted further with sodium monochloroacetate, they produce silicone betaines [24,25] (Fig. 9). If the epoxide ring is reacted with tertiary amines in the presence of acids (e.g., acetic acid), the silicone quaternaries are obtained [30,31] (Fig. 10).

Silicone betaines and silicone quaternaries are excellent additives in hair and skin care preparations [51]. More highly substantive than the silicone betaines, the silicone quaternaries show good compatibility with anionic surfactants, hence are suitable as additives in shampoos. Silicone betaines and silicone quaternaries are good antistatic agents. In comparison to pure organic polymeric quaternary compounds, silicone quats and betaines give hair a silky silicone gloss and feel. Unlike their organic counterparts, these silicone polymers do not have the disadvantage of building up on the hair.

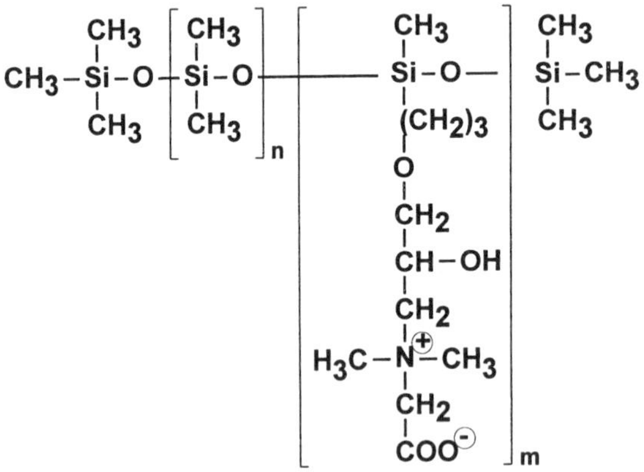

FIG. 9 Polybetaine polysiloxane copolyol.

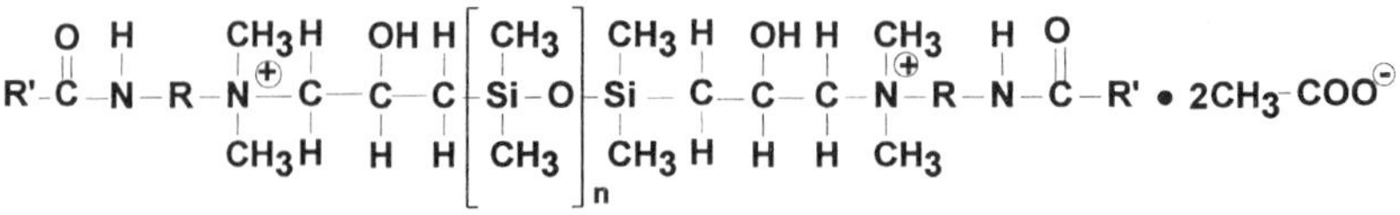

FIG. 10 α, ω—Quaternized polysiloxane [INCI (CTFA): quaternium-80]: R′ = Cocos.

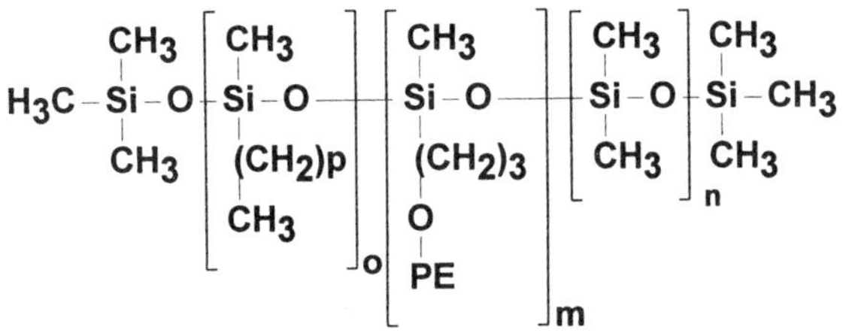

FIG. 11 Polyalkyl polyether polysiloxane copolymer [INCI (CTFA): cetyl dimethicone copolyol]: PE = $(\text{—}C_2H_4O)_x\,(C_3H_6O)_y$ —H.

Shampoo formulas can be further optimized for after-feel and conditioning (i.e., static control) by incorporating small amounts of a polyalkyl polyether polysiloxane—cetyl dimethicone copolyol (Fig. 11) and α,ω-quaternized polysiloxane—quaternium-80 [30,31]. Compared to standard two-in-one shampoos, these products give better rinse and less severe buildup problems.

The silicone quaternary compounds are also good conditioning agents. They are exceptionally mild to skin and eyes, unlike their organic counterparts. Previous studies [2,4,9,41,52] have discussed the substantivity of these compounds to hair substrates. The high cost of silicone quaternaries as sole conditioning agents often precludes their use. However, as small adjunctants in traditional conditioning bases at levels of 0.2–0.4%, they give dramatic improvements in conditioning and improved tactile properties. At similar use levels in shampoos, they are very effective conditioning agents.

Interestingly, by combining diquaternary silicone compounds with copolyols, one can develop good bases for hair conditioners. The apparent synergism noted between the diquaternaries and the copolyols allows for low usage levels and corresponding cost control. The combinations are more costly than an organic conditioner, but they are very mild and can be formulated over a wide pH range.

Another class of conditioning agents are the thiosulfate-modified siloxanes, often referred to as Bunte salts after the German chemist who first synthesized thiosulfate esters [53].

Anionic siloxanes are synthesized by the reaction of polysiloxanes with either $NaHSO_3$ or $Na_2S_2O_3$ [21,22]. Although the corresponding sulfonates have found limited application in cosmetic preparations, the thiosulfates are a subject of ongoing application research (Fig. 12). Silicone thiosulfates can be cross-linked via S—S bonds by reaction with thioglycolic acid, sodium sulfide, or cysteine on the hair. The cross-linking reaction permits the use of silicone thiosulfates in cold wave products as thin-film-forming agents. In permanent wave products, mixed S—S bonds to human hair might be possible. This could lead to wash-resistant conditioning.

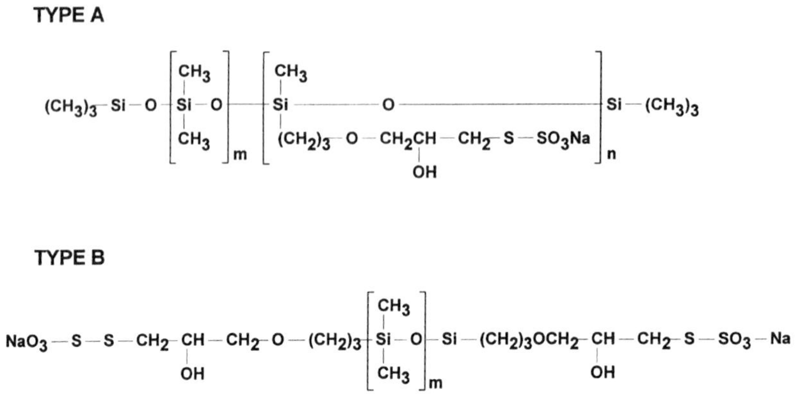

FIG. 12 Polysiloxane Bunte salts.

The routes of synthesis for these polymers have been described [2,4,9,41,52]. A recent study [52] has shown that the thiosulfate-modified siloxanes can be powerful conditioning agents on hair. Hair that had been damaged by bleaching was used in the studies. This hair showed a typical damaged structure, with the scales lifting partially off the surface of the hair. After treatment, the hair shaft is smooth, with the scales lying flat on the surface. In addition, tactile differences were noted.

The untreated hair is rough and dull, while the treated hair feels smooth, exhibiting better combability and gloss. The hair surface carries predominantly anionic charges, but because of the amino groups of the amino acids there are also cationic centers present [54]. Silicone Bunte salts can be absorbed by electrostatic forces alone. The adsorption is further enhanced by the polymeric character of these Bunte salts.

There is, however, an alternative mechanism that can explain the substantivity of dimethicone thiosulfates. This mechanism is related to the reactivity of the Bunte salt group (Fig. 13). This activity was used earlier in the development of reactive dyes that carried Bunte salt groups. These dyes could be fixed onto cellulosic or woolen textiles by using sodium sulfide as an additive [55]. The first two reactions in Fig. 13 are between Bunte salt groups and sodium sulfide. The final two lines show the reactions of Bunte salts with mercapto-functional reagents (e.g., thioglycolic acid).

During the reaction, the anionic thiosulfates are converted into mixed disulfides. These mixed disulfides may rearrange and thus form the thermodynamically more stable symmetrical disulfides [56], as shown in the fourth reaction equation of Fig. 13. In these reactions, the hydrophilic dimethicone thiosulfate becomes hydrophobic, loses its water solubility or dispersibility, and precipitates from the

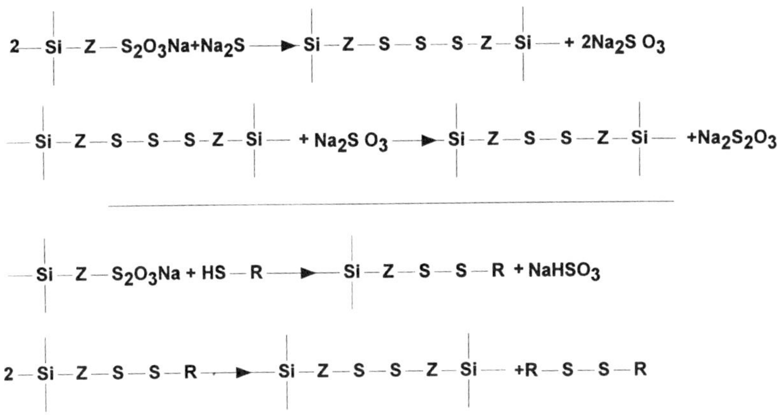

FIG. 13 Formation of disulfides from dimethicone thiosulfates: examples of R include —CH_2CO_2H (thioglycolic acid), and —CH_2CHCO_2H (cysteine).

solution. If a polyfunctional comb-like branched dimethicone thiosulfate undergoes this reaction, a cross-linked insoluble polymer will be formed.

These polymers, especially the type A polysiloxane bunte (Fig. 12), can be used in shampoos and conditioners for the treatment of damaged hair. The shampoos and conditioners should be formulated with a pH range of 7–8. This polymer, because of its hydrophilic nature and ability to create shine on damaged hair, is gaining acceptance in the ethnic hair care market.

A series of sensory tests was conducted to judge the response of hair tresses to the various organomodified siloxanes. The sensory tests were performed by trained evaluators, selected for their ability to differentiate between undamaged and damaged hair tresses. The treated hair tresses were judged both wet and dry. The following examinations were carried out with the wet hair tresses [57]:

- detangling of the hair immediately after application of the test solution
- combability
- feel

The following examinations were carried out after the hair tresses had been carefully dried in an air-conditioned room (23°C, 55% relative humidity):

- combability
- feel
- gloss of the dry hair

A rating scale of 0 to 4 was used (4, excellent; 3, very good; 2, good; 1, fair; 0, poor).

The sensory judgment was carried out by the test panel in sets of tests with not more than seven hair tresses. Two of these hair tresses were used as a positive and a negative comparison and were identified to the testers. The other hair tresses to be examined were coded with random numbers. For every test, one of the five tresses served as a control; it was treated with a quaternized cellulose (polyquaternium-10). The results are presented in Table 9.

Recently, new categories of silicone surfactants have been synthesized [58,59]. One new category is the silicone phosphobetaines. Silicone phosphobetaines have unique properties because they contain:

- a fatty portion
- an ionizable phosphate group
- a cationic quaternary nitrogen portion

However, unlike the naturally occurring phospholipids or the other phosphobetaines, these molecules contain a covalently bonded silicone moiety. Since there are two types of dimethicone copolyol compounds, comb and terminal, there are

TABLE 9 Judgment from the Sensory Tests

	Detangling, wet	Combability, wet	Wet feel	Combability, dry	Dry feel	Gloss
Quaternium-80 ($N = 10$)	2.6	2.6	2.4	3.2	2.8	2.0
Quaternium-80 ($N = 20$)	2.6	2.4	2.2	2.6	2.4	1.8
Quaternium-80 ($N = 80$)	2.6	2.2	2.4	2.8	3.0	2.6
Dimethicone propyl PG-betaine	2.4	2.8	2.2	2.6	2.6	2.2
Dimethicone copolyol D	2.4	2.4	2.0	3.2	3.0	2.4
Dimethicone copolyol G	2.2	2.2	2.0	3.2	2.8	2.6
Cetyl dimethicone copolyol	2.4	2.4	2.0	2.8	2.8	1.8
Dimethicone/sodium PG-propyldimethicone thiosulfate copolymer	2.8	2.0	2.2	2.6	2.8	2.2
Dimethicone/disodium PG-propyldimethicone thiosulfate copolymer	1.8	1.4	2.0	3.0	2.8	2.2

two substitution patterns observed in silicone phosphobetaines [26,27]. The difference is in location of the organofunctional group within the silicone molecule, and in the resulting three-dimensional structure, which in turn affects the surfactant properties of the compound (Fig. 14).

The presence of the polyalkylene oxide in the silicone polymer results in an inverse cloud point in aqueous solutions. Inverse cloud point phenomena are well known and are observed in nonionic surfactants. The inverse, or upper, cloud point is defined as a temperature above which the solute compound has minimal solubility in water and other polar solvents. If heat is applied to an aqueous solution of a surfactant at its inverse cloud point, the surfactant will become insoluble, and the solution will turn milky. Since it is no longer in solution above this point, the silicone polymer has minimal water solubility and maximum substantivity to hair, skin, and fiber at temperatures at or above its inverse cloud point.

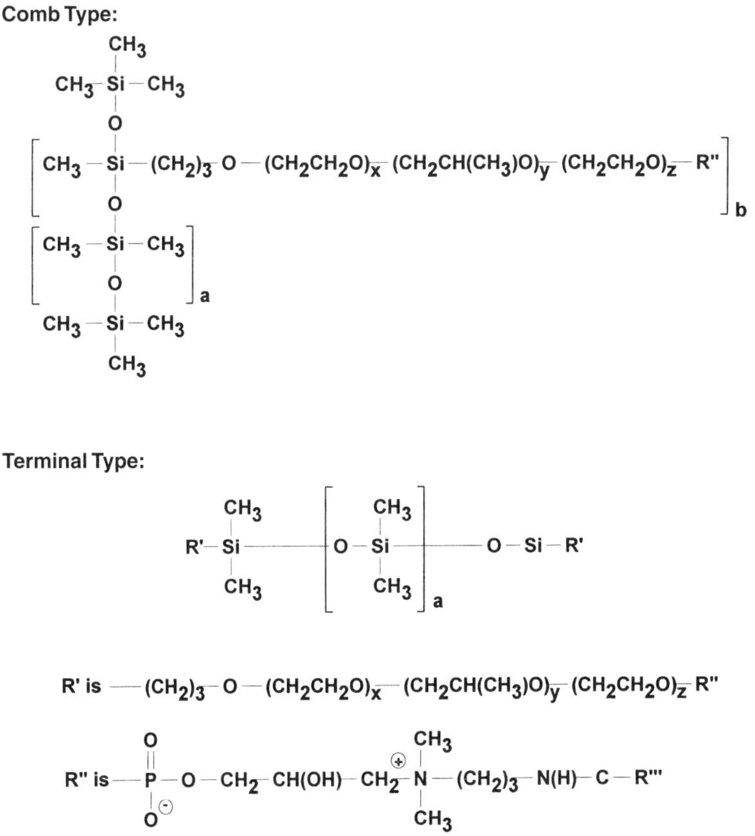

FIG. 14 Structures of comb- and terminal-type silicone phosphobetaines: R‴ is derived from a fatty acid.

Silicone phosphobetaines are amphoteric surfactants, which means that they can exist with a positive or negative charge, or both, in aqueous solutions. The charge on the molecule depends on the pH of the solution.

This silicone polymer, by virtue of its unique pendant group, is highly foaming and nonirritating to the eyes and skin; it deposits on fiber surfaces to form effective surface-modifying finishes. Therefore, these compounds could be well suited to applications in the personal care market.

Because of the variety of modifications possible, molecules prepared by this technology can be water soluble, insoluble, or dispersible. It should also be noted that the selection of a silicone polymer having a high molecular weight will result in a silicone phosphobetaine that has predominantly silicone characteristics, while the selection of a low molecular weight silicone polymer will result in a molecule with relatively little silicone character. This allows the selection of molecules for specific applications and formulation properties. This is true for other silicone polymers also.

Silicone phosphobetaines incorporate the desirable properties of silicone (mildness and substantivity to hair and skin) and the surface active properties of the phosphobetaine in one molecule. The ability of these materials to form sheets as well as micelles is significant because it may allow for the development of efficacious emulsions, delivery systems for skin care and liquid crystals.

Carboxysilicone polymers have also been developed. A typical structure is given in Fig. 15. These polymers can be complexed with a quaternary compound to form complex quats [33,34]. The carboxysilicone complexes offer substantially lower irritancy, good water solubility, and good hair rewettability after treatment,

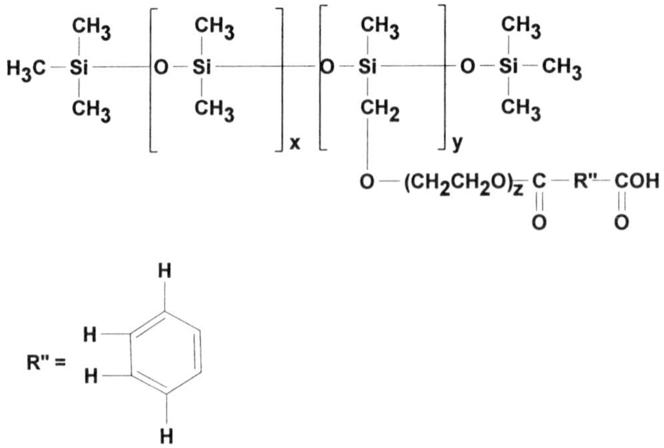

FIG. 15 General structure of carboxysilicone polymers.

while still providing the softness characteristic of the traditional quaternary incorporated in the complex. Because the complex can offer greater compatibility with anionic surfactants, formulators have far greater latitude in the preparation of hair or skin conditioning products. Formulators can incorporate hydrocarbon quaternary carboxysilicone compounds into a variety of personal care products easily, without the use of elaborate thickening systems or the need for emulsification/homogenization.

It is possible to synthesize silicone polymers with different modifications along the siloxane polymer chain. This process forms compounds with differing solubility preferences on different portions of the same molecule. The polyalkyl polyether polysiloxane copolymers (Fig. 11), which are very effective emulsifiers for water-in-oil emulsions, are an example.

Water-in-oil emulsion characteristics are reviewed in Chapter 8. The water-in-oil formulations produced by the polyalkyl polyether polysiloxane copolymers offer new avenues of exploration for the product development chemist. Since the emulsions formed can be cold-processed, these copolymers allow for the incorporation of heat-sensitive materials. The formulas are water resistant and in many cases waterproof on the skin. This polymer is ideal for use in barrier products, drug delivery systems, and sun protection products.

One class of silicone polymers that is of increasing interest consists of the polyalkyl polysiloxane copolymers. These are lipophilic emollient polymers (Fig. 16).

It is known that polyalkyl-modified polysiloxanes are compatible with emollients, waxes, and other silicone polymers [60]. They are excellent additives for improving the gliding properties of emulsions. Investigations were made to determine the maximum compatibility of such products with ester oils and liquid paraffin to obtain spreading and lubricity data. Mixtures of polyalkyl polysiloxane copolymers of different structures and molecular weights were combined with emollients such as isopropyl myristate, octyl stearate, and caprylic/capric triglycerides, and then compared by means of a laboratory in vitro spreading test [4,61].

The polyalkyl polysiloxanes are good lubricants for pigments. Since they reduce the cohesive forces of esters, by combining them with esters, one can

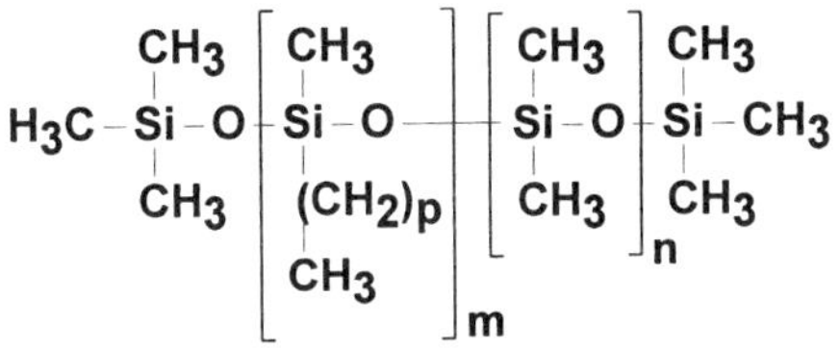

FIG. 16 Polyalkyl polysiloxane copolymer.

ensure that the surface of the pigment will be more evenly coated. If a good surface coat is applied onto a pigment, less energy is needed to mill the pigment to a smaller size. This property is important in pigmented personal care products because the size of the pigment particle is directly related to wear time and color development. The incorporation of these alkyl dimethicones into either the pigment grinding medium or the formulation base reduces the level of pigments needed for color development and increases the wear time or longevity of the product on the skin. Incorporating an alkyl dimethicone into an oil phase will allow the emulsion to spread easily and uniformly on the skin. This can be important to a formulation not only for product aesthetics, but to facilitate the uniform release of active compounds from a treatment base.

Alkyl dimethicones have good lubrication properties and can function to improve emolliency and skin feel. In hair conditioning emulsions, they can also contribute to gloss and a good combing force reduction.

A standard sunscreen oil phase was prepared to test the effectiveness of the silicone polymers for spreading enhancement. The gelatin spreading test was then used to evaluate these oil phases. Table 10, which gives the increase of the spreading area with the above-mentioned polyalkyl silicone copolymer, Table 10 reveals increases of 1.4 to 2.6 compared with the oil phase without the copolymer. This enhancement also affects the spreading and even distribution of sun filter agents and can lead to an increase in the sun protection factor in formulations. A growing trend in the industry is to formulate water-in-oil emulsions for sun protection products. Placing the active ingredients in the external oil phase coupled with an alkyl polysiloxane such as cetyl dimethicone can maximize the sun protection

TABLE 10 Spreading Rate Enhancement of an Oil Phase with Cetyl Dimethicone

Oil phase	Enhancement (%)
Cetyl dimethicone copolyol	7.0
Octyl methoxycinnamate	25.0
Benzophenone-3	10.0
Emollient	58.0
Cetyl dimethicone	0.5
Type of emollient in test formula	**Factor of spreading rate enhancement with cetyl dimethicone**
Caprylic/capric triglycerides	1.4
Cetearyl octanoate	2.4
Octyl palmitate	2.5
Isopropyl myristate	2.6
Decyl oleate	2.6

response of a compounded sunscreen. Sunscreen emulsions using an organomodified silicone emulsifier have tested at a higher sun protection factor (SPF) level with lesser amounts of chemical actives. If a silicone wax, such as cetyl dimethicone is included, improvements in the spreading coefficients of the actives are increased and SPF values are enhanced. Independent testing laboratories have confirmed that organomodified silicone emulsions can produce elevated SPF values. For example, an SPF value of 22 can be achieved with as little as 3% active [61] (Table 11). While this elevation in SPF can mean reduced raw material costs and increased efficacy for the manufacturer, the technology also results in less product sensitivity.

The enhanced spreading of emollients and actives can also be demonstrated in a formulated sunscreen product. A formula was developed with a total oil phase of 30%. A 3% level of octyl methoxycinnamate was used as the sole active in the base. The level of cetyl dimethicone was varied from 0.0 to 2.0%. A water-in-oil emulsifier based on polyglyceryl-4 isostearate–cetyl dimethicone copolyol, and hexyl laurate as a cosolvent, was used in the formula. Figure 17 gives the results

TABLE 11 SPF Enhancement of Water-in-Oil Emulsions with Silicone Emulsifiers

		SPF	
Sun filter	Level (%)	Expected	Obtained
Octyl dimethyl *P*-aminobenzoic acid	1.4	2–3	12
Octyl methoxycinnamate	3.0	5–6	18–22

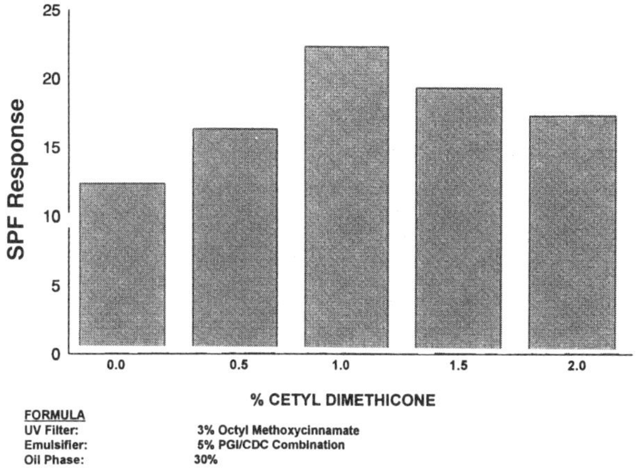

FIG. 17 Influence of cetyl dimethicone on SPF response.

of this study. The effects of cetyl dimethicone are maximized at a usage level of 1% in the base. Once the enhanced spreading effect had been achieved, it was determined that increasing the spreading agent's level in the formula gave no further deposition or formula leveling benefits for the skin. Increasing the level does not further enhance the SPF response. Other studies made with alkyl-modified siloxanes elicited similar profiles for maximizing the spreading effects of esters, oils, and formulated emulsions.

It has been observed that when a personal care product that includes a silicone polymer is formulated, once an effect has been achieved—whether it be conditioning, gloss spreading, or other cosmetic application—use performance benefits; the addition of further amounts of the silicone polymer does not seem to further boost the effects. Thus, the efficiency of silicone polymers can be demonstrated. Typically the use level of a properly selected polymer is low and can be classified as a minor adduct in a formulated product.

Human and environmental safety are critical requirements of ingredients for personal care use. Considerable data have been published on silicones. The methyl polysiloxanes, in particular, are extensively used in personal care [1,2,4,9]. Based on expert panel reviews [62,63], these methyl polysiloxanes are considered to be safe cosmetic ingredients as presently used.

The environmental impact of methyl silicones also has been studied in recent years [64,65]. It is necessary to consider the low molecular weight species separately from the high molecular weight species because of the environmental partitioning differences. Low molecular weight methyl silicones are too volatile to bioaccumulate; instead they partition to the atmosphere, where they undergo oxidative degradation. The partial oxidates have decreased lipophilicity. They complete the oxidation process to form water, silicic acid, and carbon dioxide. These materials have short atmospheric lifetimes, hence have no potential to deplete stratospheric ozone, nor can they contribute to the formation of ozone in the urban atmosphere [66,67].

High molecular weight methyl silicones partition to the sludge fraction in wastewater treatment plants. When the sludge is used as a fertilizer, these materials chemically degrade to water, silicic acid, and carbon dioxide. There is no immediate effect on landfills or agricultural soil. It is generally accepted that methyl silicones and silicone polymers have no known negative environmental effects.

V. CONCLUSION

The silicone industry has offered a wide selection of organomodified siloxanes to the personal care industry. Silicone technology offers formulators ingredients that can improve personal care products in several significant ways. Conventional products such as protective oils and lotions can be more aesthetically pleasing. Silicone additives can improve clarity, spreadability, and smoothness while reducing the tackiness of formulations. The aesthetic qualities of silicone formulation also suggest distinctive new formulations.

Hydrophobic derivatives find applications in emulsions or nonaqueous preparations. Hydrophilic nonionic and ionic products find uses in aqueous as well as anhydrous systems. In skin care products, improved silky feel is obtained by the addition of a suitable silicone compound. The desired properties in hair treatments are improved wet and dry combability, conditioning, antistatic effects, and improved gloss, combined with a silky feel. Silicone polymers in pharmaceutical and cosmetic bases can produce superior emulsions, contribute to barrier protection of the skin, and maximize the delivery of active ingredients to the skin.

Silicone polymers have low toxicity profiles, and their inclusion in many personal care products has resulted in formulation having better overall toxicity profiles.

The market for silicone-based polymers will continue to grow. For formulators and marketers of personal care products, the flexibility and utility of these silicones will make these polymers valuable additives for innovative and functional products for the consumer.

REFERENCES

1. A. DiSapio, C. Fry, and D. Zellner, *Preprints 17th IFSCC Congress*, Yokahama, 1992, p. 334.
2. D. Floyd, in *Cosmetic and Pharmaceutic Applications of Polymers* (C. Gebelein et al., eds.), Plenum Press, New York, 1991, p. 49.
3. D. Floyd, B. Macpherson, and K. Jenni, *Preprints 19th IFSCC Congress*, Sydney 1996.
4. D. Floyd and K. Jenni, *CRC Handbook on Polymers*, CRC Press, Boca Raton, FL, 1998.
5. A. DiSapio, *Cosmet. Toiletries, 102*, 102 (March 1987).
6. A. DiSapio, *Drug Cosmet Ind.* May 1994, p. 29.
7. A. Disapio, *Soap, Cosmet. Chem. Spec.* September 1994, p. 50.
8. A. Disapio and P. Fridd, *Preprints 15th IFSCC Congress 1*, London, 1988, p. 89.
9. D. Schaefer, *Preprints 15th IFSCC Congress 1*, London, 1988, p. 103.
10. M. Starch and C. Krosic, *Cosmet. Techno.* November 1982, p. 20.
11. W. Noll, *Chemistry and Technology of Silicones*, Academic Press, New York, 1968.
12. B. Grüning and G. Koerner, *Tenside Surf. Detergents 26*, 312 (1989).
13. D. Schaefer, *Tensides Surf. Detergents. 27*, 154 (1990).
14. M. Rosen, *J. Am. Oil Chem. Soc. 49*, 293 (1972).
15. R. Meals, in *Encyclopedia of Chemical Technology*, 2nd ed., Vol. 18, 1969, p. 221.
16. I. Yilgör, J. Riffle, and J. McGrath, in *ACS Symposium Series 282*, American Chemical Society, Washington, DC, 1985, Chapter 14.
17. U.S. Patent 5,149,765 to Siltech,
18. U.S. Patent 4,960,845 to Siltech,

19. U.S. Patent 5,296,434 to Siltech,
20. U.S. Patent 4,717,498 to Mclntyre,
21. U.S. Patent 3,392,182 to Goldschmidt,
22. U.S. Patent 4,537,595 to Goldschmidt,
23. U.S. Patent 5,073,619 to Siltech,
24. U.S. Patent 4,609,750 to Goldschmidt,
25. U.S. Patent 4,654,161 to Goldschmidt,
26. U.S. Patent 5,091,493 to Siltech,
27. U.S. Patent 5,237,035 to Siltech,
28. U.S. Patent 5,098,979 to Siltech,
29. U.S. Patent 5,153,294 to Siltech,
30. U.S. Patent 4,833,225 to Goldschmidt,
31. U.S. Patent 4,892,166 to Goldschmidt,
32. U.S. Patent 5,196,499 to Siltech,
33. U.S. Patent 5,248,783 to Siltech,
34. U.S. Patent 5,296,434 to Siltech,
35. U.S. Patent 5,070,171 to Siltech,
36. U.S. Patent 5,070,168 to Siltech,
37. German Patent 1,595,730 to Goldschmidt,
38. German Patent 3,133,869 to Goldschmidt,
39. German Patent 2,588,074 to Goldschmidt.
40. W. Davis and D. Jones, *Polym. Prepr. 11*, 447 (1970).
41. W. Wolfes, *Preprints 14th IFSCC Congress*, *1*, Barcelona, 1986, p. 473.
42. S. Wendell and A. DiSapio, *Cosmet. Toiletries 98*, 103 (1983).
43. M. Garcia and J. Diaz, *J. Soc. Cosmet. Chem. 27*, 379 (1976).
44. K. Yahagi, *Int. J. Cosmet. Sci. I 13*, 221 (1991).
45. Y. Suzuki and K. Yahagi, *Preprints 31st SCCJ Congress*, Tokyo, 1991, p. 1.
46. A. Niak, J. Vives Rego, and J. Cot, *Preprints 14th IFSCC Congress*, *2*, Barcelona, 1986, p. 993.
47. P. Busch, *Arztl. Kosmetol. 19*, 270 (1989).
48. H. Hoffges, D. Hollenberg, and K. Wisotzki, *Seifen Oele Fette Waechse 115*, 615 (1989).
49. W. Newmann, G. Cohen, and C. Hayes, *J. Soc. Cosmet. Chem. 24*, 773 (1973).
50. D. Floyd, H. Leidreiter, B. Sarnecki, and U. Maczkiewitz, *Preprints 17th IFSCC Congress*, Yokahama, 1992, p. 297.
51. H. Meyer, *Cosmet. Toiletries Manuf. Suppl. XV*, 5 (January 1990).
52. B. Grüning and H. Leidreiter, *Seife Oele Fette Waesch. 3*, 117 (1992).
53. H. Bunte, *Chem. Ber. 7*, 646 (1874).
54. K. Ohbu, *Colloid Polym. Sci. 264*, 798 (1986).
55. O. Sam, *J. Soc. Dyers Colori. 80*, 416 (1964).
56. V. Bell, *Proc. Int. Wolltextil—Forschungkonferenz 3*, 595 (1976).
57. H. Leidreiter, K. Jenni, and C. Jorbandt, *Conf. Proc. In-Cosmet.* Barcelona, 1994, p. 149.
58. A. J. O'Lenick Jr. and C. S. Sitbon, *Cosmet. Toiletries 111*, 67 (April 1996).
59. J. Imperante, A. J. O'Lenick Jr., and J. Hannon, *Cosmet. Toiletries*, *109*, 81 (March 1994).

60. U.S. Patent 4,275,101 to RCA.
61. D. Floyd, B. Sarnecki, and B. Macpherson, *Soap, Perf. Cosmet. 69*, 26 (March 1996).
62. Cosmetic Ingredient Review, *J. Am. Coll. Dermatol. 1*, 4 (1982).
63. Cosmetic Ingredient Review, *J. Am. Coll. Dermatol. 10*, 1 (1991).
64. C. Frye, *Soap Cosmet. Chem. Spec.* August 1983, p. 33.
65. G. Chandra, A. Disapio, C. Frye, and D. Zellner, *Cosmet. Toiletries 109*, 63 (March 1994).
66. R. Atkinson, W. Carter, and S. Aschmann, Final Report from Statewide Air Pollution Research Center, University of California, Riverside, to Dow Corning Corp., November 1992.
67. W. Carter, J. Pierce, I. Malkina, and D. Luo, Final Report from Statewide Air Polution Research Center, University of California, Riverside, to Dow Corning Corp., October 1992.

8

Silicone Surfactants: Emulsification

BURGHARD GRÜNING and ANDREA BUNGARD Surfactant Division,
Th. Goldschmidt AG, Essen, Germany

I. FUNDAMENTALS

Silicone-based emulsifiers have gained considerable market importance over the last two decades, and especially within the last 10 years. Without any doubt, this product group of organomodified silicones represents a very modern class of emulsifiers.

Applications of silicone surfactants to obtain colloid stabilization (e.g., stabilization of pigments) have gained little interest. We note, however, a few examples of treatments of titanium dioxide, which are mostly carried out with unmodified polydimethylsiloxanes, to enhance the oxide's dispersion in plastics and paint resins. Other applications, such as the stabilization of coal slurry, have been considered [1].

Silicone emulsifiers are polymeric or oligomeric surfactants that possess more than one hydrophilic and more than one hydrophobic functional group. In contrast to low molecular weight surfactants, these multifunctional emulsifiers attach to the interface with several segments. The energy of adsorption is the sum of the interactions of all segments, which thus is considerably higher than the energy of individual, of "monomeric" surfactants. In other words, these multifunctional polymeric emulsifiers adsorb more strongly at the interface. They can be used in low concentrations and are very efficient in generating highly stable emulsions. They will not desorb from the interface and migrate through the bulk phase because of their firm adsorption to the interface.

The success of silicone emulsifiers is certainly related to their special silicone character, which differs in many respects from that of organic low molecular weight and other polymeric emulsifiers. The reasons for these differences may be summarized under three headings:

The *chemistry of organomodified silicones* offers great versatility for the flexible synthesis of well-defined, customized oligomeric or polymeric molecules.

The *physicochemistry of silicones* causes the silicone part in silicone surfactants to contribute hydrophobic as well as oleophobic properties to the molecule (since, unlike mineral oils and natural oils, silicones are not soluble in water or in organic oils).

The *high flexibility of the silicone polymer chain*, especially of the polydimethylsiloxane chain, enables the molecules of rather high molecular weight to achieve optimal arrangements at the borderline between different phases in a relatively short time scale.

A. Chemistry of Silicone-Based Emulsifiers

Since the chemistry of silicone surfactants is described extensively earlier in this book [2], a short survey to cover the most important aspects with respect to emulsifiers will be sufficient here. The chemistry of silicone surfactants has also been described in review articles [3,4].

Silicone-based emulsifiers are oligomeric or polymeric mostly linear silicone molecules, that are modified by means of organic residues to add to the hydrophobic and oleophobic silicone substituents, which are compatible with aqueous or oily media. The synthesis is carried out in two basic steps.

First, in the equilibration reaction, the silicone backbone is synthesized, and second, in the modification reaction(s), the organic residues are added to the silicone backbone. Starting materials for the synthesis of the silicone backbone in the equilibration are usually silicone oils, functional siloxanes, and low molecular weight silicones like hexamethyldisiloxane, which deliver the end groups of the

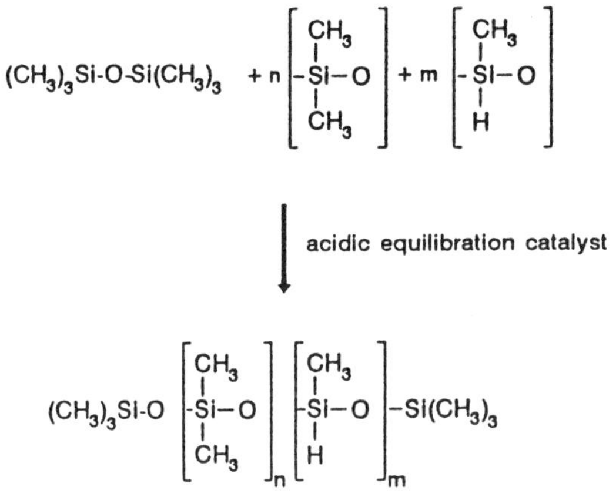

FIG. 1 Siloxane with reactive Si—H groups, comblike structure.

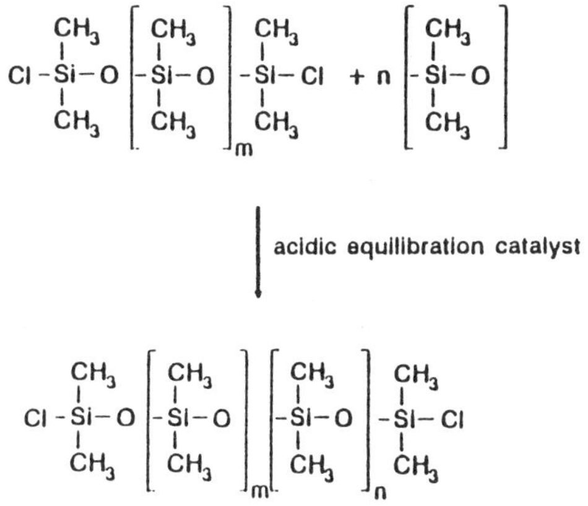

FIG. 2 Siloxane with reactive Si—Cl groups, linear structure.

polymer chain. As examples, Figs. 1 and 2 show the synthesis of a Si—H functional siloxane and a Si—Cl functional one, respectively.

The equilibration reaction can produce functional silicone backbones that are well defined with respect to molecular weight and functionality. Silicone chemistry thus offers a tool that is unique in polymer chemistry for the customized synthesis of functional polymeric raw materials on a production scale.

The modifying organic residues can be bound to the silicone by two fundamentally different linkages: the Si—O—C linkage can be obtained by reaction of hydrogen- or chlorosiloxanes with hydroxyfunctional reagents (Fig. 3). This link-

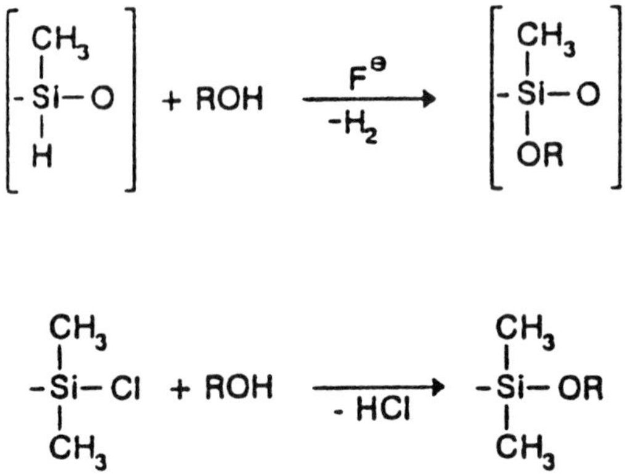

FIG. 3 Organomodified siloxanes with Si—O—C linkage.

age, however, is potentially hydrolytically unstable, so the resulting products can be used only in nonaqueous systems. In the case of emulsions this will certainly be an exception.

Organic residues bearing the Si—C linkage are bound to the siloxane in a hydrolytically stable way. This kind of product is usually made by the platinum-catalyzed hydrosilylation reaction of hydrogen siloxanes with α,β-unsaturated reagents. The organo group most frequently used for modification of siloxanes with respect to emulsifiers is the polyoxyalkylene residue. Typical examples for α,β-unsaturated reagents of this group are allyl polyethers (Fig. 4) [5]. The polyethers usually consist of oxyethylene and oxypropylene units, which can vary in their proportion. To achieve hydrophilic properties, oxyethylene units are predominantly used.

The structures of the organomodified siloxanes can be either comb like or linear; the linear compounds, in turn, are further subdivided into α,ω-functional (ABA) or bipolar linear (AB) types (Fig. 5). In the linear types, a rather long uninterrupted dimethylsiloxane segment can be achieved, and each molecule has a defined mono- or difunctionality, whereas the functionality in the comblike structured type is ruled by statistics.

There are silicone-based emulsifiers of each type; only the comblike structured silicone polyethers however, are commercially important as emulsifiers. From a basic hydrogen siloxane structure, mixed modified polyether and alkyl-substituted siloxanes can be prepared [6]. In these molecules water-soluble polyether residues and oil-soluble hydrocarbon residues are united by a siloxane backbone, which itself is neither water nor oil soluble. It can be assumed that these molecules will arrange at the water–oil interface in a double comblike manner (Fig. 6).

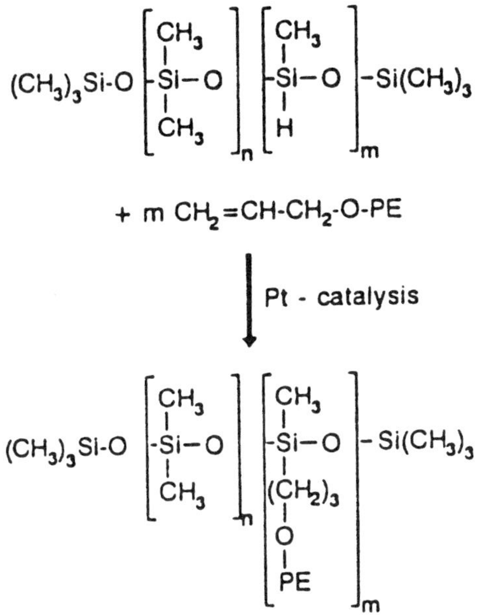

FIG. 4 Synthesis of silicone polyethers by hydrosilylation (Si—C -linkage); PE, polyether (e.g., polyoxyalkylene glycol monoalkyl ether).

Polydimethylsiloxane–polyoxyalkylene copolymers and polyalkylmethylsiloxane–polyoxyalkylene copolymers represent the two most important groups of silicone-based emulsifiers. The literature also cites many examples of organomodified siloxanes that are said to have emulsifying properties, carrying ionic functions. Examples are given in Table 1 [7–17].

The nonionic, organomodified silicones are predominantly used to stabilize water-in-oil emulsion systems. Nevertheless, there are also examples of oil-in-water emulsions. Ionic organosilicone types are recommended to be used as oil-in-water emulsifiers only.

Of commercial importance so far are solely nonionic organomodified silicones, which are used for water-in-oil emulsions of different kinds.

B. Selected Physicochemical Aspects

Polydimethylsiloxanes (PDMS) are not compatible with aqueous media. They are hydrophobic and contribute the desired hydrophobic properties to the silicone surfactant (i.e., the silicone-based emulsifiers). If the molecular weight of a given PDMS is high enough, which usually is the case if it consists of more than six to eight methylsiloxy units, that PDMS also is not compatible with mineral oils or

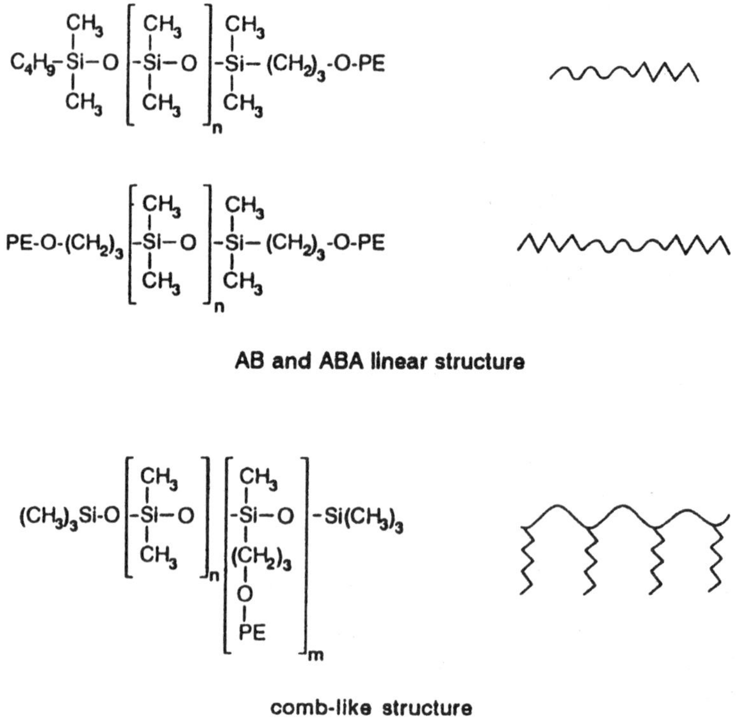

FIG. 5 Linear and comblike basic structures of organomodified siloxanes.

more polar oils (e.g., ester oils, natural fats or oils). So the polydimethylsiloxane part contributes both hydrophobic and oleophobic properties to the molecule. This property clearly differentiates the silicone-based emulsifiers from hydrocarbon based materials.

The surface tension of PDMS depends on its molecular weight [18], increasing from 15.7 mN/m for hexamethyldisiloxane to 20–21 mN/m for medium and high molecular weight oligomers and polymers. These values are generally considerably lower than those of organic oils, which are typically 30–35 mN/m. Their low surface tension together with their incompatibility enables the silicones to act as the hydrophobic part in surfactants in aqueous and also in organic media. If the silicone is combined with oil-soluble hydrocarbon residues (e.g., in long chain alkyl-substituted siloxanes), these products are surface active in organic oils. For example, the surface tension of 2-ethylhexyl palmitate is reduced by an alkyl-substituted siloxane (molecular weight ≈6000) from 30 mN/mm to approximately 25 mN/m (Fig. 7).

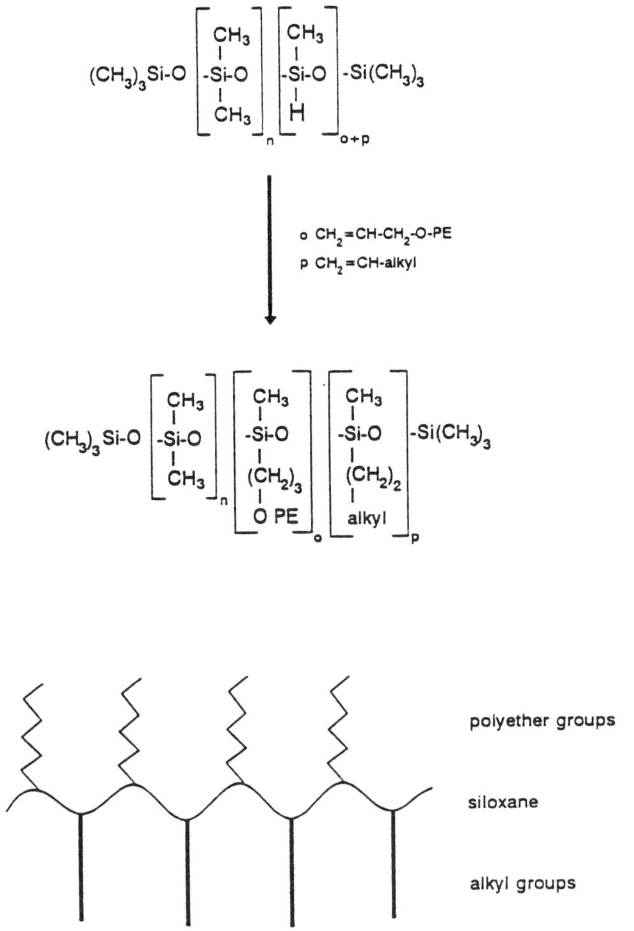

FIG. 6 Hydrophilically and oleophilically organomodified silicones; PE, polyether (e.g., polyoxyalkylene glycol monoalkyl ether).

The long chain hydrocarbon residues are important factors in the achievement of solubility and balanced compatibility of organomodified silicones with the oil phase. Where long chain alkyl residues are missing, no specific interactions with hydrocarbon-type oil phases take place. Interfacial tension measurements in decane–water systems containing water-soluble silicone polyether copolymers illustrate the absence of interaction between silicone and hydrocarbon [19]. Although in this study silicone surfactants of medium molecular weight ($\approx$4500) were effective in reducing the surface tension of the water phase to approximately

TABLE 1 Ionic Silicone Derivatives

Derivative	Structure[a]	Ref.
Silicone sulfosuccinates	$-\mathrm{Si}-\mathrm{R}-\mathrm{O}-\mathrm{CO}-\mathrm{CH}(\mathrm{SO_3Na})-\mathrm{CH_2}-\mathrm{CO_2}-\mathrm{Na}$	7,8
Silicone sulfonates	$-\mathrm{Si}-\mathrm{R}-\mathrm{O}-\mathrm{CH_2}-\mathrm{CH}(\mathrm{OH})-\mathrm{CH_2}-\mathrm{SO_3}-\mathrm{Na}$	9
Silicone sulfates	$-\mathrm{Si}-\mathrm{R}-\mathrm{O}-\mathrm{CH_2}-\mathrm{C}(\mathrm{C_2H_5})(\mathrm{CH_2}-\mathrm{OSO_3NH_4})-\mathrm{CH_2}-\mathrm{OSO_3}-\mathrm{NH_4}$	10
Silicone taurates	$-\mathrm{Si}-\mathrm{R}-\mathrm{O}-\mathrm{CH_2}-\mathrm{CH}(\mathrm{OH})-\mathrm{CH_2}-\mathrm{NH}-(\mathrm{CH_2})_2\mathrm{SO_3Na}$	11
Silicone phosphates	$-\mathrm{Si}-\mathrm{R}-\mathrm{O}-\mathrm{P}(=\mathrm{O})(\mathrm{OH})-\mathrm{ONa}$	12
Silicone thiosulfates	$-\mathrm{Si}-\mathrm{R}-\mathrm{O}-\mathrm{CH_2}-\mathrm{CH}(\mathrm{OH})-\mathrm{CH_2}-\mathrm{SSO_3Na}$	13
Silicone betaines	$-\mathrm{Si}-\mathrm{R}-\mathrm{O}-\mathrm{CH_2}-\mathrm{CH}(\mathrm{OH})-\mathrm{CH_2}-\mathrm{N}^{\oplus}(\mathrm{CH_3})_2-\mathrm{CH_2CO_2^{\ominus}}$	14
Silicone sulfobetaines	$-\mathrm{Si}-\mathrm{R}-\mathrm{O}-\mathrm{CH_2}-\mathrm{CH}(\mathrm{OH})-\mathrm{CH_2}-\mathrm{N}^{\oplus}(\mathrm{CH_3})_2-\mathrm{R'}-\mathrm{CH_3CO_2^{\ominus}}$	15
Quaternary silicone compounds	$-\mathrm{Si}-\mathrm{R}-\mathrm{N}^{\oplus}(\mathrm{CH_3})_2-(\mathrm{CH_2})_3\mathrm{SO_3^{\ominus}}$	16,17

[a] R, alkylene; R′, amidoalkyl.

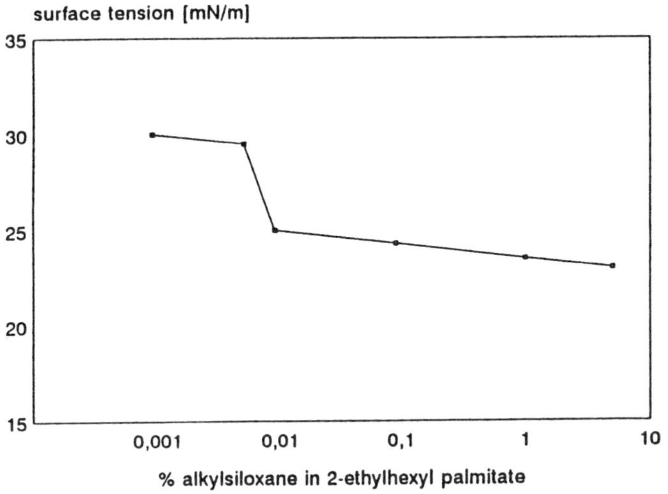

FIG. 7 Surface tension of solutions of long chain alkyl-modified silicone in 2-ethylhexyl palmitate at 25°C.

30 mN/m, they were not very effective in reducing the interfacial tension between the aqueous phase and decane, where only remarkably high values of 8–9 mN/m were achieved. It is suggested that the lack of long chain alkyl groups in the investigated silicone polyether copolymers prevents the molecules and the decane phase from interacting.

It can be understood that silicone emulsifiers carrying long chain alkyl residues are preferentially used for hydrocarbon- or fat-containing emulsions, whereas polydimethylsiloxane polyether emulsifiers without long alkyl chains are predominantly used for emulsions based on silicone oil.

Because of the specific properties of the silicone backbone the hydrophile–lipophile balance (HLB) system is difficult to apply to silicone emulsifiers. This is easy to understand when we recall that the HLB system was originally developed for fatty alcohol ethoxylates and later transferred to emulsifiers and surfactants of other chemical classes. Calculations based on the measurement of critical micelle concentrations give, however, some idea of the hydrophobic behavior of the silicone part. It was found that each methyl and methylene group of the modified silicone surfactant contributes as much as a methylene group in hydrocarbon-based surfactants, whereas the Si—O part of the molecule behaves indifferently, bringing in neither hydrophilic nor hydrophobic contributions. These results combined with the simple equation of Griffin [20] can be applied to a typical water-in-oil emulsifier, polyhexadecylmethylsiloxane–polyoxyalkylene copolymer (PCMS-POA), with a molecular weight of 10,000–15,000. Thus an HLB value of approximately 4 is calculated. In comparative emulsifying tests with the same emulsifier, an experimental HLB value was found in the range of 4 to 6 [21], which corresponds fairly well to the calculated value.

C. Flexibility of the Silicone Polymer Chain

The extraordinarily high flexibility of the polydimethylsiloxane chain distinguishes these polymers fundamentally from other hydrocarbon-based polymers and can be regarded as a reason for many special physicochemical properties of silicones, which are important with respect to emulsifiers. A characteristic property of polymers, strongly influenced by this high flexibility, is the glass transition temperature (T_g) [22]. The T_g of PDMS is 146 K, which is 2 K lower than that of polyethylene [23] in spite of the methyl groups, which cause more steric hindrance than the hydrogen atoms in the polyethylene. The unique flexibility results from the bonding relations in the PDMS backbone, since there is almost no rotational barrier along the Si—O—Si-bonds (Table 2) [22,24,25].

The superior flexibility of the backbone enables the PDMS, and certainly to a great extent also the organomodified siloxane derivative, to arrange itself in an optimal way more easily and more rapidly than would be possible for a hydrocarbon or other organic polymer backbone. This is especially significant if, like an emulsifier, the molecule is required to arrange at an interface.

A comparison of interfacial tensions in oil-in-water systems of polydodecylmethylsiloxane–polyoxyalkylene copolymer (PLMS-POA) and polyglycerol diisostearate [26] shows that the silicone-based emulsifier reduces the interfacial tension at lower concentrations (on a molar basis) than the organic emulsifier: for example, an interfacial tension of 25 mN/m is reached with the silicone derivative at approximately 0.03 μmol/L and with the polyglycerol ester at about 1.7 μmol/L.

TABLE 2 Characteristic Data on Si—O—Si Bonds

Property	Data	Ref.
Si—O bond length	0.165 nm	22
Si—O—Si bond angle	130 ± 10°	24
Rotation barrier	0.8 kJ/mol	25

The greater flexibility of the silicone backbone presumably is of major importance and gives rise to the better performance of the silicone.

II. APPLICATION OF SILICONE EMULSIFIERS

A. General

The most important application of silicone emulsifiers is in water-in-oil emulsions, which are intended for use in cosmetic formulations. The emulsions may be either antiperspirant formulations or creams and lotions for skin care and sun protection. There are additionally some technical applications such as in polishing formulations or emulsion polymerization. The kind of oil phase—whether it consists of volatile or nonvolatile silicone oils or of organic hydrocarbons, ester, or natural based oils—generally is critical for the choice of silicone emulsifier: in water-in-silicone emulsions, polydimethylsiloxane–polyoxyalkylene copolymers (PDMS-POA) are used, whereas in water-in-organic-oil emulsions the emulsifier contains additional long chain alkyl groups. The long chain alkyl groups (e.g., dodecyl or hexadecyl groups) create important interactions with the organic oil phase; they are of no or minor importance if the oil phase is a silicone fluid. In this case the compatibility of the silicone part of the emulsifier with the continuous phase will be crucial. The results of A. Harashima et al. [27] illustrate this further: polydimethylsiloxane–polyether copolymers of linear ABA and AB structures (A = polyoxyalkylene, B = polydimethylsiloxane) and comblike branched structure (see Fig. 5) were compared with regard to their ability to stabilize O/W and W/O emulsions of volatile or nonvolatile silicone fluids. The authors found that the emulsifying ability of the branched type is inferior to that of the linear types and tentatively attributed this result to a lower compatibility of the polydimethylsiloxane portion of the branched type to the silicone phase. Because of the branched structure, the undisturbed polydimethylsiloxane segments in this type are shorter than the linear types.

Although linearly built emulsifiers may have advantages in some applications, commercial silicone emulsifiers generally have the comblike structure, which has proved to be the most versatile for achieving optimized products providing high performance. In any event, ternary copolymers consisting of dimethylsiloxane–polyoxyalkylene and long chain alkyl groups can be synthesized only in the comblike manner (see Fig. 6).

The most important application of silicone emulsifiers is the stabilization of water-in-oil emulsions. The oil may be either a silicone fluid or an organic oil. Examples involving oil-in-water and nonaqueous emulsions are much less frequent. Because of the strong adsorbance of the silicone emulsifiers, more complicated systems can be also realized, where it is important that the emulsifier be prevented from migrating and interfering with other surfactants. Examples of such systems are multiple emulsions such as the water-in-oil-in-water type.

B. Water-in-Silicone-Oil Emulsions

Water-in-silicone-oil emulsions have gained considerable importance in applications that require the special properties of silicones—in particular, the excellent spreading and film-forming properties, hydrophobicity, gloss, dry nonsticky feel, and volatility (low molecular weight, cyclic silicones). Prominent examples are cosmetic antiperspirant emulsions and polish formulations.

Antiperspirants contain highly concentrated aqueous aluminum chlorohydrate or aluminum/zirconium chlorohydrate solutions. By emulsifying these solutions in cyclic dimethylsiloxanes, one can transfer them into an easily applicable form. The siloxanes facilitate application and reduce the stickiness of the chlorohydrate. They evaporate without giving a chilling effect. Other nonvolatile silicone oils may also be added, which will result in a residual dry, nonsticky film. Hydrophobic water-soluble polydimethylsiloxane–polyoxyalkylene copolymers of relatively high molecular weight are preferred as emulsifiers. They are used in typical concentrations of 1–3 wt% [28]. To achieve emulsions with increased stability, organic water-in-oil emulsifiers like polyglycerol fatty acid esters may additionally be used [29]. The patent literature gives further hints on how to obtain antiperspirant emulsions with improved stability by using a combination of volatile silicone fluid, alkanoic acids, and waxy esters in the oil phase [30] or by using organic emulsifiers of both low and high HLB value in addition to the silicone emulsifier [31].

Of further importance for the stability of these emulsions is their increased viscosity, which of course is also desired to facilitate the application of the formulation on the skin. The viscosity can be regulated quite easily by the phase volume ratio of aqueous and oil phase. Liquid formulations of a viscosity of 500–2000 mPa·s consist of 70–75 wt% aqueous phase, whereas with 80–85 wt% aqueous phase, nonflowable gel-like emulsions are obtained. Addition of propylene glycol to the aqueous phase equalizes the refractive indices of both phases and thus the formulation becomes transparent. Typical antiperspirant formulations are given in Table 3 [32,33].

Polish formulations in the form of water-in-silicone emulsions can be obtained in ways similar to those described for antiperspirants. An important effect of silicones in this application is to create easily spreading films that also transport other polishing oils and waxes. Typically these polishes contain abrasive material in the water phase and, in addition to silicone fluid, other solvents like mineral oil and alcohols in the oil phase. A. Kasprzak describes emulsions that also contain aminofunctional siloxanes in the oil phase [34]. They are stabilized by a combi-

TABLE 3 Typical Antiperspirant Formulations

	Formulation composition (wt%)	
	Liquid antiperspirant	Gel-like antiperspirant
Emulsifier[a]	10.5	10
Volatile silicones	16.3	7
Polyglyceroleate	1.5	—
Aluminum chlorohydrate (50% solution in water)	38.2	50
Water	33.5	17
Propylene glycol	—	16

[a]Consists of 1 part PDMS-POA and 9 parts of cyclic polydimethylsiloxanes.

nation of a PDMS-POA (molecular weight 2500) and an oil-soluble polyoxyethylene nonylphenyl ether (HLB 8.6).

Increasing the molecular weight of the emulsifier is an effective means of achieving emulsions with further improved stability. Thus with an emulsifier having an approximate molecular weight, of 30,000, one can obtain emulsions of water-in-volatile silicone fluid that contain ethanol or other water-soluble (potentially destabilizing) alcohols in the aqueous phase [35]. These emulsions contain, like the emulsions described above, a second oil-soluble organic emulsifier.

The molecular weight of the emulsifier can be further increased by slight cross-linking (e.g., with divinylsilanes during the addition reaction of polyoxyalkylene residues to the Si—H functional silicone backbone by hydrosilation) [36]. The cross-linking reactions, via hydrosilylation of vinyl groups bound to polydimethylsiloxane derivatives, can be driven even further, leading to continuous solidified silicone phases in which aqueous droplets are dispersed. To prepare these dispersions, a water-in-silicone emulsion is first formed, and stabilized by a PDMS-POA. This way pharmaceutical delivery devices can be produced [37].

C. Water-in-Organic-Oil Emulsions

As pointed out before, compatibility of the silicone polyoxyalkylene copolymer with organic oils is achieved or increased by the attachment of long chain alkyl groups to the silicone backbone. The replacement of dimethylsiloxy units by alkylmethylsiloxy units may be partial or complete. Referring to Fig. 6, the index n in the formula may even become zero. Both kinds of copolymer (i.e., with and without dimethylsiloxy units) are effective emulsifiers. Examples of fatty alkyl groups used are the dodecyl or hexadecyl residues.

Emulsifiers of this kind for the stabilization of water-in-oil emulsions were first described in the 1980s [38–40]. The materials themselves, silicone polyoxyalkylene copolymers [5] and long chain alkyl-substituted silicone polyoxyalkylene copolymers [41], were invented earlier and were regarded as useful mainly for dispersible waxes. Based on this knowledge, later patents were filed that deal with an improved production procedure using higher saturated alcohols as solvent [42]. Others deal with related structured silicone copolymers:

alkyl groups being attached to the silicone backbone by ether groups [43],
limited degree of polymerization and excluding dimethylsiloxy units [44] or
increased molecular weight by slight cross-linking with difunctional agents during the hydrosilylation reaction [45,46]

1. Technical Applications

Although by far the most applications of silicone surfactant emulsifiers are in the cosmetic field, there are also some interesting technical applications.

In inverse emulsion polymerization of acrylic monomers, silicone emulsifiers can be used to obtain smooth water-in-oil emulsions consisting of 20–25% external oil phase and 75–80% internal aqueous phase. With 3% emulsifier, smooth emulsions can be easily produced without applying high shear homogenization. Upon completion of the polymerization, the emulsion is inverted by nonylphenol ethoxylates to isolate the polymer [47].

In the drilling of oil and gas wells, completion fluids are used that are brine-in-oil emulsions, which must be stable at high temperatures and high pressures. Calcium chloride, calcium bromide, or zinc bromide are used to obtain brines with high specific gravity to improve pressure control. These conditions require emulsifiers of high performance, which are met by the silicone-based emulsifiers or mixtures of silicone emulsifiers and organic emulsifiers [48].

2. Cosmetic Applications

Cosmetic emulsions have to meet high stability standards. This means not only long-term stability at ambient temperatures but also stability in freeze–thaw cycles and at higher temperatures, often above 50°C, which is especially important for emulsions used in sun protection products. Generally the emulsions also contain ingredients like water-miscible alcohols or polar oils, which tend to have a destabilizing effect. For these reasons emulsions that are based on well-known conventional organic emulsifiers, such as glycerol mono/dioleate or sorbitan mono/dioleate, are not sufficiently stable and need additional oil-soluble waxes. These specially adopted waxes raise the viscosity of the oil phase by building liquid crystalline gel-like structures, thus immobilizing the lipid phase. The adverse effect of the waxes is directly related to the high viscosity of the oil phase: the emulsions spread only slowly and create a certain kind of stickiness and tenacity; these typical qualities of W/O creams, however, usually are not acceptable to consumers. Furthermore the liquid crystalline phases are temperature sensitive and disappear above their melting temperature, thus limiting the temperature stability of these emulsions.

Water-in-oil systems have distinct advantages for personal care products. When such products are applied, the external oil phase comes directly into contact with the skin and forms immediately a continuous film. After the water has evaporated, a uniform oil film is left on the skin, which protects the epidermis from external attack and keeps the skin flexible and hydrated.

Silicone emulsifiers received considerable interest as soon as they appeared on the market because these potent polymeric emulsifiers offer a way of achieving water-in-oil emulsions that do not need additional stabilizing waxes. Thus easily spreading light emulsions can be realized which show the excellent properties of the water-in-oil formulations combined with elegant application behavior, such as easy spreading and rapid penetration, of oil-in-water emulsions.

A silicone emulsifier consisting of dodecylmethylsiloxy units and methyl(polyoxyalkylene)siloxy units (polydodecylmethysiloxane–polyoxyalkylene copolymer, PLMS-POA) was developed mainly for the stabilization of water-in-mineral-oil emulsions [49]. Besides mineral oil, the oil phase may contain to some extent volatile silicones or polar natural oils like jojoba oil. With only 2% silicone emulsifier, stable emulsions can be obtained. If more polar oils (e.g., natural triglycerides) are integrated into the formulation, coemulsifiers like glyceryl sorbitan isostearate or sorbitan sesquioleate are needed. For further illustration, two examples are given in Table 4 [49]. The creams are produced at room temperature by adding the previously prepared water phase to the oil phase with continuously agitation, followed by homogenization.

Another silicone-based emulsifier is constituted of three different building blocks, namely dimethylsiloxy, cetylmethylsiloxy, and methyl(polyoxyalkylene)siloxy units (polyhexadecylmethylsiloxane–polyoxyalkylene copolymer, PCMS-POA). Besides alkyl and polyoxyalkylene residues, this emulsifier contains polydimethylsiloxane segments. The emulsifier was originally developed for the stabilization both of water-in-silicone fluid and water-in-mineral-oil emulsions [21]. In follow-up investigations, the outstanding emulsifying abilities of this product were gradually elucidated, the first manifestation being the possibility of stabilizing emulsions based on pure natural triglycerides. Later the quality of the emulsifier was further illustrated by its potential to form special emulsion types like the following:

emulsions that maintain stability at high temperature
nonaqueous emulsions
multiple emulsions
emulsions containing liposomes

TABLE 4 Water-in-Oil Emulsions with PLMS-POA

Phase	Ingredient	Content (wt%)
	Formulation 1	
Oil	PLMS-POA	2.00
	Mineral oil	20.00
	Trioxypropylene myristyl ether	0.50
Water	Sodium chloride	2.00
	Water	up to 100
	Formulation 2	
Oil	Mineral oil	15.0
	Mink oil	2.0
	Avocado oil	2.0
	Glyceryl sorbitan isostearate	1.5
	PLMS-POA	2.0
	Myristyl myristate	2.0
Water	Sodium chloride	2.0
	Glycerol	2.0
	Water	up to 100

Source: Ref. 49.

Water-in-oil emulsions, where the oil phase consists solely of natural or synthetic triglycerides, are not easy to stabilize. The silicone emulsifier PCMS-POA permits the production of soft, easy-to-apply, highly stable emulsions [50]. The effectiveness of the silicone emulsifier is clearly demonstrated in a comparison with common organic emulsifiers. Some of the results are shown in Table 5 [50]: these emulsions contain 25% oil phase (isopropyl stearate 9.2%, decyl oleate 9.2%, jojoba oil 4.6%, and emulsifier 2.0%); the water phase included small amounts of electrolyte (NaCl, 0.4%), propylene glycol (1.8%), and preservative (0.2%). Cold and heat stability of the emulsions was tested. The stability at low temperatures was determined in freeze–thaw cycles, and the heat stability was simply tested by storage. The results clearly demonstrate the superiority of the silicone emulsifier.

In a more detailed comparison between the two polymeric emulsifiers triglycerol trioleate (TGTO) and PCMS-POA, again the excellent efficacy of the silicone emulsifier was elaborated [51]. The work included measurements of interfacial tensions with a spinning drop or sessile drop tensiometer in water–oil systems containing 7.5% emulsifier in the oil phase. Tables 6 and 7 give interfacial tensions and stability data of two formulations, respectively. Whereas lower interfacial tensions are obtained with the polyglycerol-based emulsifier, the more stable emulsions are clearly obtained with the silicone emulsifier. There is obviously no cor-

TABLE 5 Comparison of Organic W/O Emulsifiers with the Silicone Emulsifier PCMS-POA

	Stability[a]				
		Heat			
Emulsifiers[b]	Cold (°C)[c]	Time (months)	Temp (°C)	Time (weeks)	Temp (°C)
LAN + GMO (1:1)	1 × –5	1	20	<1	45
GMO	5 × –5	3	20	<1	45
TGTO	5 × –5	2	20	<1	45
SMO	5 × –5	3	20	<1	45
Cas-7EO	1 × –5	1	20	<1	45
MEODO + EODO (1:1)	1 × –5	1	20	<1	45
PCMS·POA + TGTO	5 ×–15	4	45	2	60
PCMS·POA	5 × –25	6	45	4	60

[a]Gives most demanding conditions under which emulsion stayed stable.
[b]LAN, lanolin alcohol; GMO, glycerol mono/dioleate; TGTO, triglycerol trioleate; SMD, sorbitol mono/dioleate; Cas-7EO, polyoxyethylene-(7)-castorwax; MEODO + EODO, mixture of methyloxyethylene-(22)-oxydodecene-(7) copolymer and oxyethylene-(22)-oxydodecene-(7) copolymer.
[c]Freeze–thaw cycles.
Source: Ref. 50.

TABLE 6 Interfacial Tension and Stability

		Heat stability days at				
Oil[a]	Interfacial tension at 25°C (mN/m)[b]	70°C	60°C	50°C	40°C	Freeze–thaw cycles: –15°C to +20°C
		Emulsifier TGTO				
PAR	<2.0	4	10	10	20	0
IPIS	ND	0	1	3	20	0
DOL	<0.3	0	0	0	1	0
D5	—[c]	0	0	0	0	0
		Emulsifier PCMS-POA				
PAR	3.2	>30	>30	>30	>30	>10
IPIS	ND	20	25	30	>30	>10
DOL	1.6	25	25	>30	>30	>10
D5	3.3	10	10	10	>30	1

[a]PAR, paraffin oil; IPIS, isopropyl isosterate; DOL, decyl oleate; D5, decamethylcyclopentasiloxane.
[b]ND, not determined.
[c]Not miscible.
Source: Ref. 51.

TABLE 7 Composition of Formulations in Table 6

	Content (wt%)	
Ingredient	PCMS-POA formulation	TGTO formulation
PCMS-POA	2.0	—
TGTO		3.0
Oil	19–21	23–38
NaCl	0.4	—
$MgSO_4$	—	0.4
Glycerol	—	3.0
Water	to 100	to 100
Content of oil phase[a]	With PCMS-POA (wt%)	With TGTO (wt%)
PAR	21	23
IPIS	19	25
DOL	20	34
D5	19	38

[a]See note a, Table 6.
Source: Ref. 51.

relation between the ability of the emulsifier to reduce the interfacial tension and its ability to stabilize the emulsion. The higher molecular weight of the silicone copolymer compared to the polyglycerol ester (≈15,000 vs. ≈1000) may be one important reason for the superior performance of the former. The emulsions need a small amount of electrolyte, which causes a considerable increase in cold and heat stability. In emulsions based on organic emulsifiers, 0.25–0.50% magnesium sulfate gives the best effects, mainly with respect to heat stability. In emulsions based on silicone emulsifiers, optimal stability is achieved with 0.8–2% sodium chloride. Table 8 correlates sodium chloride content and stability [51]. Further investigations looked into an even lower optimum sodium chloride content: 0.5%. The electrolyte presumably affects the hydratization of the polyoxyalkylene residues in a way that is advantageous for emulsion stability.

The oil phase volume of the formulations shown in Table 7 differs depending on the kind of emulsifier and the oil. It has been adapted to achieve comparable viscosity and optimal stability in each of the formulations. The organic emulsifier produces considerable differences in oil phase volume depending on the oil, whereas with the silicone emulsifier the volumes are comparable and in the range of 19–21%.

The silicone emulsifier PCMS-POA is able to stabilize emulsions with 50–80% internal phase volume: that is, with medium to very high water content. Three

TABLE 8 Stability of PCMS-POA Emulsions Correlated with Sodium Chloride Content

NaCl (%)	Stability at 60°C (days)	Freeze–thaw stability: –20°C to +20°C (cycles)
0.0	0	0
0.1	3	1
0.2	7	6
0.4	30	>10
0.8	>30	>10
1.2	>30	>10
2.0	>30	>10
3.0	>30	7
4.0	10	0
5.0	5	0

Source: Ref. 51.

parameters determine the viscosity of W/O emulsions. These are the content and the particle size of dispersed aqueous phase, and the viscosity of the continuous oil phase. The viscosity of the emulsion increases with the content of internal aqueous phase and inversely with its particle size, because interactive forces between water droplets rise with the increasing packing density. The particle size mainly affects the stability and should not be greater than 2 μm in creams and 3 μm in lotions. The volume of the internal phase and the viscosity of the oil phase are the parameters of choice to adjust the viscosity of the emulsion. Thus, for example, in the preparation of W/O creams, oils low in viscosity can be used without viscosity-enhancing additives as long as the optimal phase volume ratio is chosen.

There are no clear correlations between chemical structure, molecular weight, or polarity of the oil and emulsion viscosity. Emulsions of linear polydimethylsiloxanes or triglycerides of unsaturated C_{18} fatty acids, for instance, show distinctly higher viscosities than corresponding formulations containing fatty acid esters of univalent alcohols, branched hydrocarbons of similar viscosity, or cyclic siloxanes. Obviously there are specific interactions of the emulsifier with the respective oil.

To achieve optimal stability, a certain volume range of both phases has to be considered, which depends on the kind of oil phase, consisting of oil, stabilizing waxes, and emulsifier. Figure 8 plots the effects of these variables for a simple system consisting of PCMS-POA, decyl oleate, water, and sodium chloride [21]. Lower volumes of oil phase give creamy emulsions, in which the disperse phase tends to coalesce and separate. Higher volumes result in emulsions low in viscosity, which suffer from a gradual aggregation of water droplets leading to oil separation.

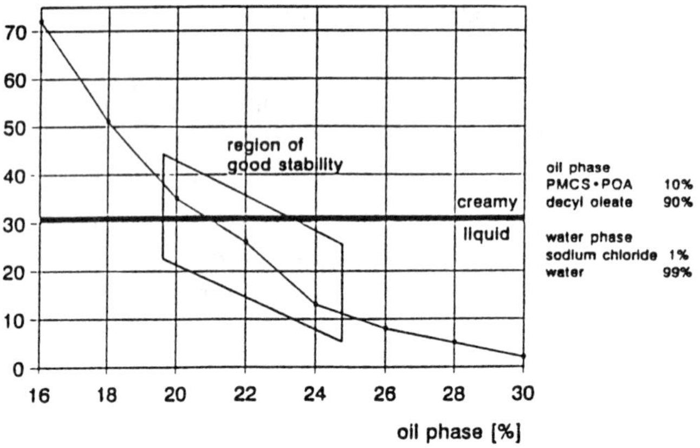

FIG. 8 Influence of phase volume on emulsion viscosity. (From Ref. 21.)

3. Cold Stability

The ability of the silicone emulsifier PCMS-POA to create stable emulsions consisting only of water, sodium chloride, oil, and emulsifier without the incorporation of any stabilizing waxes allows us to investigate the influence of the solidifying point of the oils and the oil phases on the cold stability of emulsions. Corresponding experiments have not been possible before because high melting waxes had to be used in combination with organic emulsifiers to achieve sufficient stability at ambient conditions. In a thorough investigation using 17 oils with a broad range of melting points, a relationship between the solidifying point of the oil phase and the cold stability of the emulsions was clearly demonstrated [52]. The results are shown in Table 9. A linear correlation is, however, not possible. This study measured the pour point, the temperature at which a liquid loses its flowability, instead of the solidifying point. The cold temperature stability was determined by means of freeze–thaw cycles, in which the freeze temperature was decreased at every cycle by 5°C. Thus in the first cycle the emulsion is cooled to –5°C, in the second to –10°C. All emulsions solidify at temperatures of –10°C or slightly below, because the aqueous phase crystallizes. This, however, does obviously not affect the stability of the emulsion. For comparison, glycol-in-oil emulsions having the following composition were prepared:

2% emulsifier
22% paraffin oil or isopropyl isostearate
76% polyethylene glycol 200

TABLE 9 Cold Stability of Emulsions and Pour Point of Their Oil Phases

Oil	Oil phase pour point (°C)[a]	Cold stability (°C)[b]
Isopropyl isostearate	–29	–30
Ethylhexyl isostearate	–30	–30
Isooctadecyl isononanoate	–30	–30
Paraffin oil, 30 mPa·s	–23	–30
Decamethylcyclopentasiloxane	<–30	–30
Propylene glycol dioleate	–19	–25
Caprylic/capric acid triglyceride	–19	–25
Isostearyl isostearate	–17	–30
Oleyl oleate	–9	–20
Decyl oleate	–8	–25
Isopropyl myristate	–6	–10
Cetyl/stearyl isooctanoate	–4	–10
Ethylhexyl stearate	+3	–15
Isopropyl palmitate	+10	–5
Jojoba oil	+7	–25
Octamethylcyclotetrasiloxane	+5	–5
Isopropyl stearate	+17	–5

[a]Oil phase: 94% oil + 6% emulsifier.
[b]Emulsions: 19–29% oil phase + 81–71% aqueous phase; aqueous phase: 99.5% water + 0.5% NaCl.
Source: Ref. 52.

These emulsions, which can also be stabilized by PCMS-POA, increase in viscosity only when the temperature is lowered, but they do not solidify down to –20°C.

4. Emulsifying Methods

Since additional waxes often are not needed to stabilize silicone emulsifier based emulsions, there is no need to heat the oil or water phase to dissolve the high melting waxes before the emulsification process. Thus, instead of the conventional methods, where either both phases or at least the oil phase must be heated in advance, silicone emulsifiers allow the preparation of emulsions in a cold/cold process, which saves energy and time and does not stress temperature-sensitive ingredients [21,49]. First a preemulsion is prepared by slowly pouring the aqueous phase into the oil phase containing the emulsifier, with continuous stirring. If the water phase is added too quickly, or if the oil is added to the water, instable emulsions are formed. The desired degree of dispersion is adjusted afterwards by intensive stirring or by means of treatment with a homogenizer or a high shear mixer. These methods will produce long-term stable emulsions in which droplet sizes of 2–3 μm have been determined by microscopy. Care has to be taken, however, not to overhomogenize the emulsions, because they might then undergo an increase in droplet size, leading to a viscosity decrease during storage.

5. Formulations with High Content of Emulsion Stressing Compounds

Active ingredients that are integrated in emulsions for pharmaceutical or cosmetic reasons often have detrimental effects on emulsion stability. Some ingredients are to some extent surface active and hinder the adsorption of emulsifier at the interface. Others may have a negative effect on the hydratization of the hydrophilic groups. Table 10 surveys substances that may cause stress if emulsified with PCMS-POA [51]. Although a broad range of water-soluble ingredients can be emulsified, it is worth mentioning that emulsions containing aluminum chlorohydrate solutions and alkaline solutions with cyclic siloxanes as external phase can-

TABLE 10 PCMS-POA Based Emulsions: Maximum Concentration of Stressing Compounds

	Maximum concentration (wt%)	
Stressing compound	Formulation 1[a]	Formulation 2[b]
Sodium cumolsulfonate	4	—
Sodium lactate (pH 7)	3	—
Sodium salicylate (pH 7)	3	5
Aluminum chlorohydrate	20	—
Urea	30	30
N-Methyl acetamide	20	20
Nicotinic acid methyl ester[c]	20	10
Lactic acid (pH 2)	1	—
Citric acid (pH 2)	0.7	—
Alkali hydroxide	—	—
Propylene glycol	>40	—
Glycerol	>40	—
PEG 200[d]	>40	—

[a]PCMS-POA, 2 wt%, paraffin oil, 18.0 wt%; NaCl, 0.5 wt%; tested substance, x wt%, water (79.5 wt% – x wt%).
[b]PCMS-POA, 2 wt%, decamethylcyclopentasiloxane, 18.0 wt%; NaCl, 0.5 wt%; tested substance, x wt%; water (79.5 wt% – x wt%).
[c]Partly dissolved in oil phase.
[d]Polyoxyethylene glycol ether (molecular weight 200).
Source: Ref. 51.

not be prepared by these recipes. The emulsions can tolerate a high content of polyethylene glycol, which can be even further increased up to 100% polyethylene glycol (0% water); an example was mentioned above. Hydrotropes, organic α-hydroxy acids, or nicotinic acid methyl ester, for example, can be integrated as well in emulsions. The stability of emulsions containing stressing compounds can be further improved by the use of hydrocolloids like hydroxyethylcellulose or carboxymethylcellulose in the aqueous phase [53].

D. Nonaqueous Oil-in-Oil Emulsions

Stable emulsions of polyethylene glycol in paraffin oil or in isopropyl isostearate have been mentioned in earlier chapters. These are only two examples of unusual nonaqueous emulsions that can be stabilized with silicone emulsifiers. Further examples with glycerol fatty acid esters in the oil phase are given in Table 11 [51]. One of the phases in all these emulsions is quite hydrophilic, and this property is enhanced by adding small amounts of water and sodium chloride. To achieve sufficient stability, the emulsions further require small amounts of waxes, such as microcrystalline paraffin wax or glycerol trihydroxystearate, in addition to the silicone emulsifier. Similarly, polyol-in-silicone emulsions can also be obtained [26].

Examples of completely different but also unusual systems are emulsions/dispersions wherein one phase (presumably the internal) consists of silicone oil and the other of lipophilic hydrocarbon-based surfactants (e.g., distearyldimethyl ammonium methylsulfate, ethoxylated fatty alcohols), which in some cases may also contain lipophilic esters, mineral oil, or polyethylene glycol [54]. The prepa-

TABLE 11 Composition (wt%) of Some Nonaqueous Emulsions[a]

	Formulation				
Ingredient	1	2	3	4	5
Glycerol trioleate	26.0	26.0	—	—	—
Propylene glycol	67.0	—	67.0	—	—
Water	2.7	—	2.7	—	2.7
Sodium chloride	0.3	—	0.3	—	0.3
Glycerol	—	70.0	—	70	—
Glycerol tricaprylate/caprinate	—	—	26.0	26.0	—
Paraffin oil	—	—	—	—	26.0
Polyethylene glycol (200)	—	—	—	—	67.0

[a]All formulations contain PCMS-POA, 2.0 wt%; microcrystalline paraffin wax, 1.2 wt%; and glycerol trihydroxystearate, 0.8 wt%.
Source: Ref. 51.

rations are useful as fabric conditioning compositions. The emulsions do not need long-term stability because they solidify after preparation and will be molten again when used. These mixtures are stabilized by silicone emulsifiers. Since the emulsification problem is not very fastidious, the nature of the silicone-based emulsifier is not critical. Similar to water-in-silicone-oil emulsions, solely polyoxyalkylene-substituted polydimethylsiloxanes are effective as emulsifiers. Depending on the kind of surfactant–hydrocarbon phase, however, also silicones carrying long chain alkyl residues or silicones carrying both alkyl and polyether substituents may be suitable to stabilize oil-in-oil emulsions of these types.

E. Oil-in-Water Emulsions

While there is extended knowledge on the stabilization of water-in-oil emulsions by silicone emulsifiers, examples of the reversed type are rare. Ionically modified silicones have frequently been described to be suitable as oil-in-water emulsifiers. Examples were listed in Table 1. There are, however, no cases known to the authors in which substituted silicones of these types are actually used for this purpose.

Examples of oil-in-water emulsions refer exclusively to nonionic dimethylsiloxane polyoxyalkylene copolymers. Chemically they are similar to the water-in-oil emulsifiers, but they differ from those by having a considerably higher hydrophilicity, which can be achieved by a sufficiently high degree of grafting of the polydimethylsiloxane backbone with polyoxyethylene substituents. In a systematic study Si—H-functional polydimethylsiloxanes were grafted with undecylenic acid polyoxyethylene esters that differed in molecular weight [55]. The effective emulsifiers of this study were based on Si—H-functional silicones that contain at least 50% dimethylsiloxy groups to contribute hydrophobicity to the molecule. The hydrophobicity of the decene spacer groups alone is not sufficient. The grafted copolymers of highest molecular weight, with a high degree of substitution and longest polyoxyethylene side groups, produced the most stable emulsions. Thus dodecane-in-water emulsions were obtained, containing 0.1% emulsifier. They were stable for more than 30 days, but there was a slight increase in droplet size, on average, 10–15 μm diameter. A polydimethylsiloxane–polyoxyethylene copolymer, which turned out to be one of the best of the investigated series, is characterized by a backbone having a molecular weight of 2050 and a polyoxyethylene residue content of 84 wt%.

The long chain polyoxyalkylene residues provide essential steric stabilization to the emulsions, which in contrast to W/O emulsions contain only 30% or even less of the dispersed phase.

The authors also determined the effect of the copolymers on surface and interfacial tension. Like others [51], they found no correlation between effectiveness in reducing interfacial tension and emulsion stability.

Similar to other nonionic ethoxylated surfactants, silicone polyethers show inverse solubility in water [56,57]. Aqueous solutions of these surfactants can be characterized by their cloud point, which is the temperature above which the solution separates into two phases. This behavior of polydimethylsiloxane–polyoxyalkylene copolymers can be used to adjust the coagulation temperature of latex emulsions [58]. Thus, for example, approximately 2% of a slightly hydrophobic silicone polyether is added to a natural latex emulsion of 60% solid matter to achieve a coagulation temperature of 45–48°C. An example of a suitable silicone polyether is given in Fig. 9.

Concentrated defoamer emulsions, which can easily be diluted with water before use, are formulated by emulsifying hydrophobic organomodified silicones, such as alkoxy- or polyether-substituted silicones, using hydrophilic silicone polyethers in water [59]. Moreover, the emulsions also contain fine particulate silica. An emulsion may typically consist of:

38%	end chain ethoxy-substituted polydimethylsiloxane	(oil phase)
2%	silica	(oil phase)
30%	silicone polyether	(emulsifier)
30%	water	(water phase)

The required amounts of emulsifier are noticeably high. Since, however, the emulsifier partly also acts as defoamer, this is no drawback.

F. Multiple Emulsions

The fascinating properties of multiple emulsions, which may be of the oil-in-water-in-oil (O/W/O) or water-in-oil-in-water (W/O/W) type, have attracted recurring interest, in particular when the protection of sensitive ingredients or a controlled release of active substances is required [60–62]. Like other polymeric surfactants, silicone-based emulsifiers are especially suited to stabilize this kind of emulsion because their polymeric nature permits them to be adsorbed strongly at the interface. Two kinds of interface have to be stabilized, which are in the case of O/W/O emulsions that between the inner oil phase and the water phase and that between the water phase and the continuous oil phase. A migration of the emulsifiers from one interface to the other would lead to destabilization and further to inhomogeneity or transformation into a two-phase system. A scheme of an O/W/O emulsion is shown in Fig. 10.

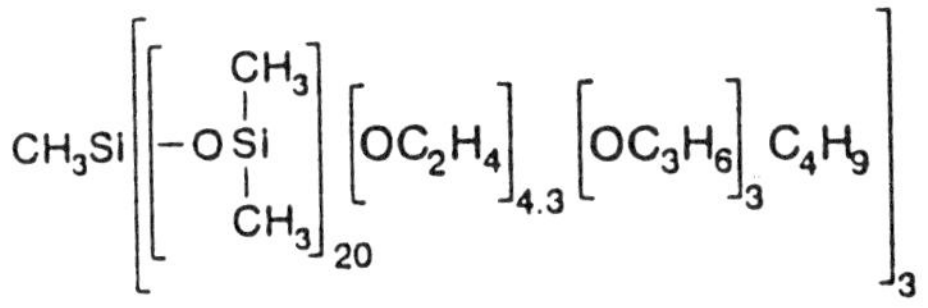

FIG. 9 Polyoxyalkylene-substituted siloxane as coagulant.

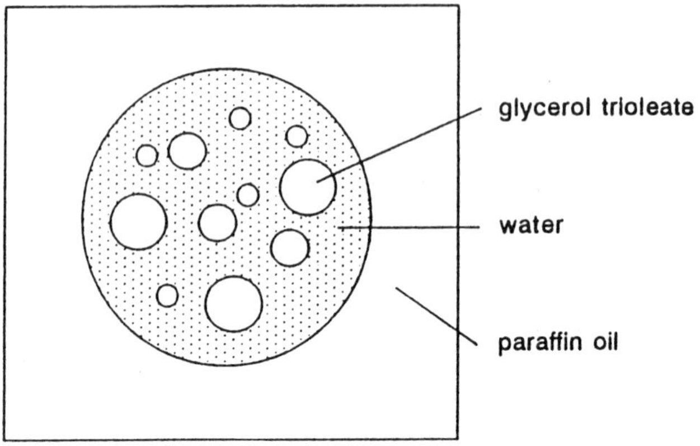

FIG. 10 Schematic representation of an O/W/O emulsion. (From Ref. 51.)

Stability for this type of multiple emulsion can be achieved by two polymeric emulsifiers: a hydrophilic polyacrylate copolymer, which carries lipophilic alkyl and hydrophilic polyoxyethylene groups for stabilization of the O/W interface and PCMS-POA [51]. The HLB values of the emulsifiers, which can be determined experimentally, should be above 10 for the hydrophilic emulsifier and below 6 for the hydrophobic emulsifier [63]. The composition of the multiple emulsion is given in Table 12. The emulsion is made in a two-step procedure: first high shear energy is used to prepare an O/W emulsion (droplet size below 0.8 μm), and in the second step the emulsion is stirred into the oil phase containing the silicone emulsifier. This kind of multiple emulsion has proven stable for at least 6 months at 20 and 45°C.

TABLE 12 Composition of an O/W/O Emulsion

Phase	Ingredient	Concentration (wt%)
External oil phase, 28 wt%	Paraffin oil	24
	PCMS-POA	2
	Paraffin wax	1.3
	Glycerol trihydroxystearate	0.7
Water phase, 54.1 wt%	Water	53.6
	Sodium chloride	0.5
Internal oil phase, 17.9 wt%	Glycerol trioleate	15
	Polyacrylate copolymer	2.9

Source: Ref. 51.

A different approach to O/W/O emulsions is based on the preparation of an O/W precursor emulsion by the paste method [64]. Into aqueous solutions containing 25% emulsifier, a fourfold amount of vegetable oil is emulsified with high shear stirring to give a transparent paste, which is diluted with water to 67% oil phase. The result is a fine O/W emulsion. With this well-known method, stable emulsions with hydrophilic nonionic surfactants such as poly-(25)-oxyethylene monostearyl ether or ethoxylated-(60)-castor oil can be obtained. In the second step the O/W preemulsion is further diluted to the finally desired amount of water, whereupon this emulsion is immediately poured with gentle stirring into the external oil phase, consisting of emulsifier PCMS-POA, paraffin wax, and paraffin oil. The final products comprise 2.0% of lipophilic W/O emulsifier and 0.65% of hydrophilic O/W emulsifier.

The preparation of W/O/W emulsions follows in principle a very similar process, which preferentially is also a two-step procedure. However, one-step procedures are mentioned in the literature, as well [65]. The high emulsifying power of the silicone-derived emulsifiers, their strong adsorption to the W/O interface, and the pronounced elasticity of the interfacial film formed by the emulsifiers again are the basis for the stability. These qualities result in a high tolerance toward other surfactants that must be used to stabilize the outer O/W interface. Virtually the only precondition required of these latter surfactants is to be sufficiently hydrophilic, with a preferred HLB value of at least 13 [66]. Nevertheless their emulsifying activity must be supported by water-soluble, high molecular weight, alkyl-modified polyacrylates [64]. The hydrophilic, ionic surfactants can even be omitted if the external phase is gelified by suitable polymers such as alkyl-modified polyacrylates [67].

The primary, thoroughly homogenized W/O emulsion should be a highly viscous liquid. To prepare the final multiple emulsion, it is combined with the premixed external water phase; and in the following step with gentle stirring, the W/O/W system is formed. The surfactants that stabilize the O/W interface must provide self-emulsification of the W/O emulsion. This can be effected, for instance, with ionic surfactants such as laurylamidopropyl betaine or sodium lauryl ether sulfate and also with lauryl glucoside in concentrations of 0.1–1%. A composition of a W/O/W emulsion is given in Table 13 [64].

After the preparation of the emulsion, an increase in viscosity can be observed which comes to an equilibrium after 24 h. Thus, for example, the viscosity of an emulsion rises from approximately 1000 mPa·s initially to approximately 16,000 mPa·s after 24 h. The authors explain this phenomenon plausibly by citing an osmotic process: water diffuses from the external phase through the oil into the internal phase, where the electrolyte content is higher; migration of electrolyte to the external phase is not observed. After equilibrium viscosity has been reached, the emulsion remains stable in appropriate tests at 40°C for at least 3 months. The final composition of the emulsion will be different from that given in Table 13

TABLE 13 Composition of a W/O/W Emulsion

Phase	Ingredient	Concentration (wt %)
Internal water phase, 37.5 wt%	Water	37.2
	Sodium chloride	0.3
Oil phase, 12.5 wt%	Paraffin oil	11.5
	Silicone emulsifier (W/O) 1.0	
External water phase, 50 wt%	Water	49.6
	Acrylate thickener 0.1	
	O/W-emulsifier	0.3

Source: Ref. 64.

because the volume of the internal water phase will be higher than the initial values, and that of the external water phase less.

If the emulsions are further diluted with water for photographic purposes, the droplets continue to grow and the oil phase thins to films less than 500 nm thickness (Fig. 11), indicating the high stability and elasticity of the interfacial emulsifier films. The size of the W/O droplets was determined from photomicrographs to be between 5 and 25 μm and that of the enclosed water droplets between 1 and 4 μm.

The very hydrophilic silicone emulsifiers described earlier (Sec. II. E), can be used to stabilize the outer O/W interface [68]. In combination with the hydrophobic water-in-oil-in silicone emulsifiers, emulsions could be obtained that were characterized by an average W/O droplet size of 4.5 μm. In contrast to the W/O/W emulsions described above, these emulsions undergo some leakage of electrolyte from the internal water phase, which is detected by conductivity experiments.

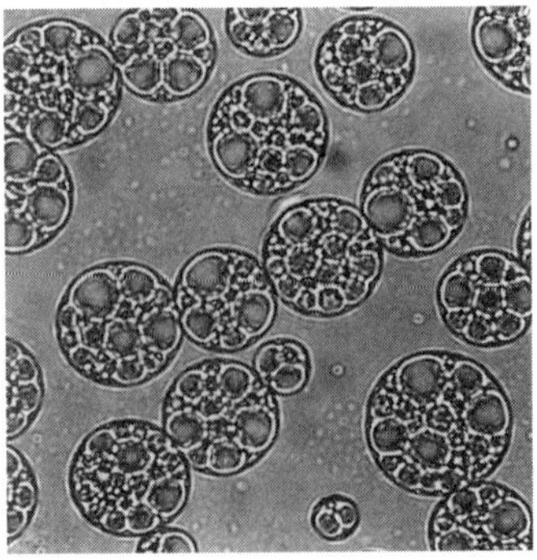

FIG. 11 Photomicrograph of a diluted W/O/W emulsion.

Like the O/W/O type, the W/O/W triple emulsions offer a versatile tool for the formulation of cosmetic or pharmaceutical preparations to deliver sensitive active ingredients in a protected form. It is, even possible to integrate two water-soluble substances that are incompatible with each other (because they may react) in one formulation: for example, $Fe^{II}Cl_2$ in the internal phase and $K_3Fe^{III}(CN)_6$ in the external phase. This emulsion is very stable and does not show any change during storage. Only when it is broken, which can be achieved by the addition of *n*-butanol, the iron salts react to give the well-known Prussian blue [64]. This example is illustrative and somewhat spectacular.

Other interesting examples of long-term stable emulsions show the opportunities afforded by the W/O/W system, especially in cosmetic applications. Sensitive ingredients like dihydroxyacetone or panthenol can for instance be effectively protected against oxidation by dissolving them in the inner water phase.

With the W/O/W technology based on silicone emulsifiers, it is possible to create emulsions that consist of 20–25% oil phase like normal O/W emulsions and show the desired high viscous flow behavior of lotions or creams without using the typical waxes. The waxes build up liquid crystalline structures in the continuous aqueous phase, thus providing high viscosity and stability. The waxes, however, also affect the application behavior of the cream or lotion. They often generate some kind of unpleasant stickiness when the emulsion is rubbed into the skin. By avoiding the use of waxes, the W/O/W technology can be used to obtain extremely light emulsions with elegant application properties. The viscosity of these emulsions is generated just by the volume of the internal phase of more than 75%. The oil content in this kind of emulsion can be reduced, even to much lower values than 20%. Thus with only 6% oil phase, stable emulsions are still achieved. Here again the outstanding emulsifying power of silicone emulsifiers is exemplified.

G. Further Developments

The strong, almost irreversible adsorption to the interface of the polymeric silicone-based W/O emulsifiers offers the opportunity to create emulsion formulations that contain all kinds of surfactant-stabilized systems but do nothing to destroy their integrity. Further examples beyond the multiple emulsions are liposome-containing formulations [69] and emulsions containing liquid crystals. Thus transmission electron microscopy was used to prove the existence of liposomes consisting of phospholipids in a W/O emulsion stabilized by a silicone emulsifier [70]. In these investigations the existence of liposomes in W/O emulsions was shown the first time, and in addition the long-term stability of liposomes in this formulation was proven (because the emulsion had been aged for several months before examination). With the appropriate choice of immiscible liquids, emulsions can be formed which contain up to five different liquid or liquid crystalline phases [60].

With the existing polymeric silicone-derived water-in-oil emulsifiers, a group of the most effective highly specialized surfactants is at the disposal of the formulator to obtain emulsions that meet high stability standards. Certainly the steadily increasing interest in this versatile tool will result in new ideas for innovative formulations to serve the future demands of the consumer. Also, on the O/W emulsifier side, there are promising developments that may lead to new products having an efficacy level comparable to that of the known W/O emulsifiers.

REFERENCES

1. M. J. Owen, in *Silicon-Based Polymer Science* (J. M. Zeigler and F. W. G. Fearon, eds.), American Chemical Society, Washington, DC, 1990, p. 707.
2. This volume, Chapter 7.
3. S. A. Snow, L. J. Petroff, V. Cobb, M. Stanga, R. A. Ekeland, G. Legrow, and R. Thimineur, Chemical Specialties Manufacturers 53rd midyear meeting, Chicago, 1967; published in *Manuf. Chem Aerosols News*, August 1967, p. 55.
4. B. Grüning and G. Koerner, *Tenside Surf. Detergents 26*, 3 (1989).
5. L. A. Haluska, U.S. Patent 2,846,458 to Dow Corning Corp. (1956).
6. R. L. McKellar, U.S. Patent 3,427,271 to Dow Corning Corp. (1996).
7. A. R. L. Colas and F. A. D. Renauld, U.S. Patent 4,777,277 to Dow Corning Corp. (1988).
8. B. D. Maxon, U.S. Patent 4,717,498 to Mclntyre Chemical Co. (1987).
9. B. Kanner and R. A. Pike, U.S. Patent 3,507,897 to Union Carbide Corp. (1966).
10. E. L. Morehouse, U.S. Patent 3,997,580 to Union Carbide Corp. (1972).
11. E. L. Morehouse, German Patent 1,921,872 to Union Carbide Corp. (1968).
12. A. J. O'Lenick Jr., U.S. Patent 5,070,171 to Siltech Inc. (1990).
13. B. Grüning, U. Holtschmidt, and G. Koerner, German Patent 3,323,881 to Th. Goldschmidt AG (1983).
14. K. Hoffmann, H. J. Kollmeier, and R. D. Langenhagen, German Patent 3,417,912 to Th. Goldschmidt AG (1985).
15. W. N. Fenton, M. J. Owen, and S. A. Snow, European Patent Application 276,114 to Dow Corning Corp. (1988).
16. W. G. Reid, U.S. Patent 3,389,160 to Union Carbide Corp. (1968).
17. D. Schaefer and M. Krakenberg, German Patent 3,719,086 to Th. Goldschmidt AG (1987).
18. W. Noll, *Chemie und Technologie der Silicone*, Verlag Chemie, Weinheim/Bergstralsse, 1968, p. 404.
19. M. Gradzielski, H. Hoffmann, P. Robisch, W. Ulbricht, and B. Grüning, *Tenside Surf. Detergents 27*, 336 (1990).
20. W. C. Griffin, *J. Soc. Cosmet. Chem. 5*, 249 (1954).
21. P. Hameyer, *Seifen-Oele-Fette-Waechse*, *116*, 392 (1990).
22. M. J. Owen, in *Silicon-Based Polymer Science* (J. M. Zeigler and F. W. G. Fearon eds.), American Chemical Society, Washington, DC, 1990, pp. 705.
23. W. A. Lee and R. A. Rutherford, in *Polymer Handbook* (J. Brandrup and E. H. Immergut, eds.), Wiley, New York, 1975, pp. III–139.

24. W. Noll, in *Chemie und Technologie der Silicone*, Verlag Chemie, Weinheim/Bergstrassee, 1968, p. 261.
25. A. Zombeck, Advanced Technology Conference, Barcelona, 1994.
26. G. H. Dahms and A. Zombeck, *Cosmet. Toiletries 110*, 91 (1995).
27. A. Harashima, R. Mikami, H. Kondo, A. Sasaki, and T. Hamachi, 17th IFSCC Congress, Yokohama, 1992.
28. A. A. Zotto, R. J. Thimineur, and W. J. Raleigh, U.S. Patent 4,988,504 to General Electric Co. (1987).
29. R. P. Gee and J. W. Keil, U.S. Patent 4,122,029 to Dow Corning Corp. (1997).
30. W. J. Keil, U.S. Patent 4,265,878 to Dow Corning Corp. (1979).
31. W. J. Keil, U.S. Patent 4,268,499 to Dow Corning Corp. (1979).
32. J. Roidl, *Parfuem, Kosmet. 67*, 232 (1986).
33. J. Smith, L. Madore, and S. Fuson, *Drug Cosme. Ind. 157*, 46 (1995).
34. K. A. Kasprzak, U.S. Patent 4,218,250 to Dow Corning Corp. (1978).
35. M. S. Starck, U.S. Patent 4,311,695 to Dow Corning Corp. (1979).
36. S. F. Rentsch, European Patent Application 529,847 to Dow Corning Corp. (1992).
37. L. M. J. Aguadisch and F. S. Rankin, European Patent Application 281,236 to Dow Corning Ltd. (1988).
38. J. W. Keil, U. S. Patent 4,532,132 to Dow Corning Corp. (1983).
39. J. W. Keil, European Patent Application 125,779 to Dow Corning Corp. (1984).
40. W. Wolfes, R. Hüttinger, H. J. Kollmeier, R. D. Langenhagen, and A. Walter, German Patent 3,436,177 to Th. Goldschmidt AG (1986).
41. R. L. McKellar, U.S. Patent 3,427,271 to DOW Corning Corp. (1966).
42. P. L. Brown, U.S. Patent 4,520,160 to Dow Corning Corp. (1983).
43. S. Ichinohe, European Patent Application 459,705 to Shin Etsu Chemical Co. (1991).
44. W. J. Raleigh and R. J. Thimineur, U. S. Patent 5,401,870 to General Electric Co. (1993).
45. C. B. Bahr, P. Y. Lo, A. W. Lomas, and D. J. Romenesko, U.S. Patent 5,136,068 to Dow Corning Corp. (1989).
46. C. B. Bahr, A. W. Lomas, P. Y. Lo, and D. J. Romenesko, European Patent Application 298,402 to Dow Corning Corp. (1987).
47. K. Plochocka and J. C. Chuang, U.S. Patent 5,216,070 to ISP Investment Inc. (1991).
48. D. J. Romenesko and H. M. Schiefer, U.S. Patent 4,381,241 to Dow Corning Corp. (1981).
49. D. G. Krysik and J. v. Reeth. *Drug* Cosmet. *Ind. 146*, 28 (1990).
50. P. Hameyer, *Seifen-Oele-Fette-Waechse 117*, 214 (1991)
51. B. Grüning, P. Hameyer, and C. Weitemeyer. *Tenside Surf. Detergents 29*, 78 (1992).
52. P. Hameyer, *Seifen-Oele-Fette-Waechse 118*, 600 (1992).
53. J. Bara and M. Mellul, World Patent Application 93/14742 to L'Oréal (1992).
54. T. J. Taylor and S. Q. Lin, European Patent Application 544,493 to Unilever (1991).
55. Y. Sela, S. Magdassi, and N. Garti. *Colloid Polym. Sci. 272*, 684 (1994).
56. S. C. Vick, *Soap, Cosmet. Chem. Spec. 60*, 36 (1984).
57. H. J. Kollmeier, *World Surfactants Congress Proceedings* Munich, 1984, vol. IV, p. 195.
58. G. Sinn, H. Hornig, and W. Simmler, German Patent 1,268,828 to Farbenfabriken Bayer AG (1961).

59. M. Keup and R. Sucker, German Patent 4,343,185 to Th. Goldschmidt AG (1993).
60. A. Bevacqua, K. Lahanas, J. Cohen, and G. Cioca, *Cosmet. & Toiletries 106*, 53 (1991).
61. M. De Luca, P. Rocha-Filho, J. Grossiord, A. Rabaron, C. Vantion, and M. Seiller, *Int. Cosmet. Sci. 13*, 1 (1993).
62. S. Raynal, J. L. Grossiord, M. Seiller, and D. Clausse, *J. Controlled Release 26*, 129 (1993).
63. B. Grüning, P. Hameyer, and C. Weitemeyer, German Patent Application 4,206,732 to Th. Goldschmidt AG (1992).
64. P. Hameyer and K. R. Jenni, 18th International IFSCC Congress, Venice, 1994.
65. J. Nielsen, M. Piatkiewics, A. Müller, P. Kröpke, and S. H. Gohla, 18th International IFSCC Congress, Venice, 1994.
66. B. Grüning, P. Hameyer, and C. Weitemeyer, German Patent Application 4,322,174 to Th. Goldschmidt AG (1992).
67. J. F. Nadaud and L. Sebillotte, World Patent Application 94/01073 to L'Oréal (1992).
68. Y. Sela, S. Magdassi, and N. Garti. *Colloids Surfaces A: Physicochem. Eng. Aspects 83*, 143 (1994).
69. D. T. Floyd and K. R. Jenni, CRC Press, Boca Raton, FL, *Handbook on Polymers*, 1998.
70. C. C. Müller-Goymann, *Parfuem, Kosmet. 73*, 452 (1992).

9

Use of Organosilicone Surfactants as Agrichemical Adjuvants

DONALD PENNER Department of Crop and Soil Sciences, Michigan State University, East Lansing, Michigan

RICHARD BUROW* Dow Corning Corporation, Midland, Michigan

FRANK C. ROGGENBUCK Department of Crop and Soil Sciences, Michigan State University, East Lansing, Michigan

I. INTRODUCTION

Agrichemical adjuvants are added to agricultural spray tank mixes to function as wetters, spreaders, stickers, defoamers, and so on. In 1983 the Weed Science Society of America defined the word "adjuvant" as "any substance in a herbicide formulation or added to the spray tank to improve herbicidal activity or application characteristics." Organosilicone adjuvants have been used for many years as defoamers, but this chapter focuses, on the use of organosilicone surfactants as activator adjuvants: that is, adjuvants that enhance the activity of the active ingredient in the spray solution.

*Retired.

The evaluation of organosilicones for enhancing herbicide activity was first reported by Jansen in 1973 [1]. The late 1980s saw a burgeoning of interest in the development and evaluation of organosilicone adjuvants for enhancing herbicide activity. This coincided with the development of large numbers of postemergence herbicides and a shift to reduced-till or no-till agriculture, which is much more dependent on postemergence herbicides. The organosilicone adjuvants have been found to dramatically enhance the activity and rainfastness of some postemergence herbicides, particularly the salts of weak acids such as isopropylamine salt of glyphosate (Roundup herbicide), sodium salt of bentazon (Basagran herbicide), and sodium salt of acifluorfen (Blazer herbicide) [2–5].* Figure 1 [6] shows that the organosilicone adjuvant Sylgard 309 Silicone Surfactant increased the efficacy of acifluorfen herbicide compared with acifluorfen with a conventional crop oil concentrate adjuvant or no adjuvant.† Figure 2 [6] shows that the efficacy of the acifluorfen/Sylgard 309 solution was still very high, even when 2.54 cm of simulated rainfall was applied to the plants 15 min after the herbicide application. The addition of organosilicone adjuvant to the spray solution made the application "rainfast."

Organosilicone adjuvants have also been used with the foliar application of organic and inorganic fungicides [7,8], insecticides [9], nutrients [10], and growth regulators [11,12]. Efficacy has been shown with specific representatives of each of these classes of agrichemicals.

II. CHEMISTRY OF ORGANOSILICONE ADJUVANTS

Most of the organosilicone surfactants used as agrichemical adjuvants are silicone polyethers (also called siloxylated polyethers or silicone polyether copolymers) having the following general structure:

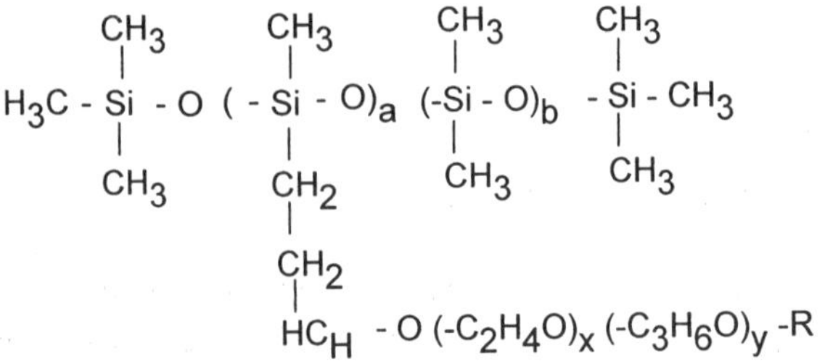

*Roundup is a trademark of Monsanto Corporation; Basagran and Blazer are trademarks of BASF Corporation.

†Sylgard is a trademark of Dow Corning Corporation.

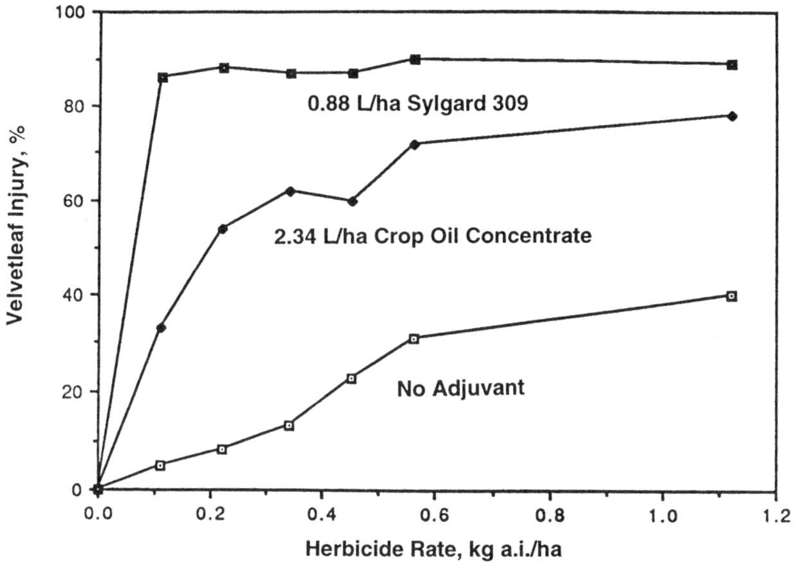

FIG. 1 Visual velvetleaf injury as a function of adjuvant and acifluorfen rate in the absence of rainfall. (Adapted from Ref. 6.)

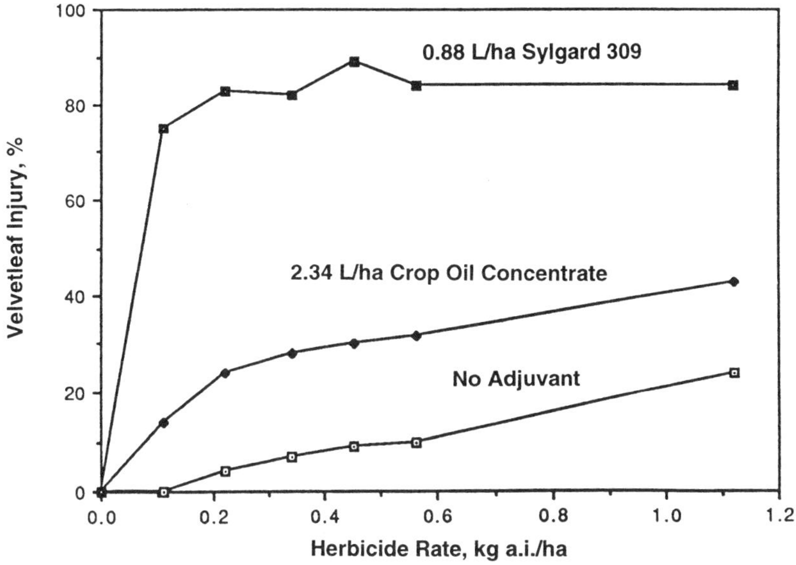

FIG. 2 Visual velvetleaf injury as a function of adjuvant and acifluorfen rate with 2.54 cm simulated rainfall. (Adapted from Ref. 6.)

TABLE 1 Organosilicone Adjuvants

Product name	Manufacturer or distributor	Principal functioning components	Recommended use range (%)
Action 99	Universal Coop./ Countrymark/ Growmark	Polyalkylene, modified heptamethyltrisiloxane plus nonionic surfactant	0.188–1
Break-Thru	Goldschmidt Chemicals	100% polyether–polymethylsiloxane copolymer	0.05–0.1
Century	Precision Labs, Inc.	100% polyether–polymethylsiloxane copolymer	0.09–0.20
Dow Corning 211	Dow Corning Corp.	Heptamethyltrisiloxane polyether	0.06–1
Dyne-Amic	Helena Chemical	Blend of polyalkylene oxide modified polydimethylsiloxane, nonionic emulsifiers, and methylated soybean oil	0.5
Eth-N-Gard	WILFARM L.L.C.	Heptamethyltrisiloxane polyether and ethylated seed oil	0.25–0.5
Excel 2000	Coastal Chemical Corp.	Polyether–polymethylsiloxane copolymer	0.125–0.25
Excel 2000 50 DF	Coastal Chemical Corp.	Proprietary siloxane copolymer	0.25
Freeway	Loveland Industries	Silicone polyether copolymer and alcohol ethoxylates	0.09–1
Galactic	Custom Chemicides	Polydimethylsiloxane, nonionic surfactant	0.06–1
Impact	Jay-Mar, Inc.	Proprietary blend of polyalkylene oxide modified heptamethyltrisiloxane and nonionic surfactant	0.188–1
Inforce	United Suppliers/ Chemorse Ltd.	Silicone polyether copolymer plus nonionic surfactant	0.125–1
Kinetic	Helena Chemical	Blend of polyalkylene oxide modified polydimethylsiloxane and block copolymers of ethylene and propylene oxide	0.09–0.5

TABLE 1 Continued

Product name	Manufacturer or distributor	Principal functioning components	Recommended use range (%)
Kinetic DC	Helena Chemical	Blend of polyalkylene oxide modified dimethylsiloxane, nonionic surfactants, and polymerized ethoxylates	0.25
Kinetic HV	Helena Chemical	Polyalkylene oxide modified polymethylsiloxane and block copolymers of ethylene and propylene oxide	0.25–1
Matrixx	Coastal Chemical Corp.	Polymethylsiloxane copolymer and polyethoxy ethers	0.25–0.5
Motion	Loveland Industries	Silicone polyether copolymer and alcohol ethoxylates	0.09–1
Peerless	Custom Chemicides	Organosilicone, vegetable oil, poly-fatty acid esters, polyethoxylate, alkylaryl phosphate esters	0.05–0.25
Silkin	Terra International	Polyalkylene oxide modified heptamethyltrisiloxane	0.188–1
Sil-Fact	Drexel Chemical Co.	Blend of organosilicone surfactant and alcohol ethoxylates	0.09–0.5
Sil-Fact HV	Drexel Chemical Co.	Blend of organosilicone surfactant and alcohol ethoxylates	0.09–1
Silwet Energy	Brewer International	Polyalkylene oxide, modified polymethyl-siloxane	0.09–0.188
Silwet L-77	OSi Specialties/Loveland Industries, Helena Chemical	Silicone–polyether copolymer 100%	0.05–0.25
Sun Energy	Brewer International	Methylated sunflower oil, organosilicones, and NIS	0.25–1
Sylgard 309	Dow Corning Corp/ Wilbur-Ellis Co., WILFARM, L.L.C.	Heptamethyltrisiloxane polyether	0.06–1

Source: Adapted from Ref. 14.

where $a = 1$–100, $b = 0$–50, $x = 4$–30, $y = 4$–30, and R = H, CH_3, or $COCH_3$. The x and y values for polyether chain length are mean values. Biological evaluation of many organosilicone surfactants in this family has shown that the lower molecular weight silicone polyethers produce the greatest enhancement of herbicide activity. The trisiloxane surfactant, wherein $a = 1$, $b = 0$, $x = 7$, and $y = 0$, has been found to be an optimum structure, particularly for enhancement of water-soluble postemergence herbicides [13].

Organosilicone surfactants of the foregoing structure are in commercial use as agricultural adjuvants. These surfactants are also used in adjuvant blends in combination with organic surfactants and/or vegetable oils. A representative listing of commercially available organosilicone adjuvants and organosilicone adjuvant blends is shown in Table 1 [14].

III. EFFECT ON SURFACE TENSION AND SPRAY DROPLET SPREADABILITY

Organosilicone surfactants such as Silwet L-77 and Sylgard 309 reduce the equilibrium surface tension of water to about 21 mN/m.* Knoche et al. [15] reported that Silwet L-77 reduced the equilibrium surface tension of water to 20–22 mN/m at a concentration of 0.01%. Figure 3 shows equilibrium surface tension versus concentration in water for Sylgard 309 and a conventional nonionic organic surfactant [16]. Equilibrium surface tension decreases as the concentration of organosilicone surfactant increases to about 0.007%, which is the critical micelle concentration (cmc). Solutions at concentrations above the cmc are turbid. Depending on the chemicals in an agricultural spray solution, organosilicone adjuvants can lower the dynamic surface tension to values between 21 to 25 mN/m.

The ability of organosilicone surfactants to dramatically reduce the surface tension of water leads to rapid spreading on hydrophobic surfaces [7,17]. Water droplets or agrichemical spray solutions containing organosilicone surfactants spread completely over leaf surfaces in a matter of seconds, and the droplet contact angle is zero. This rapid spreading is sometimes called "superwetting" or "superspreading." Organosilicone adjuvants such as Silwet L-77 and Sylgard 309 are often referred to as "superwetters" or "superspreaders" [18].

Figure 4 shows leaves of velvetleaf (*Abutilon theophrasti* Medicus), which were photographed 25 s after being sprayed with Blazer herbicide with the following adjuvants: 0.25% nonionic surfactant (Fig. 4a), 1.0% crop oil concentrate (Fig. 4b), and 0.25% Sylgard 309 silicone surfactant (Fig. 4c). Figure 4c shows the superwetting that results from the addition of a silicone adjuvant to an agricultural spray solution.

---

*Silwet is a trademark of the Osi Specialties, a division of Witco Corporation.

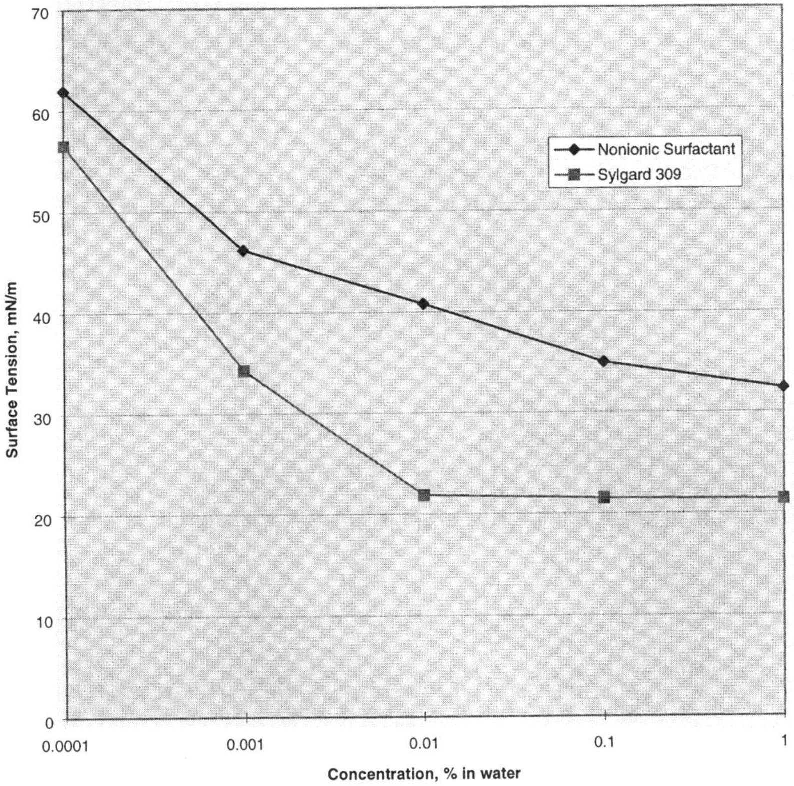

FIG. 3 Equilibrium surface tension of Sylgard 309 and nonionic surfactant. (From Ref. 16.)

(a) (b) (c)

FIG. 4 Wetting and spreading of acifluorfen herbicide on velvetleaf with various adjuvants: (a) nonionic surfactant, (b) crop oil concentrate, and (c) Sylgard 309. (Photographs courtesy of Dow Corning Corp.)

Some investigators believe that the T shape of the molecule is responsible for superspreading [19]. The T shape reputedly leads to "molecular zippering action," whereby the surfactant molecules at the leading edge of the spreading droplet roll over mechanically in a sort of tractor-tread motion. Others have argued instead of the T shape being the cause of superspreading, the superspreading results from the combination of low dynamic surface and interfacial tension [20,21]. The presence of a preexisting water layer on the surface may also be a factor [22].

The effectiveness of any adjuvant in enhancing spray droplet spread is affected by the pesticide in the spray solution and by the nature of the surface to which the droplet is applied [23]. Leaf surfaces of various weeds differ enormously [24], and droplet response may differ significantly from one weed species to another. In general, organosilicone adjuvants spread rapidly (superwet) on virtually all weed species.

Stevens [25] considers that the spreading ability of the organosilicone adjuvants has been a major incentive for their development as adjuvants. It could be anticipated that enhancing the spreading of a contact fungicide or insecticide spray solution might enhance its biological activity. Indeed, there are numerous reports in the literature of enhancement of fungicide and insecticide performance resulting from the addition of organosilicone adjuvants.

The efficacy enhancement of herbicides is somewhat different. It has been found that the spread of the herbicide spray droplet as influenced by adjuvants may have no direct correlation to the efficacy of the adjuvant [26]. Roggenbuck et al. [27] studied the relationship between leaf position in velvetleaf plants, herbicide absorption, and the spread of spray droplets containing Sylgard 309. The authors found a negative relationship between spread and herbicide absorption; That is, the leaves with the best spreading had the least absorption. The enhancement of herbicide uptake is discussed in greater depth in a later section.

For cost reasons, organosilicone surfactants are often blended with organic nonionic surfactants. Adjuvant blends containing organosilicones may retain most of the capability to reduce surface tension and enhance spray droplet spreadability but do not retain the efficacy of the pure organosilicone in enhancing herbicide activity or rainfastness [13,28]. The enhancement achieved appears to correlate directly to the concentration of organosilicone adjuvant in the blend.

IV. PESTICIDE ACTIVITY ENHANCEMENT

A. Concentration of Adjuvant Required for Activity

When used in concentrations sufficient to produce rapid spreading, organosilicone surfactants enhance the activity of fungicides and insecticides [7,9]. Figure 5 shows the ability of Sylgard 309 to enhance the efficacy of sulfur fungicide (S) in control of powdery mildew (*Erysiphe graminis*) in wheat [8]; it does this by facil-

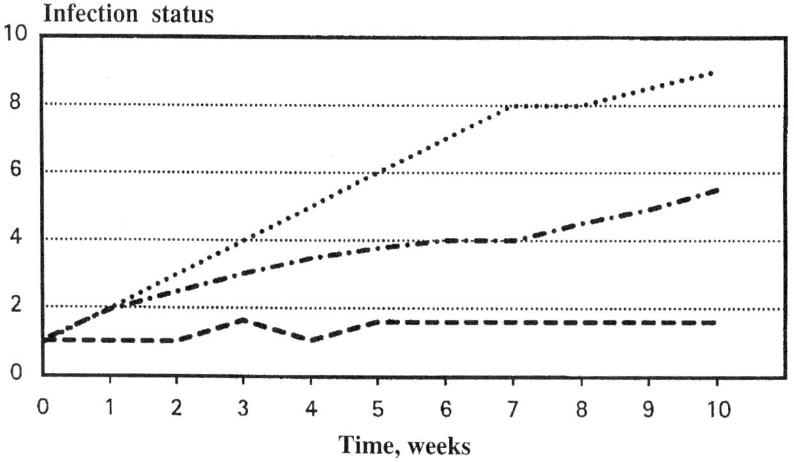

FIG. 5 Effect of Sylgard 309 on the performance of sulfur fungicide on wheat: dots, control, dots and dashes, sulfur alone; dashes, S plus organosilicone surfactant. Scale: 1, no infection; 9, maximum infection. (From Ref. 8.)

itating spread of the sulfur dispersion to completely cover and protect the leaf. Silwet L-77 and Sylgard 309 product labels recommend using 0.03–0.06% in fungicide and insecticide spray solutions. This concentration is sufficient to achieve good enhancement without risking phytotoxicity (i.e., injury) to sensitive crops.

Where a chemical must be absorbed into plant tissue, a higher concentration is required than is needed for rapid spreading. Data presented by Stevens et al. [29] indicate that ^{14}C-deoxyglucose absorption into bean leaf increased with Silwet L-77 concentration, as shown in Table 2. Very little ^{14}C-deoxyglucose is absorbed at

TABLE 2 ^{14}C-Deoxyglucose Absorption in 10 min by Bean (*Vicia faba* L.) Leaves

Silwet L-77 concentration (%)	^{14}C-Deoxyglucose absorbed (%)
0	4
0.1	12
0.2	18
0.5	49

Source: Ref. 29.

Silwet L-77 concentrations up to 0.2%, but the amount absorbed increases rapidly as Silwet L-77 concentration is increased from 0.2% to 0.5% in the spray solution.

Buick et al. [30] studied the absorption of ^{14}C-triclopyr herbicide by bean (*Vicia faba* L.) leaves. A concentration of 0.25% of Silwet L-77 provided absorption of the ^{14}C-triclopyr significantly greater than 0.1%, although the surface tension of the spray solution reached a minimum of 22.8 mN/m at 0.05%.

Roggenbuck et al. [31] evaluated the efficacy of a structured series of organosilicone adjuvants for efficacy in enhancing Na–acifluorfen activity on velvetleaf. Activity more than doubled as the adjuvant concentration increased from 0.18% through 0.55%. Figure 6 shows the increase of herbicide activity with organosilicone adjuvant concentration. The equilibrium surface tension data are shown on the same graph in support of the investigators' conclusion that herbicide activity enhancement and surface tension are not directly related.

B. Influence of pH on Activity

The effect of pH on the stability of organosilicones has been studied extensively by Knoche et al. [12,15,32]. The organosilicone Silwet L-77 was observed to be stable at spray tank pH values of 6–8 [15,33], but degradation occurred at pH values of 3–5 and 9–10. The degradation reaction appeared to follow first-order kinet-

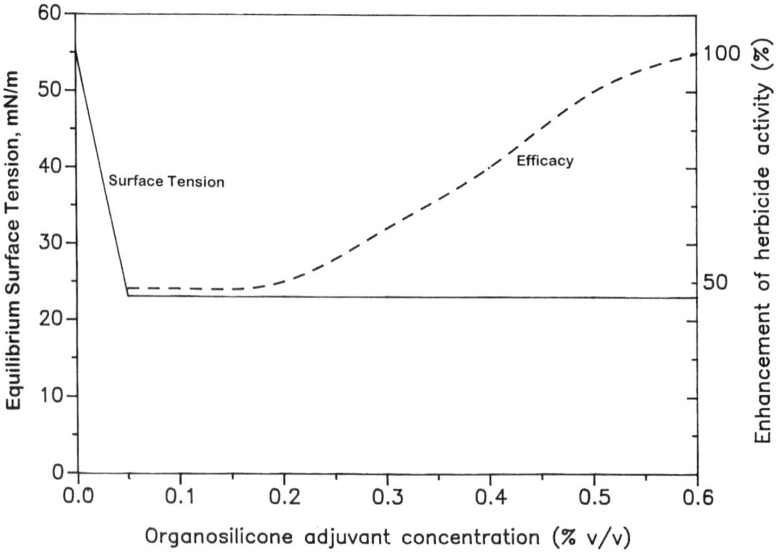

FIG. 6 Schematic diagram of the effect of Sylgard 309 concentration on equilibrium surface tension and on efficacy enhancement of acifluorfen herbicide. (From Ref. 34.)

ics and was considered to involve the hydrolytic cleavage of the siloxane bond [15]. The rate of Silwet L-77 degradation at pH 3.0 increased as the temperature increased from 15°C to 35°C and decreased as the concentration in the spray solution increased from 0.025% to 0.4% [15]. Knoche et al. [12] attributed the greater stability of Silwet L-77 in well water versus deionized water to the greater buffering capacity of the well water at pH 2.5.

The degradation of Silwet L-77 exhibited a lag phase at concentrations above the critical micelle concentrations. The duration of the lag phase was linear with lag phase (L) = $-0.42 + 12.41 \times$ concentration (%) with an r^2 of 0.982 [15]. The degradation process as a function of low pH is readily observed as a change in the turbidity, appearance of the solution, and the change in the surface tension of the solution.

Studies have shown that organosilicone adjuvants vary somewhat in their response to pH. Knoche et al. [15] evaluated six different organosilicones. Table 3 shows the change in surface tension values at several intervals after preparation of pH 3.0 solutions containing the different surfactants. The organosilicones Silwet L-7001 and Silwet L-7602 showed no change in stability as measured by changes in surface tension up to 24 h after preparation of the pH 3.0 solution [15]. Unfortunately, these two surfactants did not exhibit the low surface tensions required for rapid wetting.

In a study of acifluorfen formulated as Blazer herbicide, spray solutions were buffered to pH 4,7, and 9 and held various lengths of time before spraying [34]. Sylgard 309 was included in each spray solution at 0.375%. At that high concentration of organosilicone surfactant, the performance of the spray solutions was unchanged for 72 h, except for the pH 4 solution, which was stable for only 24 h. Results are presented in Table 4.

TABLE 3 Changes in Surface Tension in Organosilicones Over Time

	Surface tension (mN/m) at 24°C at pH 3.0 for three solution ages		
Organosilicone at (0.1%)	0 h	3 h	6 h
Silwet L-77	22.6	51.9	55.5
Silwet L-7001	36.5	36.5	36.5
Silwet L-7602	34.3	33.6	33.9
Silwet L-7607	24.3	29.0	32.8
Sylgard 309	23.2	28.8	54.2
Dow Corning 211	22.6	50.4	58.7

Source: Ref. 15.

TABLE 4 Visual Injury (%) from Sprays Stored Before Spraying[a]

		Storage time (h)			
Blazer treatment rate	pH	0	6	24	72
0.0336 kg/ha	4	60	66	58	25
	7	61	64	74	73
	9	60	65	64	66
0.336 kg/ha	4	100	98	89	78
	7	100	98	100	100
	9	100	98	100	100

[a]Evaluated 7 days after treatment
Source: Ref. 34.

In many agricultural applications of spray solutions, the optimal concentration of the organosilicone is greater than 0.1% and the time elapsed from spray preparation to spray deposition and drying on the leaf surface is less than 3 h. The organosilicone surfactant should remain stable and effective in most agricultural spray solutions for that period of time. The indiscriminate addition of spray acidifiers should be avoided, however, as they may cause rapid loss of adjuvant efficacy.

C. Basis for the Efficacy of Organosilicone Adjuvants

The use and mode of action of adjuvants for herbicides has been reviewed by Kirkwood [35], but the author does not mention the organosilicone adjuvants. The action or efficacy of organosilicone adjuvants in agrichemical sprays is not readily explained by a single mode or basis. At least two separate bases can be identified.

1. Enhanced spread of active ingredient over the plant surface. The rapid spreading character of the superspreader organosilicone adjuvants is useful in the application of contact fungicides, insecticides, and herbicides, where thoroughness of coverage of the plant is critical to the activity of the active ingredient.

2. Rapid wetting, low droplet angles, and improved retention. Rapid leaf wetting and low droplet contact angles have been related to the low equilibrium surface tension of spray solutions containing organosilicone adjuvants, while spray drop retention has been found to correlate with low dynamic surface tension [36].

Baylis and Hart [37] have argued that enhanced spread of the active ingredient on the plant surface to the site of active ingredient action can be considered a mechanism or mode whereby the superspreader organosilicone adjuvant can exert its action. The plant species specificity observed by Field et al. [38] and Baylis and Hart [37] for herbicide activity enhancement with Silwet L-77 could be expected from the diversity of plant cuticles occurring in nature [24].

Adjuvant concentrations greater than that necessary to reach the critical micelle concentration appear to be necessary for maximum reduction of the contact angle for Silwet L-77 [39]. Knoche [32] suggests that the concentration-dependent wetting is related to surfactant adsorption at the leaf–spray droplet interface, which reduces the active surfactant concentration in the spray solution. The studies of Lin et al. [40] indicate that for superspreading to occur with the organosilicone surfactants, the surface requires a minimum level of moisture. The superspreading characteristic of certain organosilicone adjuvants is critical for this basis of efficacy.

V. ENHANCED PENETRATION OF THE PESTICIDE INTO THE PLANT

The plant cuticle is considered to be the prime barrier to the penetration into plants by pesticides. The wax component of the cuticle may present the greatest barrier [41–43]. The foliar absorption of herbicides is believed to occur by diffusion through the cuticle [44]. The rate of diffusion is a function of the concentration gradient across the cuticle, the diffusion coefficient of the herbicide in the nonpolar components of the cuticle, the thickness of the cuticle, and the partition coefficients of the herbicide between the external environment of the cuticle–cuticle and the cuticle–subtending aqueous interface [45]. The ultrastructure of the various components of the cuticle may also influence adsorption.

Absorption via diffusion through the cuticle is greater when herbicides are applied in low spray volumes because the droplets are more concentrated as they dry, and the concentration gradient across the cuticle is steeper [46,47]. Movement of the herbicide away from the cuticle in the leaf interior also affects the concentration gradient [48]. Currier and Dybing [49] speculated that if adjuvants reduced the surface tension sufficiently, the spray solution could move through the stomatal aperture, and this would enhance foliar absorption. The surface tension of the spray droplets would need to be reduced to less than 30 mN/m [50]. Even though the spray solution passed through the stomatal aperture, the herbicide or pesticide would still need to diffuse through the cuticle lining the substomatal cavity [51]. Guard cells and the basal cells of trichomes have been recognized as sites of rapid absorption [52,53].

Stevens [25] has proposed that the organosilicone adjuvants enhance herbicide absorption primarily by facilitating stomatal infiltration. He argues that the lowering of the surface tension to values near 22 mN/m [30] facilitates passage of the spray solution through the stomatal aperture [50]. This argument is based on experiments [29] in which ^{14}C-deoxyglucose treatments were applied to bean (*Vicia faba* L.) leaves that had been excised 1.5 before treatment. The level of herbicide absorbed in the excised leaves was greatly reduced in comparison to the level absorbed in leaves on whole plants. Stevens [25] points out that excising the leaves closes the stomata, and argues that, the greatly reduced herbicide absorption proves that herbicide solution enters the leaves by stomatal infiltration. This argument is advanced without considering that photosynthesis stops when a leaf is excised, and that immediately following excision, the leaf begins to dehydrate, and transport of nutrients and herbicides through the cuticle ceases. In other words, herbicide absorption is greatly reduced in an excised leaf primarily because all its processes are interrupted, not because its stomata close.

If stomatal infiltration is the mechanism for rapid absorption, then the concentration of any adjuvant that reduces the surface tension to the required low value (≈22 mN/m) for rapid spreading should produce rapid absorption. But it has been determined that not all organosilicone adjuvants that lower the surface tension to this level are equally effective in enhancing herbicide absorption [5]. Additionally, the low concentrations of organosilicone adjuvant required to reduce the surface tension to about 22 mN/m are not sufficient to provide good absorption. While adjuvant concentrations above the 0.007% cmc produce surface tensions in the 22 mN/m range, absorption is almost nil at these low concentrations. The data show that concentrations of 0.2–0.4% are required to achieve maximum levels of absorption [29,31]. This suggests lack of a direct relationship between surface tension reduction/increased spreading and enhanced herbicide absorption.

It seems more likely that the adsorption enhancement seen with the use of the organosilicone adjuvants involves greatly enhanced cuticular penetration of the pesticide. Roggenbuck et al. [27] proposed this as the primary mechanism and stated the belief that stomatal infiltration as proposed by Stevens [25,54] makes only a negligible contribution to the enhanced herbicide absorption.

Supporting this hypothesis, studies conducted by Roggenbuck et al. [55] showed that the organosilicone adjuvant Sylgard 309 was absorbed very rapidly by velvetleaf, with 75% being taken up in 15 min. The herbicides acifluorfen and bentazon were absorbed at essentially the same rate as the adjuvant. Evidently, the adjuvant and the herbicide were absorbed together, and the adjuvant may have acted as a cosolvent. Stock and Holloway [56] refer to this phenomenon as interactive facilitation. Schonherr and Bauer [57] stress the necessity of matching the cuticle penetration velocities of the adjuvant and the active ingredient. Rapid initial herbicide absorption can be explained by rapid cuticular penetration of both adjuvant and herbicide, with the adjuvant acting as a cosolvent. This rapid absorption can be followed by a period of much slower absorption, wherein the herbicide remaining on the surface (after all the adjuvant has been absorbed) is slowly absorbed through the cuticle. This pattern of herbicide absorption was observed by Gaskin [58]. Rapid herbicide absorption explains the "rainfastness" reported by various researchers [2,5,28,39] and shown in Fig. 2. Further, Roggenbuck et al. [27] observed a negative correlation between herbicide adsorption and stomatal density, providing further evidence that organosilicone adjuvants exert their action by enhancing cuticular penetration. Cuticular penetration appears to be the most likely explanation for the observed results.

VI. CONCLUSION

The number of organosilicone surfactants and blends listed in Table 1 testifies to the considerable popularity that organosilicone adjuvants have gained in a short period of time. The dramatic "superwetting" characteristic of the organosilicones has undoubtedly been a factor in their performance and in their market acceptance. Their use with postemergence herbicides has been further aided by the shift in farming practices to no-till and minimum-till management, which requires increased use of postemergence herbicides. The organosilicone adjuvants have shown the greatest efficacy with weakly anionic herbicides applied for control of certain broadleaf weeds, and with foliar-applied insecticides and fungicides. Further exploration of organosilicone chemistry will likely lead to the development of organosilicone adjuvants able to enhance efficacy with a broader range of pesticides against an even wider range of pest targets.

REFERENCES

1. Jansen, L. L., *Weed Sci. 21*:130–135 (1973).
2. Reddy, K. N., and M. Singh, Organosilicone adjuvant effects on glyphosate efficacy and rainfastness. *Weed Technol. 6*:361–365 (1992).
3. Reddy, K. N., and M. Singh, Organosilicone adjuvants increased the efficacy of glyphosate for control of weeds in citrus (*Citrus* spp.). *HortScience 27*(9):1003–1005 (1992).
4. Roggenbuck, F. C., L. Petroff, and D. Penner, Use of silicone adjuvants to increase activity and rainfastness of Na-acifluorfen, IPA-glyphosate, and Na-bentazon. *Proc. North Central Weed Control Conf. 43*:82 (1988).
5. Roggenbuck, F. C., L. Rowe, D. Penner, L. Petroff, and R. Burow, Increasing postemergence herbicide efficacy and rainfastness with silicone adjuvants. *Weed Technol. 4*:576–580 (1990).
6. Roggenbuck, F. C., L. Rowe, D. Penner, R. F. Burow, R. A. Ekeland, and L. J. Petroff, Use of silicone adjuvants to increase activity and rainfastness of acifluorfen. *Proceedings of the Brighton Crop Protection Conference, Weeds*, 1989, pp. 219–224.
7. Green, C. F., and D. A. Green, Relocation of fungicides to stem bases of winter wheat using organosilicone copolymers in relation to disease suppression. *Third International Symposium on Adjuvants for Agrochemicals*, Cambridge, 1992.
8. Demes, H., M. Gaudchau, and R. F. Burow, Role of organosilicone surfactants in enhancing the performance of inorganic fungicides. *Pestic. Sci. 38*:278–280 (1993).
9. Adams, A. J., J. S. Fenlon, and A. Palmer, Improving the biological efficacy of small droplets of permethrin by the addition of silicone-based surfactants. *Ann. Appl. Biol. 112*:19–31 (1988).

10. Leece, D. R., and J. F. Dirou, Organosilicone and alginate adjuvants evaluated in urea sprays foliar-applied to prune trees. *Commun. Soil Sci. Plant Anal. 8*:169 (1977).
11. Greenberg, J., S. P. Monselise, and E. E. Goldschmidt, Improvement of gibberellin efficacy in prolonging the citrus harvest season by the surfactant L-77. *J. Am. Soc. Hortic. Sci. 112*:625 (1987).
12. Knoche, M., H. Tamura, and M. J. Bukovac, Stability of the organosilicone surfactant Silwet L-77 in growth regulator sprays. *HortScience 26*:1498–1500 (1991).
13. Burow, R. F., D. Penner, F. C. Roggenbuck, and R. M. Hill, Relationship of organosilicone adjuvant structure and phase behaviour to activity enhancement of acifluorfen and glyphosate. *Proceedings of the Fourth International Symposium on Adjuvants for Agrochemicals*, Melbourne, Australia, 1995 (FRI Bulletin 193, pp. 54–59).
14. Kapusta, G., *A Compendium of Herbicide Adjuvants.* Southern Illinois University, Carbondale, 1996, pp. 21–22.
15. Knoche, M., H. Tamura, and M. J. Bukovac, Performance and stability of the organosilicone surfactant L-77: Effect of pH, concentration, and temperature. *J. Agric. Food Chem. 39*:202–206 (1991).
16. Dow Corning Internal Report, Midland, MI.
17. Policello, G. A., G. J. Murphy, P. J. G. Stevens, and W. A. Forster, Dynamic surface tension effects on spray droplet adhesion of organosilicones. *Proceedings of the Brighton Crop Protection Conference,* Weeds, 1993.
18. He, M., R. M. Hill, Z. Lin, L. E. Scriven, and H. T. Davis, Phase behavior and microstructure of polyethylene trisiloxane surfactants in aqueous solution. *J. Phys. Chem., 97*:8820–8834 (1993).
19. Goddard, E. D., and K. P. A. Padmanabhan, A mechanistic study of the wetting, spreading, and solution properties of organosilicone surfactants, in *Adjuvants for Agrichemicals* (C. L. Foy, ed.), CRC Press, Boca Raton, FL, 1992, pp. 373–383.
20. Hill, R. M., M. He, H. T. Davis, and L. E. Scriven, Comparison of the liquid crystal phase behavior of four trisiloxane superwetter surfactants. *Langmuir 10*:1724–1734 (1994).
21. Hill, R. M., and R. F. Burow, Why organosilicone adjuvants spread, in *Pesticide Formulations and Applications Systems*, ASTM STP 1328, Vol. 17, American Society for Testing and Materials, Philadelphia, 1997.
22. Zhu, X., W. G. Miller, L. E. Scriven, and H. T. Davis, Superspreading of water–silicone surfactant on hydrophobic surfaces. *Colloids Surf. A. 90*:63–78 (1994).
23. Penner, D., The impact of adjuvants on herbicide antagonism. *Weed Technol. 3*:227–231 (1989).
24. Harr, J., R. Guggenheim, G. Schulke, and R. H. Falk, *The Leaf Surface of Major Weeds.* Sandoz Ltd., Basel, 1991, 132 p.
25. Stevens, P. J. G., Organosilicone surfactants as adjuvants for agrochemicals. *Pestic. Sci. 38*:103–122 (1993).
26. Wanamarta, G., D. Penner, and J. J. Kells, Identification of efficacious adjuvants for sethoxydim and bentazon. *Weed Technol. 3*:60–66 (1989).
27. Roggenbuck, F. C., R. F. Burow, and D. Penner, Relationship of leaf position to herbicide absorption and organosilicone adjuvant efficacy. *Weed Technol. 8*:582–585 (1994).

28. Sun, J., C. L. Foy, and H. L. Witt, Effect of organosilicones surfactants on rainfastness of primisulfuron in velvetleaf (*Abutilon theophrasti* Medicus). *Proc. Northeast. Weed Sci. Soc. 49*:34 (1995).
29. Stevens, P. J. G., R. E. Gaskin, S. O. Hong, and J. A. Zabkiewicz, Contributions of stomatal infiltration and cuticular penetration to enhancements of foliar uptake by surfactants. *Pestic. Sci. 33*:371–382 (1991).
30. Buick, R. D., R. J. Field, A. B. Robson, and G. D. Buchan, A foliar uptake model of triclopyr, in *Adjuvants for Agrichemicals* (C. L. Foy, ed.), CRC Press, Boca Raton, FL, 1992, pp. 87–99.
31. Roggenbuck, F. C., L. Rowe, D. Penner, R. Burow, R. Ekeland, and L. Petroff, Comparison of statistical methods for evaluating silicone adjuvants for Na-acifluorfen, in *Adjuvants for Agrichemicals* (C. L. Foy, ed.), CRC Press, Boca Raton, FL, 1992, pp. 411–421.
32. Knoche, M., Organosilicone surfactant performance in agricultural spray application: A review. *Weed Res. 34*:221–239 (1994).
33. Policello, G. A., R. E. Rackle, and G. J. Murphy, Formulation considerations for Silwet L-77. *Proc. South. Weed Sci. Soc. 44* (1991).
34. Dow Corning–Michigan State University Study.
35. Kirkwood, R. D., Use and mode of action of adjuvants for herbicides: A review of some current work. *Pestic. Sci. 38*:93–102 (1993).
36. Anderson, N. H., D. J. Hall, and D. Seaman, Spray retention effects of surfactants and plant species. *Aspects Appl. Biol. 14*:233–243 (1987).
37. Baylis, A. D., and C. A. Hart, Varying responses among weed species to glyphosate–trimesium in the presence of an organosilicone surfactant. *Proceedings of the Brighton Crop Protection Conference*, *Weeds*, 1993.
38. Field, R. J., N. N. Dobson, and L. J. Tisdell, Species-specific sensitivity to organosilicone surfactant-enhancement of glyphosate uptake, in *Adjuvants for Agrichemicals* (C. L. Foy, ed.), CRC Press, Boca Raton, FL, 1992, pp. 423–432.
39. Field, R. F., and N. G. Bishop, Promotion of stomatal infiltration of glyphosate by an organosilicone surfactant reduces the critical rainfall period. *Pestic. Sci. 24*:55–62 (1988).
40. Lin, Z., R. M. Hill, H. T. Davis, and M. D. Ward, Determination of wetting velocities of surfactant superspreaders with the quartz crystal microbalance. *Langmuir 10*: (1994).
41. Baker, E. A., and M. J. Bukovac, Characterization of the components of plant cuticles in relation to the penetration of 2,4-D. *Ann. Appl. Biol. 67*:243–253 (1971).
42. Bukovac, M. J., Some factors affecting the absorption of 3-chlorophenoxy-α-propionic acid by leaves of the peach. *Proc. Am. Soc. Hortic. Sci. 87*:131–138 (1965).
43. Hull, H. M., Leaf structure as related to absorption of pesticides and other compounds, in *Residue Research*, Vol. 31 (F. A. Gunther and J. D. Gunther, eds.), Springer-Verlag, New York, 1970, pp. 1–155.
44. Price, C. E., Penetration and translocation of herbicides and fungicides in plants, in *Herbicides and Fungicides—Factors Affecting Their Activity* (N. R. McFarlane, ed.), Chemical Society Special Publication 29, The Chemical Society, London, 1976, pp. 42–66.

45. Price, C. E., A review of the factors influencing the penetration of pesticides through plant leaves, in *The Plant Cuticle* (D. F. Cutler, K. L. Alvin, and C. E. Price, eds.), Linnaean Society Syrup. Ser. 10, Academic Press, London, 1982, pp. 237–252.
46. Al-Khatib, K., D. R. Gesly, and C. M. Boerboom, Effect of thifensulfuron concentration and droplet size on phytotoxicity, absorption, and translocation in pea (*Pisum sativum*). *Weed Sci. 42*:482–486 (1994).
47. Penner, D., F. C. Roggenbuck, R. F. Burow, L. J. Petroff, and B. Thomas, Determination of whether dosage or concentration is the critical factor for efficacy of Sylgard 309 organosilicone adjuvant, *Proceedings of the 43rd International Crop Protection Symposium*, Ghent, 1991.
48. Richard, E. P. Jr., and F. W. Slife, *In vivo* and *in vitro* characterization of the foliar entry of glyphosate in hemp dogbane (*Apocynum cannabinum* L.). *Weed Sci. 27*:426–433 (1979).
49. Currier, H. B., and C. D. Dybing, Foliar penetration of herbicides—Review and present status. *Weeds 7*:195–213 (1959).
50. Schonherr, J., and M. J. Bukovac, Penetration of stomata by liquids: Dependence on surface tension wettability and stomatal morphology. *Plant Physiol. 49*:813–819 (1972).
51. Foy, C. L., Review of herbicide penetration through plant surfaces. *J. Agric Food Chem. 12*:473–476 (1964).
52. Sargent, J. A., The penetration of growth regulators into leaves. *Annul Rev. Plant Physiol. 16*:1–12 (1965).
53. Linskens, H. F., Wilkinen, and A. L. Stoffers, Cuticle of leaves at the residue problem. *Residue Rev. 8*:136 (1965).
54. Stevens, P. J. G., R. E. Gaskin, S. O. Hong, and J. A. Zabkiewicz, Pathways and mechanisms of foliar uptake as influenced by surfactants, in *Adjuvants for Agrichemicals* (C. L. Foy, ed.), CRC Press, Boca Raton, FL, 1992, pp. 385–398.
55. Roggenbuck, F. C., D. Penner, R. F. Burow, and B. Thomas, Study of the enhancement of herbicide activity and rainfastness by an organosilicone adjuvant utilizing radiolabelled herbicide and adjuvant. *Pestic. Sci. 37*:121–125 (1993).
56. Stock, D., and P. J. Holloway, Possible mechanisms for surfactant-induced foliar uptake of agrochemicals. *Pestic. Sci. 38*:165–177 (1993).
57. Schonherr, J., and H. Bauer, Modelling penetration of plant cuticle by crop protection agents and effects of adjuvants on the rate of penetration. *Pestic. Sci. 42*:185–208 (1994).
58. Gaskin, R. E., Effect of organosilicone surfactants on the foliar uptake of herbicides: Stomal infiltration versus cuticular penetration. *Fourth International Symposium on Adjuvants for Agrochemicals*, 1995 (FRI Bull No 193, pp. 243–248).

10
Polymer Surface Modifiers

ISKENDER YILGÖR Department of Chemistry, Koç University, Istanbul, Turkey

I. INTRODUCTION

Polymers with controlled surface properties have been receiving increased attention for over two decades [1–3]. This is mainly due to the critical roles played by the surface characteristics of polymeric materials in their overall performance in many diversified fields. Some of these applications include adhesives (improved wetting and adhesion), paints and coatings (surface finish, coefficient of friction, weatherability, scratch and chemical resistance), specialty films (controlled adhesion/release, barrier properties), membranes (permeability and selectivity), textile and industrial fibers (stain resistance, soiling, yarn-to-yarn friction, dye reception, dissipation of electrostatic charge), and biomaterials (tissue and blood compatibility). In these and many other fields of applications it is therefore desirable to obtain well-defined and controlled bulk and surface properties in the polymeric materials.

Although the bulk properties of polymers can be carefully controlled through their chemical structures, molecular weights, and morphologies, it is often more difficult to control the surface properties for different applications of the same material. At present, for many industrial uses the preferred method of surface modification involves the posttreatment of finished products by plasma or other chemical or photochemical methods. With this approach one can successfully obtain permanently hydrophilic (polar) or hydrophobic (nonpolar) surfaces, depending on the chemical nature of the treatment, without affecting the bulk properties of the base polymer. However, such processes can be complex and costly because they require postmanufacturing steps. A simpler approach to modify the polymer surfaces is based on the blending of small amounts of specially designed multicomponent polymeric surface active additives with the base resin before or during processing. The low solubility parameters (and resulting incompatibilities with the base polymer) and low surface energies of these additives tend to cause them to phase-separate and to migrate to the polymer–air interface during and/or after processing, to provide the desired surface properties to the system. Since they are usually employed at very low levels, they do not influence the bulk properties of the base polymer. In addition to their applications in polymers, silicone-based copolymers are finding increasing use as surface-modifying additives for fillers (such as titanium dioxide or silica) to enhance their compatibility with polymers and to provide better processing conditions and a more homogeneous distribution of the fillers in organic resins.

Typical surface-modifying additives include organofunctionally terminated silicone oligomers (with reactive or nonreactive terminal groups) or copolymers of silicones (or polydimethylsiloxanes, PDMS) or fluoropolymers with organic polymers. This chapter discusses only silicone-based surface-modifying additives and their performance. Although silicones (or siloxanes) themselves are hydrophobic, through careful design of the backbone structures and compositions of the copolymeric additives it is possible to obtain either hydrophilic or hydrophobic surface properties in the resultant system. Detailed procedures for the preparation of a large number of silicone–organic copolymers that can be used as surface-modifying additives and their chemical and physical properties are provided elsewhere [4].

II. SURFACE PROPERTIES OF SILICONE-CONTAINING COPOLYMERS

Preferential segregation at surfaces and interfaces is a general phenomenon known to occur in essentially all multicomponent copolymeric systems and polymer blends [3,5,6]. The thermodynamic driving force for segregation is the difference between the solubility parameters and the surface energies of the constituents. The component of lowest surface energy is driven to the surface (or interface) to reduce the overall surface and free energies of the system as required by thermodynamics. This migration is known to be enhanced by microphase separation (incompatibility) in the bulk phase. A number of studies in the recent literature describe the theoretical aspects and experimental verification of this phenomenon in a wide variety of polymeric systems [7–13].

Surface activity and incompatibility of silicones and silicone copolymers are a direct result of the low cohesion energy densities (low solubility parameters), fairly large volumes, and excellent backbone flexibilities of PDMS. For linear, high molecular weight PDMS, surface tension is about 20–22 mN/m and the solubility parameter is about 7.3–7.5 $(cal/cm^3)^{1/2}$. These values are much smaller than those for typical organic polymers (with the exception of highly fluorinated polymers), which have surface tensions in the range of 30–50 mN/m and solubility parameters between 8 and 14 $(cal/cm^3)^{1/2}$ [14]. As a result of these large differences, silicones that are copolymerized or blended with organic polymers tend to phase-separate in bulk and also to migrate to the polymer–air interface to form low energy surfaces as favored by thermodynamics.

Surface properties and morphologies of various organofunctionally terminated PDMS oligomers and silicone-containing copolymers have been investigated [4]. Recently, Koberstein and co-workers used pendant drop tensiometry to study the influence of molecular weights and end-group structures on the surface tensions of aminopropyl-, aminopentyl-, methyl-, and hydroxyl-terminated PDMS oligomers [15]. Molecular weights of the oligomers studied (Fig. 1) were varied between 1000 and 75,000 g/mol.

The investigators observed an increase in the surface tension for methyl-terminated oligomers and a decrease for the aminoalkyl-terminated oligomers with increasing molecular weights [15]. There was no significant change in the surface tensions of hydroxyl-terminated PDMS with an increase in the molecular weight.

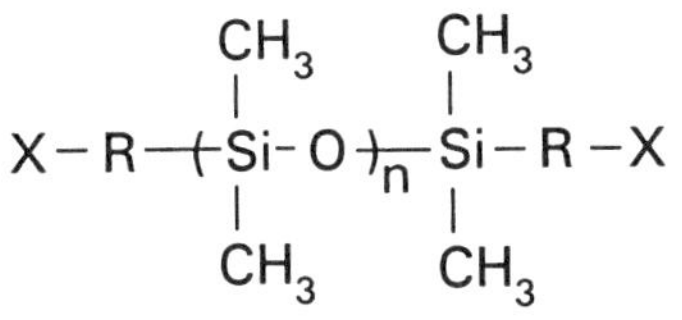

X	R
CH_3	chemical bond
OH	chemical bond
NH_2	$(CH_2)_3$ or $(CH_2)_5$

(n) 10 to 1,000

FIG. 1 Chemical structures of organofunctionally terminated PDMS oligomers.

Koberstein et al. attributed this behavior to the relative polarities of the end groups, noting that nonpolar methyl end groups are preferentially attracted to the surface, whereas highly polar (high surface energy) amine end groups are depleted from the surface. Depletion profiles of alkylamine-terminal groups from the surface, further studied by x-ray photoelectron spectroscopy (XPS or ESCA) to a maximum sampling depth of about 7 nm, were found to be largest for the lowest molecular weight oligomer and to decrease weakly with an increase in molecular weight, which is an expected behavior [16].

Syntheses and surface morphologies of linear multiblock, star block, and graft copolymers of PDMS with a wide variety of organic polymers have also been investigated. Experimental methods employed for the surface characterization of these systems include simple techniques such as contact angle measurements using water or other liquids (a technique sensitive to only the very top molecular layers of the films produced) or more elaborate, instrumental methods, which can be used to obtain quantitative information on the chemical composition and morphology of the surfaces formed and their depth profiles, ranging from about 20 Å to a few micrometers [17]. These methods include attenuated total reflectance Fourier transform infrared (ATR)-FTIR spectroscopy, angle-and energy-dependent ESCA, high resolution electron loss spectroscopy (HREELS), static secondary ion mass spectroscopy (SSIMS), low energy ion scattering spectroscopy (LEISS), forward recoil elastic spectroscopy (FRES), and neutron reflectivity. Among these techniques angle-and energy-dependent ESCA is employed most widely.

Gardella and his group recently used ESCA to investigate the surface composition and morphologies of the following copolymers: diblock and triblock polystyrene–PDMS (PS-PDMS) [18,19], multiblock and star block poly(α-methylstyrene)–PDMS (PMS-PDMS) [20], multiblock polycarbonate–PDMS (PC-PDMS) [21], and diblock nylon 6–PDMS, (N6-PDMS) [22]. Investigations on the chloroform cast films of AB, ABA, and BAB type PS-PDMS copolymers showed that outermost surface region (up to 27 Å) of every copolymer was composed of pure PDMS when PDMS fraction was high (> 0.6). PDMS surface segregation was also dependent on the architecture and segment lengths of the copolymers. A very similar observation was made for PMS-PDMS copolymers. Studies on PC-PDMS systems showed an enrichment in the surface siloxane concentration upon annealing the solvent cast films at 180°C. The surface compositions of semicrystalline N6-PDMS diblock copolymers showed a strong dependence on both the segment molecular weights and the casting solvent used in film preparation. Mixed solvents, which may control the competition between bulk crystallization of nylon 6 versus microphase separation, seemed to influence the surface morphologies in these copolymers. When the films were annealed at 235°C for 10 min, however, the surfaces became covered with a fairly thick layer (100 Å) of PDMS. Similar observations were made in the blends of semicrystalline poly(ethylene terephthalate) (PET) and PDMS copolymers [23,24].

McGrath and coworkers studied the synthesis and bulk and surface characterization of poly(methyl methacrylate)–PDMS graft copolymers [25]. These well-defined copolymers were produced by means of the macromonomer technique. Surface properties of chloroform cast films were investigated by contact angle measurements and ESCA. Depth profiling by ESCA demonstrated the formation of silicone-rich film surfaces with a gradient of composition, dependent on the molar mass of PDMS grafts. In graft copolymers with well-phase-separated bulk morphologies, the surface layer of siloxane was found to be thicker.

Kennedy, Ratner, and coworkers studied the surface properties of PDMS containing segmented polyurethanes by ESCA and SSIMS [26]. Films were prepared by solvent casting. ESCA results, as expected, showed the formation of silicone-rich surfaces. The extent of enrichment was dependent on the ratio of mixed soft segments and the casting solvent used. At a grazing angle of 80°, corresponding to a surface thickness of approximately 20 Å, ESCA showed complete coverage of the film surfaces with PDMS in copolymers containing 20 wt% of PDMS with a molecular weight of 2300 g/mol.

All these studies on multiphase silicone-containing copolymers clearly demonstrate the tendency of silicones to migrate to the surface of the films produced. Although no information is available on the kinetics of migration, which may play a significant role in industrial applications, there are strong indications that it is dependent on the amount and block length of PDMS in the system, differences in the solubility parameters of silicone and the organic segments, method of sample preparation, the solvent used, and the time and temperature of annealing.

III. SILICONE COPOLYMERS AS SURFACE-MODIFYING ADDITIVES IN POLYMER BLENDS

In light of the foregoing discussion, it is conceivable a silicone-containing copolymer blended with an organic polymer under favorable processing conditions would tend to migrate to the polymer–air surface. During this migration it would also carry the organic segments, which it is attached to, near the surface. This sequence gives rise to two important phenomena, which may advantageously be used in the surface modification of polymeric materials through blending with silicone-containing copolymers. First, the organic segments can be designed to be highly miscible with the base polymer, thus providing "anchoring sites" and, as a result, permanent surface modification to the system [4]. If pure PDMS is used as additive, its total immiscibility with the organic polymers would eventually cause it to exude from the system, thus providing only temporary surface modification. Second, organic segments in silicone copolymers can be designed to have specific functional properties (UV stabilizers, antioxidants, antistatic agents, antimicrobials, etc.), therefore, when carried to the surface, they would provide the desired properties to the products more effectively, since usually the degradation starts at the surface.

In this approach the silicone residue in the surface-modifying additive not only is essential for its surface activity but also is desirable owing to the many benefits and properties it may provide to the final system. The most characteristic property modifications of the finished products or blends are improved hydrophobicity or water repellency [27], reduced friction [28], atomic oxygen resistance [29], blood and tissue compatibility [30–32] and flame retardancy [33].

A review of the earlier studies on the use of silicone copolymers as surface-modifying additives was published in 1993 has been cited [23]. Most of the earlier work was focused on the experimental verification of surface segregation of the additives and enrichment of PDMS at the polymer–air surface. The idea of "anchoring groups" has also been identified and studied [4], demonstrating its importance in the achievement of permanent surface modification. More recent studies focused on the quantitative determination of surface composition and morphologies through the use of several newly developed instrumental techniques and the consideration of factors influencing the extent and kinetics of surface segregation. In addition, surface-modifying additives with specific functionalities, such as polymeric stabilizers have been synthesized. Some of these systems are being evaluated for possible commercial applications [34–37].

Polymer surface modification through blending offers several other advantages. Since only very small amounts of additives are used (usually 0.1–3.0 wt%), bulk properties of the base resins are not affected. Compounding and processing of the blends can be performed in solution or in an extruder (depending on the application), with conventional equipment. If melt processing is employed, silicones may also act as lubricants and processing aids and in some cases as internal mold release agents.

A. Formation of Hydrophobic Surfaces

We investigated the surface modification of high density polyethylene (HDPE) by blending with a polycaprolactone–PDMS–polycaprolactone (PCL-PDMS-PCL) triblock copolymer and a PDMS–polyurea segmented copolymer. PCL and PDMS block lengths in the triblock additive were 2000 and 3000 g/mol, respectively. Polyurea additives consisted of 90 wt% of PDMS, which had a segment molecular weight of 10,000 g/mol. Blends were prepared in a twin-screw extruder and contained 1–5% wt% of additive. Compression-molded films were annealed at 60°C for 1 h and kept under vacuum before characterization. The surface behavior of these films was studied by static water contact angle measurements. As can be seen from Table 1, although HDPE itself is fairly hydrophobic (with contact angles around 90°), addition of siloxane copolymers saturates the surfaces with PDMS and results in a substantial increase in contact angles to around 110°. How-

TABLE 1 Influence of Silicone-Based, Surface-Modifying Additives on the Processing Conditions and Surface Properties of HDPE

Sample description[a]	Torque (A)	Screw speed (rpm)	Output (g/min)	Water contact angle (degs)
Virgin HDPE	35	250	73	90
SMA				
1.0%	35	275	108	102
2.5%	30	350	143	105
5.0%	25	350	200	108
TPSU				
1.0%	35	275	132	103
2.5%	35	350	164	110
5.0%	30	350	176	110

[a]SMA, PCL-PDMS-PCL copolymer; TPSU, PDMS-polyurea copolymer.

ever, a more dramatic improvement was observed during the compounding process. The addition of silicone copolymers, which acted as slip agents, served to reduce the torque on the screws of the extruder, as a result, we were able to increase the feed rates and screw speeds, leading to improvements of up to 300% in the output. Similar results were observed for both additives as summarized in Table 1. We later extended this approach to other polyolefins, such as low density polyethylene (LDPE) and polypropylene (PP), and made similar observations, in terms of improvements in both extruder output during processing and surface properties of the final system.

Similar types of PCL-PDMS-PCL triblock copolymer with varying block lengths have been used in the surface modification of a number of organic polymers, such as aromatic polyesters, polyurethanes, epoxies, polyacrylates, and polyolefins, at various additive levels. In each case drastic improvements in the hydrophobicity of the surfaces were observed owing to enrichment by siloxanes, as determined by contact angle and/or surface energy measurements [23,24].

Surface activity of a PS-PDMS copolymer added to a PS matrix has been investigated [38]. Results obtained from wettability and contact angle measurements and from ESCA and SIMS clearly showed the complete coverage of the air–polymer surface with PDMS when 0.2–2.0 wt% of additive is used. A surprising observation, however, was the enrichment of the additive at the interface between PS and the high energy substrates (glass, stainless steel, aluminum etc.) onto which the films have been deposited. Concentration profiles (depletion) of PDMS from surface were consistent with the exponential decrease predicted by the mean field theory, but not with the predictions of scaling theory. The surfactant behavior of the additives was found to depend on the copolymer architecture and the nature of the base resin. Surface modification of PS, PMS, and PC were also subjected to ESCA study by mixing with small amounts of PDMS-containing block copolymers [39]. In all cases enrichment of PDMS at the blend surface was observed, as expected. The influences on the surface properties of the amount and type of the additive, its molecular architecture, and the type of base resin were also investigated. However, very limited results obtained on different blends were far from indicating a trend or providing satisfactory explanations to some important questions regarding the kinetics and mechanisms of surface migration and surface properties. Blends of PDMS–poly(phenylene oxide) (PDMS-PPO) copolymers and PS homopolymers were also studied [40]. Solution-prepared blends containing 0.1–5.5% wt% of additive resulted in clear films owing to the miscibility of PS and PPO. Water contact angle measurements did not show any significant difference between the lowest (0.1%) and highest (5.5%) loading of the surface-modifying additive, clearly indicating that surface saturation by PDMS was at very low levels.

An important application of silicone-based surface modifiers is in the high performance polyester or polyamide release films. Organosiloxane–polyamide triblock copolymers prepared by the melt reaction of carboxy-terminated PDMS and caprolactam, under the catalysis of sodium or lithium hydrides, were blended into polyamides in small amounts (0.1–5 wt%) as surface modifiers [41]. Modified films thus produced showed durable, low energy surfaces with low coefficients of friction and excellent bonding to glass. Recently, Mohajer and coworkers [42] also used a variety of PDMS-containing segmented copolymers for the preparation of release films based on engineering thermoplastics. They found that in addition to modifying surface properties, in nylon 6, these additives also improved some bulk properties of the blend. By use of 1–5 wt% of surface-modifying PDMS copolymer, these workers were able to obtain nylon 6 films with reduced coefficients of friction and good release, better dimensional stability, improved film strength and integrity, and improved water resistance. Such films have applications as release films for printed circuit board fabrication, sheet molding compounds (SMC), and prepregs. PDMS-containing thermoplastic, segmented polyester copolymers were also prepared and used as high strength, high slip films with improved clarity and excellent nonsticking properties [43].

B. Formation of Hydrophilic Surfaces

Although PDMS itself is totally hydrophobic, it is possible to obtain hydrophilic surface properties in blends. This can be achieved by using amphipathic siloxane copolymers, where PDMS is chemically combined with water-soluble organic blocks such as poly(ethylene oxide)glycol (PEO) or poly(2-ethyl-2-oxazoline) (POX). With this approach one can obtain water-wettable surfaces and impart antistatic or soil release properties to the resulting system. The role of silicone in such applications is to carry the additive molecule to the air–polymer interface. Under normal conditions, after processing, blend surfaces will be rich in PDMS; when these modified surfaces come into contact with water or any other polar liquids, however, the polar segments of the additive, which are just beneath the surface layer, will rearrange to reduce the interfacial tension of newly formed interface, yielding a hydrophilic surface to the system.

PEO-containing siloxane copolymers were incorporated into polyamides, which were then spun into fibers. The modified fibers of these blends have shown antistatic and soil release properties superior to those of unmodified fibers [44]. Similar additives were also used to obtain water-wettable surfaces on polypropylene webs, which are inherently hydrophobic [45]. Results of this study suggest that the overall molecular weight of the surface-modifying additive plays a critical role on surface migration and wettability. In the study described, low molecular weight additives (number-average molecular weight $M_n \approx 850$ g/mol) produced water-wettable PP surfaces; on the other hand, higher molecular weight species ($M_n \approx 3000$ g/mol) were found to be ineffective.

We used water contact angle measurements to study the surface behavior of poly(vinyl chloride) (PVC) and poly(methyl methacrylate) (PMMA) homopolymers, blended with small amounts of POX-PDMS-POX triblock copolymers (Fig. 2) [23]. Block length of each segment was 3000 g/mol and the films were cast from THF solution. No annealing was applied. Results of this study are summarized in Table 2.

As can be seen in Table 2, in all blends, when a water droplet was placed onto the surface it first showed a somewhat higher contact angle, indicating the presence of silicone. After about 3 min, however, the contact angle went down to very small values and in some cases there was complete wetting of the surface. This was a clear indication of a dramatic change in the surface structure and morphology of the modified films, which can be explained by the displacement of PDMS moieties by more polar POX segments (or simply a flipping at the surface between blocks of the copolymeric additive). As a control, blends of POX homopolymer ($M_n \approx 3000$ g/mol) with PVC and PMMA were also prepared at same additive lev-

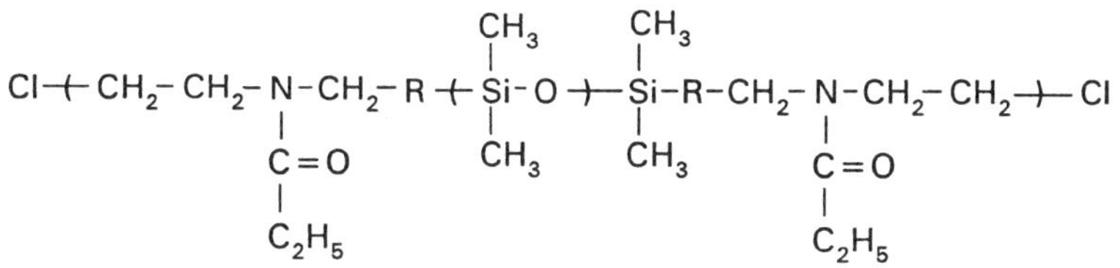

FIG. 2 Chemical structure of poly(2-ethyl-2-oxazoline)–PDMS–poly(2-ethyl-2-oxazoline) triblock copolymers: R, ethylbenzene group.

TABLE 2 Influence of Poly(2-ethyl-2-oxazoline)–PDMS–Poly(2-ethyl-2-oxazoline) Additives on the Surface Properties of PMMA and PVC, Studied by Water Contact Angle Measurements

		Water contact angle (degs)	
Sample description	Additive level (wt%)	1 min	3 min
PMMA	—	66	62
PMMA-POX control[a]	3.0	58	52
PMMA	3.0	41	37
PMMA	5.0	Wetting	Wetting
PMMA	10.0	Wetting	Wetting
PVC	—	83	77
PVC-POX Control[a]	3.0	64	60
PVC	3.0	20	15
PVC	5.0	Wetting	Wetting

[a]Additive is a POX homopolymer ($M_n \approx 3000$ g/mol).

els. Water contact angles of these blends did not indicate any change in the surface properties compared with the respective homopolymers. As given in Table 2, static water contact angles after 3 min for PVC and PMMA homopolymers were 77 and 62°; for their blends with 3 wt% of POX, these angles were 60 and 52°, respectively. This shows that in the absence of a carrier such as PDMS, which would move it to the surface, POX homopolymer added into the system is homogeneously distributed in the bulk of the polymer. These results clearly support the important role played by the silicone backbone in the additives for surface migration and surface modification in polymers.

C. Surface Active Silicone Copolymers as Polymeric Stabilizers

Silicone copolymers with UV-stabilizing allyloxytetramethylpiperidine side chains were prepared, characterized and blended with PP at 100, 250, and 500 ppm levels to compare their performance with conventional stabilizers [35]. Chemical structures of these additives are given in Fig. 3. As shown in Table 3, accelerated UV degradation studies showed that at 100 ppm additive levels there was about a 6-fold improvement in the UV stability of the system compared with unmodified systems. This went up to 14-fold when 500 ppm stabilizer was used. At the same levels, (100 and 500 ppm), conventional stabilizers provided 3.5- and 7-fold improvements, respectively. This demonstrates the effectiveness of PDMS-bound stabilizers and their superiority to conventional systems. This is a direct result of the enrichment of the silicone-based UV stabilizers on the surface of the films formed, whereas the conventional organic additives were distributed in the

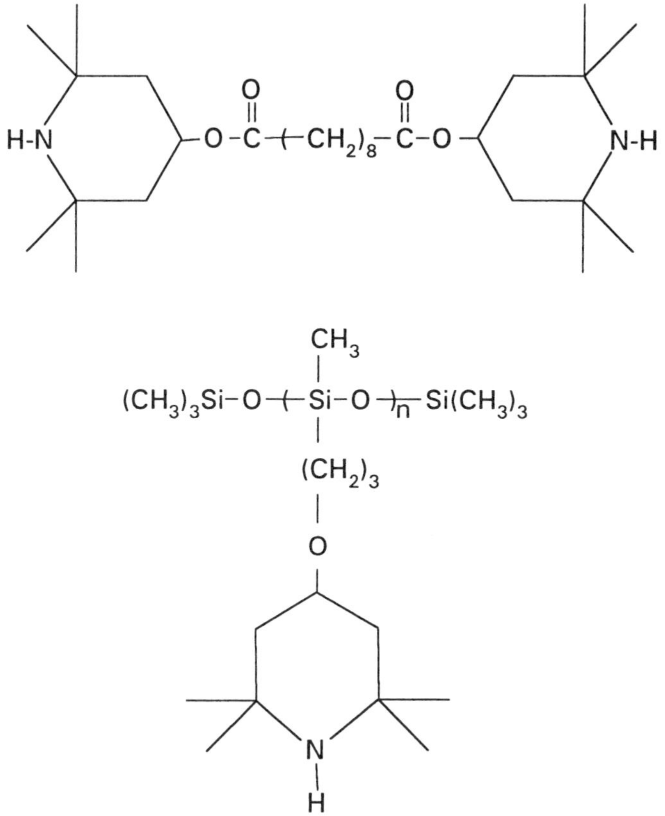

FIG. 3 Chemical structures of allyloxytetramethylpiperidine type conventional organic and silicone-based UV stabilizers.

bulk. Similar results were reported by Maycock and coworkers, who used benzotriazolyl-containing siloxane copolymers as UV stabilizers for various thermoplastic resins [37]. All these studies clearly indicate the potential use of PDMS-bound UV stabilizers in commercial applications, where performance improvements can be obtained at a reduced cost. Synthesis and characterization of various PDMS copolymers with other UV-absorbing groups were also reported [34].

Graft polysiloxanes containing sterically hindered phenolic antioxidants (Fig. 4) have also been prepared and used [36]. It was shown that because of their polymeric nature, these additives result in lower volatility, are easily blended with polyolefinic base resins, and exhibit good resistance to extraction by various solvents. When compounded with PP at low levels, they also show excellent improvement in the thermal stability of the blend.

TABLE 3 Comparison of the Effect of Conventional and Silicone-Based UV Stabilizers on Brittling Times of PP Films

	Stabilizer			
Sample	Type[a]	(ppm)	Brittling time (h)	Relative efficiency (%)
1	—	—	150	100
2	M	100	520	350
3	M	250	780	520
4	M	500	1100	730
5	P	100	850	560
6	P	250	1250	830
7	P	500	2120	1400

[a]M, conventional organic stabilizer; P, silicone-based stabilizer. Chemical structures are given in Fig. 3.
Source: Ref. 35.

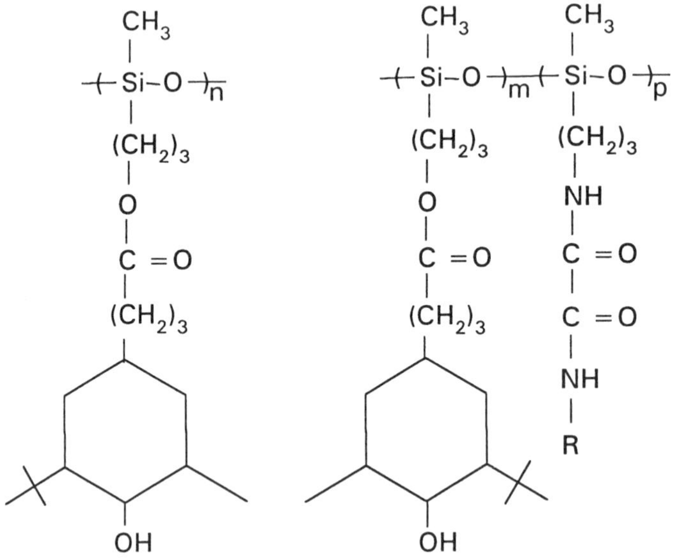

FIG. 4 Graft polysiloxanes containing sterically hindered phenolic antioxidants.

IV. CONCLUSIONS

This chapter reviewed recent developments and results of some of our original work on the use of siloxane-containing copolymers for the surface modification of organic polymers, including some new potential commercial applications. In this approach a small amount (usually 0.1–3 w%) of a specially designed PDMS-containing multiphase copolymer is blended with an organic polymer and the mixture is processed by conventional melt or solution techniques depending on the application. The resultant product shows enrichment of PDMS at the polymer–air surface. This effect is due to the low solubility parameter and low surface energies of PDMS compared with organic polymers, which result in phase separation in bulk and favor the migration of the additive to the newly formed surface. The factors that influence the kinetics of formation of modified surfaces and surface migration are not completely understood; among those identified thus far are amount and molecular weight of silicone in the copolymer; copolymer architecture; type and nature of the base polymer, its morphology (amorphous or crystalline), and transition temperatures (T_g, T_m); processing methods; and conditions and annealing temperature and time of the finished product. A number of research groups are working on better understanding of this complex relationship.

Although the use of silicone copolymers as surface-modifying additives has been mainly an academic interest for a long time, it is worthwhile to note the number of patents that have been issued recently on the development of new additives with specific functions and their possible industrial applications. Compared with conventional organic stabilizers, which are distributed mainly in the bulk of the polymers, silicone-based additives containing various stabilizing groups are concentrated near the surface of the polymers, where the degradation usually starts. Thus they provide more effective stabilization and protection to the system. In addition to the fields discussed here, biomaterials, especially those serving in implantable polyurethanes, represent another area of continuing interest for the use of silicone-based surface-modifying additives.

REFERENCES

1. F. Garbassi, M. Morra, and E. Ochiello, *Polymer Surfaces: From Physics to Technology*, Wiley, New York, 1993.
2. Y. Ikada, *Lubricating Polymer Surfaces*, Technomic Publishing, Lancaster, PA, 1993.
3. I. C. Sanchez, *Physics of Polymer Surfaces and Interfaces*, Butterworths, Boston, 1992.
4. I. Yilgör and J. E. McGrath, *Adv. Polym. Sci. 86*, 1 (1988).
5. P. Cifra, F. E. Karasz, and W. J. MacKnight, *Macromolecules 25*, 4895 (1992).
6. M. R. Vilar, A. M. Bothelho do Rego, J. Lopez da Silva, F. Abel, V. Quillet, M. Schott, S. Petitjean, and R. Jerome, *Macromolecules 27*, 5900 (1994).
7. J. J. O'Malley, R. H. Thomas, and G. M. Lee, *Macromolecules 12*, 996 (1979).
8. N. M. Patel, D. W. Dwight, J. L. Hedrick, D. C. Webster, and J. E. McGrath, *Macromolecules 21*, 2689 (1988).
9. A. N. Semenov, *Macromolecules 25*, 4967 (1992).
10. S. H. Anastasiadis, T. P. Russel, S. K. Satija, and C. F. Majkrzak, *Phys. Rev. Lett. 62*, 1852 (1989).
11. T. P. Russel, *Mater. Sci. Rep. 5*, 171 (1990).

12. G. Coulon, B. Collin, D. Ausseree, D. Chatenay, and T. P. Russel, *J. Phys. Fr. 51*, 2801 (1990).
13. J. C. Dijt, M. A. Cohen Stuart, and G. J. Fleer, *Macromolecules 27*, 3229 (1994).
14. J. Brandrup and E. H. Immergut, eds., *Polymer Handbook*, Wiley, New York, 1975.
15. C. J. Jalbert, J. T. Koberstein, R. Balaji, Q. Bhatia, L. Salvati, Jr., and I. Yilgör, *Macromolecules 27*, 2409 (1994).
16. C. J. Jalbert, J. T. Koberstein, I. Yilgör, P. Gallagher, and V. J. Krukonis, *Macromolecules 26*, 3069 (1993).
17. L. Sabbatini and P. G. Zambonin, eds., *Surface Characterization of Advanced Polymers*, VCH, New York, 1993.
18. X. Chen, J. A. Gardella Jr., and P. Kumler, *Macromolecules 25*, 6621 (1992).
19. X. Chen, J. A. Gardella Jr., and P. Kumler, *Macromolecules 25*, 6631 (1992).
20. X. Chen, J. A. Gardella Jr., and P. Kumler, *Macromolecules 26*, 3778 (1993).
21. X. Chen, H. F. Lee, and J. A. Gardella Jr., *Macromolecules 26*, 4601 (1993).
22. X. Chen, J. A. Gardella Jr., and R. E. Cohen, *Macromolecules 27*, 2206 (1994).
23. I. Yilgör, E. Yilgör, and B. Grüning, *Tenside Surf. Detergents 30*, 158 (1993).
24. I. Yilgör, W. P. Steckle Jr., E. Yilgör, R. G. Freelin, and J. S. Riffle, *J. Polym. Sci. A: Polym. Chem. 27*, 3673 (1989).
25. S. D. Smith, J. M. DeSimone, H, Huang, G. York, D. W. Dwight, G. L. Wilkes, and J. E. McGrath, *Macromolecules 25*, 2575 (1992).
26. S. C. Yoon, B. D. Ratner, B. Ivan, and J. P. Kennedy, *Macromolecules 27*, 1584 (1994).
27. J. B. Thompson and M. J. Owen, German Patent, 2,120,961 to Midland Silicones Corp. (1971).
28. E. M. Yorgkitis. C. Tran, and N. S. Eiss Jr.; T. Y. Hu, I. Yilgör, G. L. Wilkes, and J. E. McGrath, in *Rubber-Modified Thermoset Resins* (C. K. Riew and J. K. Gillham, eds.), *ACS Advances in Chemistry Ser., No. 208*, American Chemical Society, Washington, DC 1984, Chapter 10.
29. I. Yilgör, in *Adhesives, Sealants and Coatings for Space and Harsh Environments* (L. H. Lee, ed.), Plenum Press, New York, 1988, p. 249.
30. R. S. Ward, U.S. Patent, 4,675,361 to Thoratec Laboratories Corp. (1987).
31. F. Lim, C. Z. Yang, and S. L. Cooper, *Biomaterials 15*, 408 (1994).
32. R. W. Hergenrother, X.-H. Yu, and S. L. Cooper, *Biomaterials 15*, 635 (1994).
33. G. L. Gaines and D. G. LeGrand, U.S. Patent, 3,686,355 to GE Corp. (1972).
34. H. J. Riedel and H. Hocker, *J. Appl. Polym. Sci. 51*, 573 (1994).
35. S. Constanzi, L. Cassar, C. Busetto, C. Neri, and D. Gussonui, EU Patent, 343,717 to Enichem (1989).
36. C. Neri, D. Fabbri, R. Farris, and L. Pallini, *Polyolefins, Proceedings of the Eighth International Conference*, Houston, TX, 1993, pp. 428–440.
37. W. E. Maycock, R. S. Nohr, and G. J. MacDonald, U. S. Patent, 4,859,759 Kimberly Clark Corp. (1989).
38. S. Petitjean G. Ghitti, R. Jerome, P. Teyssie, J. Marien, J. Riga, and J. Verbist, *Macromolecules 27*, 4128 (1994).
39. X. Chen and J. Gardella Jr., *Macromolecules 27*, 3363 (1994).
40. R. D. Allen and J. L. Hedrick, IBM Research Report RJ 5639 (56949) (1987).
41. J. B. Thompson and M. J. Owens, U.S. Patent, 3,723,566 (1973).

42. Y. Mohajer, T.-L. H. Nguyen, V. R. Sastri, A. Degrassi, and E. L. Belfoure, PCT, Int. Publ. No. WO 91/15538 to Allied Signal Inc. (1991).
43. P. R. Ginnings, U.S. Patent, 4,496,704 to Goodyear Tire and Rubber Co. (1985).
44. P. A. Leeming, G.B. Patent 1,197,567 (1967).
45. R. S. Nohr and G. J. MacDonald, U.S. Patent, 4,923,914 to Kimberly Clark Corp. (1990).

11

Surfactant-Enhanced Spreading

T. STOEBE Eastman Kodak Company, Rochester, New York

RANDAL M. HILL Central Research and Development, Dow Corning Corporation, Midland, Michigan

MICHAEL D. WARD and L. E. SCRIVEN Department of Chemical Engineering and Materials Science, University of Minnesota, Minneapolis, Minnesota

H. TED DAVIS Dean, Institute of Technology and Department of Chemical Engineering and Materials Science, University of Minnesota, Minneapolis, Minnesota

I. INTRODUCTION

Spreading and wetting of solid substrates by liquids is a process of significant technological importance. Agents that promote effective wetting are quite useful for many industrial products, including coatings, cosmetics, agrochemicals, lubricants, and dispersants. However, the mechanism of the spreading process has remained elusive, preventing the rational design of surfactant formulations and limiting the development of these systems to empirical approaches. Studies providing insight into the mechanisms and dynamics of wetting therefore are of considerable interest.

In general, a liquid will wet a solid–vapor interface only if the spreading coefficient S is positive:

$$S = \gamma_{SV} - \gamma_{SL} - \gamma_{LV} \tag{1}$$

Where γ_{SV}, γ_{SL}, and γ_{LV} refer to the solid–vapor, solid–liquid, and liquid–vapor interfacial tensions, respectively. Surfactants can promote spreading by decreasing the liquid–vapor and/or solid–liquid interfacial tensions, causing the spreading coefficient to become more positive. For example, water has a relatively large surface tension ($\gamma_{LV} \approx 72.5$ mN/m), and wetting is inhibited over lower energy hydrophobic surfaces. Addition of surfactants can reduce this surface tension by as much as 50 mN/m, and so the resulting aqueous mixtures are much more likely to wet low energy surfaces. A positive spreading coefficient does not ensure rapid spreading, however, and this simple description cannot predict the rate (often of critical practical importance) at which a given surfactant solution will wet a particular substrate. Furthermore, solid–liquid interfacial tensions (actually, surface excess free energies) are not easily determined, which leads to uncertainties in the spreading coefficient. Furthermore, spreading is clearly a dynamic process, and dynamic interfacial tensions are probably more relevant than static values in many cases.

The need to better understand spreading dynamics has prompted experimental investigations of a number of aqueous surfactant solutions and dispersions. Early work was aimed at understanding spreading dynamics of aqueous surfactant solutions on strongly hydrophilic surfaces [1]. Recently, Zabkiewicz and Gaskin [2], Gaskin and Kirkwood [3], and Knoche et al. [4] reported that the trisiloxane surfactant denoted $M(D'E_8OMe)M$,

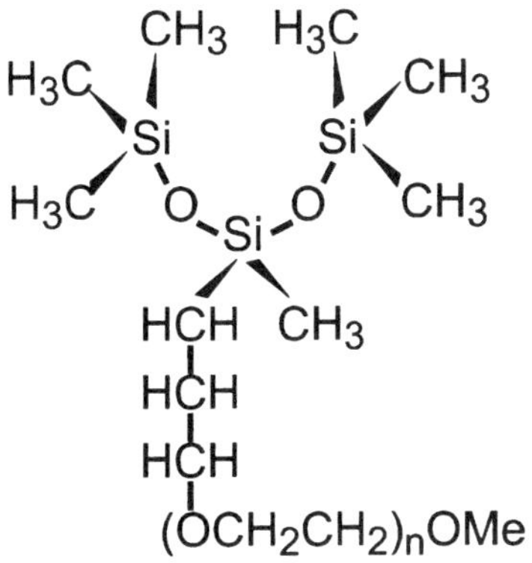

where M = $(CH_3)_3SiO$, $E_n = (OCH_2CH_2)_n$, Me = CH_3, and D′ = —Si—CH_3R, can be used as a very effective wetting agent for water-based herbicides on fairly *hydrophobic* plant leaves. Subsequent work identified other siloxane-based surfactants that greatly enhance the ability of aqueous solutions and dispersions to spread rapidly over highly hydrophobic substrates. These low molecular weight, nonionic surfactants share similar structural elements—namely, a trisiloxane hydrophobic head group coupled to a poly(oxyethylene) hydrophilic tail. The poly(oxyethylene) chain is generally polydisperse and the reported number, *n*, represents a mean value. The E_n sequence can be capped by different terminal groups, denoted explicitly as OH (hydroxyl), OMe (methoxy), or OAc (acetoxy) (see Table 1). Disiloxane, MM′E_n, and and cyclosiloxane structures, $D_3D'E_n$, $D_4D'E_n$, have also been claimed to promote rapid spreading. Note, however, that many of the surfactants discussed in this chapter form turbid (two-phase) dispersions in water, even at low concentrations, and cannot be described simply as aqueous solutions.

The use of organosilicone surfactants as adjuvants to enhance the efficacy of agrochemicals such as herbicides, fungicides, and foliage-applied fertilizer has recently been reviewed [5–7]. Such adjuvants allow agrochemicals to be used at lower levels, which is desirable for both economic and environmental reasons [5]. The reviews have focused on equilibrium and dynamic surface tensions, contact angles on hydrophobic surfaces, and spread areas relative to water (after some fixed time period), the correlation of these surface properties with greenhouse and field performance, and studies of foliar uptake and the penetration of certain agrochemicals into plant leaf surfaces. Many siloxane surfactants have been studied, but the primary interest has been in the trisiloxane surfactants, M($D'E_8$OR)M, with $n = 7$, 8, and various end caps.

The wetting efficacy of surfactants as agricultural adjuvants is often discussed in terms of spread areas relative to water—the ratio of the area of a spread droplet

TABLE 1 Surfactant Nomenclature and Structure

Surfactant[a]	Manufacturer[b]	Structure[c]
$M(D'E_4OH)M$	DCC	$(Me_3SiO)_2Si(Me)(CH_2)_3(OCH_2CH_2)_4OH$
$M(D'E_8OH)M$	DCC	$(Me_3SiO)_2Si(Me)(CH_2)_3(OCH_2CH_2)_8OH$
$M(D'E_8OMe)M$	OSi	$(Me_3SiO)_2Si(Me)(CH_2)_3(OCH_2CH_2)_8OMe$
$M(D'E_8OAc)M$	DCC	$(Me_3SiO)_2Si(Me)(CH_2)_3(OCH_2CH_2)_8OAc$
$M(D'E_{12}OH)M$	DCC	$(Me_3SiO)_2Si(Me)(CH_2)_3(OCH_2CH_2)_{12}OH$
$MDM'E_8OH$	DCC	$Me_3SiOSi(Me_2)OSi(Me)(CH_2)_3(OCH_2CH_2)_8OH$
SDS	Fisher	$CH_3(CH_2)_{11}OSO_3Na$
AOT	Fluka	$[C_4H_9CH(C_2H_5)CH_2OOCCH_2]_2CH—SO_3^-Na^+$
DTAB	Kodak	$CH_3(CH_2)_{11}N(CH_3)_3Br$
DDAB	Fluka	$(CH_3(CH_2)_{11})_2N(CH_3)_2Br$
$C_{10}E_3$	Nikko	$CH_3(CH_2)_9(OCH_2CH_2)_3OH$
$C_{12}E_3$	Nikko	$CH_3(CH_2)_{11}(OCH_2CH_2)_3OH$
$C_{12}E_4$	Nikko	$CH_3(CH_2)_{11}$ $(OCH_2CH_2)_4OH$
$C_{12}E_5$	Nikko	$CH_3(CH_2)_{11}$ $(OCH_2CH_2)_5OH$
C_8G_1	Fluka	$CH_3(CH_2)_7$-β-sD-Glucopyranoside

[a]Me, CH_3; M, Me_3SiO; D′, $Si(Me)(CH_2—$; E_n, $(OCH_2CH_2)_n$; Ac, OC(C=O)CH_3.
[b]DCC; Dow Corning Corporation, Midland, MI; OSi, OSi Specialties, Inc., Nikko, Nikko Chemicals, Tokyo.
[c]While the poly(oxyethylene) chains of the siloxane compounds are polydisperse, the dodecyl nonionic surfactants are monodisperse.

of the surfactant solution after some period of time (e.g., 45 s) to the area of a (nonspreading) droplet of water. Spread areas are much larger for certain trisiloxane surfactants than for conventional organic surfactants, polymeric organic surfactants, or fluorocarbon surfactants [8]. Although dynamic and equilibrium surface tensions are frequently discussed in this context, there is little correlation between either of these quantities and agricultural-adjuvancy.

The efficient wetting of hydrophobic leaf surfaces by aqueous formulations containing organosilicone surfactants generally has been attributed to the ability of these compounds to reduce surface tension values well below what is possible with conventional organic surfactants. (Again, however, not all surfactants dispersed in water are simple aqueous solutions.) A synergy has been claimed for mixtures of siloxane surfactants with organic surfactants [9–11]. Hill [12] reported the critical micelle concentrations for several mixtures containing siloxane surfactants and anionic, cationic, and nonionic surfactants, and the detergency properties of mixtures of a limited number of siloxane surfactants and organic surfactants. Most agrochemical formulations contain organic surfactants in addition to the siloxane surfactant, but it has been shown that some organic surfactants are antagonistic to both surface tension lowering and spreading [8,3].

Early studies of the wetting dynamics of aqueous surfactant dispersions and solutions by Ananthapadmanabhan et al. [14] and Zhu et al. [15] were performed on either Parafilm or polyethylene, which are very hydrophobic and similar in chemical composition. We refer to substrates subtending a large sessile contact angle, $\theta > 90°$ ($\cos\theta < 0$), relative to 18 MΩ water as very hydrophobic (Fig. 1a). Surfaces exhibiting smaller contact angles, $\theta < 90°$, will be distinguished as increasingly hydrophilic. The authors cited above reported that only turbid two-phase dispersions of specific siloxane surfactants exhibited rapid wetting on these surfaces. Surfactants promoting rapid spreading on these hydrophobic surfaces were described as "superspreaders." The "superspreading" behavior was suspected to be related both to the unusual molecular geometry of these surfactants and to the peculiar microstructure of the dispersed second phase [16]. Subsequent studies were performed on self-assembled monolayers, which can be used to vary systematically the substrate surface energy from highly hydrophobic to strongly hydrophilic (Fig. 1b). These studies revealed that many of the previously reported conclusions, which were based on the limited data obtained on Parafilm and polyethylene, were not valid for higher energy (more hydrophilic) substrate surfaces [17]. These observations indicated that "superspreading" was a particular exam-

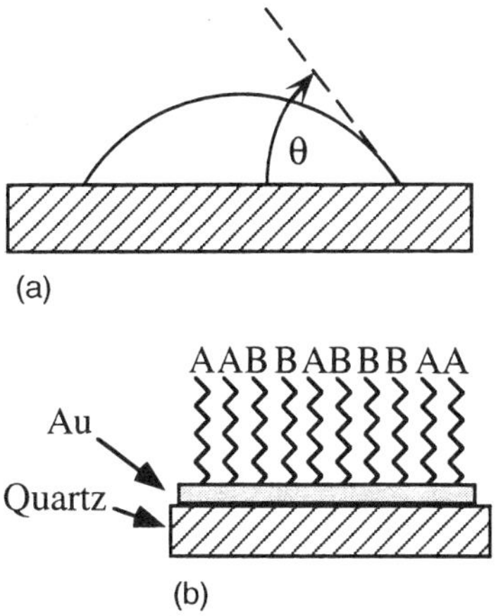

FIG. 1 (a) Schematic representation of a liquid droplet on a solid substrate and the contact angle, θ. Substrates exhibiting large values for θ ($\theta \approx 90°$, $\cos\theta \approx 0$) are termed very hydrophobic. Decreasing values of ω corresponding to substrates that are increasingly hydrophilic. (b) Schematic representation of a self-assembled alkane thiol mixed monolayer modified gold surface: A, $HS(CH_2)_{11}CH_2OH$; B, $HS(CH_2)_{11}CH_3$. The A terminal groups are very hydrophobic. Incorporation of B terminated alkane thiols yields increasingly hydrophilic substrates.

ple of a more general phenomenon, which we described subsequently as "surfactant-enhanced spreading." This term was adopted to describe rapid wetting of lower energy surfaces that were not wetted in the absence of surfactant. In addition to a series of $M(D'E_nOMe)M$ surfactants, nonionic ethoxylated dodecyl surfactants $C_{12}E_n$ [$C_{12} = CH_3(CH_2)_{11}$, $E_n = (OCH_2CH_2)_nOH$, $n = 3, 4, 5$], and glycoside surfactants C_8G_1, $C_{10}G_2$ [$C_n = CH_3(CH_2)_{n-}1$, G = β-D-glucopyranoside] [18] were also found to enhance spreading in aqueous two-phase dispersions and micellar solutions under certain conditions. These results indicated that neither the trisiloxane hydrophobe nor the poly(oxyethylene) hydrophile is peculiar to the observation of greatly increased spreading rates. Moreover, there appears to be no obvious connection between the observed aggregation behavior and microstructure of the surfactant and the enhanced spreading. Interestingly, conventional ionic surfactants failed to exhibit enhanced spreading as aqueous solutions over any of the substrates tested. Despite this new insight, the mechanism responsible for the rapid spreading has not yet been elucidated.

This chapter conveys observations we have made during recent investigations of surfactant-enhanced spreading. Section II describes results obtained on Parafilm and polyethylene substrates, Sec. III describes surface energy modification by preparation of organosulfur monolayers on solid substrates, Sec. IV introduces two experimental techniques recently used in our laboratory to obtain spreading rates, Sec. V describes spreading behavior of aqueous dispersions and solutions of trisiloxane surfactants, Sec. VI describes the spreading characteristics of aqueous solutions of ethoxylated dodecyl surfactants, Sec. VII describes the behavior of glucoside surfactants, Sec. VIII describes the spreading of ionic surfactant solutions, and Sec. IX briefly describes possible mechanisms for the observed spreading rate data.

II. WETTING RESULTS ON PARAFILM AND POLYETHYLENE

The work of Ananthapadmanabhan et al. [14] represents a pioneering effort in the investigation of the spreading behavior of several surfactant dispersions and solutions on Parafilm. Using the Wilhelmy plate technique, these researchers discovered that aqueous mixtures of $M(D'E_8OMe)M$ exhibit a break in the surface tension–surfactant concentration curve, which they attributed to the onset of aggregation near 0.007 wt%. However, the microstructure of the aggregates was not determined. Turbidity and fluorescence studies indicated that the aggregates were not small spherical micelles. Above this critical aggregation concentration, at approximately 1 wt%, the measured surface tension was 20.5 mN/m. This value is remarkably lower than that obtained when more conventional ionic or nonionic hydrocarbon-based surfactants were used, and it was presumed to be a factor in the wetting of these dispersions on even very hydrophobic substrates.

These studies were performed by depositing a 50 μL drop of 0.1 wt% aqueous surfactant dispersions and solutions on flat sheets of Parafilm and measuring the largest diametric dimension of the spreading drop after 5 min. The data, obtained in open air at room temperature, are provided in Table 2 [14]. The spreading factor was calculated as the ratio of the diameter of the surfactant-containing mixture to the diameter realized by a drop of doubly distilled water of equal volume. The siloxane surfactants were clearly the most effective wetting agents, even though perfluorinated surfactant mixtures exhibited still lower surface tensions (≈ 16 mN/m). Spreading on low energy substrates therefore is not solely dependent on surface tension, suggesting that adsorption at the liquid–vapor and liquid–solid interfaces and kinetic effects associated with surfactant diffusion are also important.

Kinetic effects have been investigated by observing the wetting of polyethylene powder and by dynamic surface tension measurements [14]. These effects were surmised from the time dependence of the surface tension immediately after the removal by suction of the polyethylene powder and its adsorbed surfactant layer. The surface of (M(D′E_8OMe)M dispersions was found to be replenished by nearly instantaneous (within 0.5 s) diffusion of molecules from the bulk phase. These results and the observed fast wetting of the polyethylene powder demonstrated the high mobility and rapid solid–liquid adsorption of these surfactants in aqueous systems.

Some of these properties may be related to the unusual molecular structure of M(D′E_8OMe)M. The trisiloxane hydrophobe has a structure quite different from those of conventional organic surfactants (a numerical comparison is given below: Sec. VI). Ananthapadmanabhan et al. [14] suggested that this structure plays a

TABLE 2 Relative Spreading Ability on Parafilm of Various 0.1 wt% Aqueous Surfactant Mixtures Measured Under Ambient Conditions

System	Spreading factor[a]	Surface tension (dyn/cm)
Water	1	72.8
M(D′E_8OMe)M	8.6	20.5
Fluorocarbon surfactant, FSA[b]	2	16.2
Nonionic surfactant, Tergitol NP-10[c]	1.7	31.1
SDS	1.2	44.3

[a]The spreading factor represents the ratio of the area covered by the surfactant mixture relative to a similar droplet of pure water after 5 min. As pointed out in the text, many surfactants form turbid (two-phase) dispersions in water, even at low concentrations.
[b]DuPont Corporation.
[c]Union Carbide Corporation.

central role in determining the rapid spreading/wetting behavior. They assert that the branched trisiloxane hydrophobe facilitates a kind of tractor-tread motion at the spreading edge, while conventional surfactants impede this motion.

Ananthapadmanabhan et al. [14] also proposed that unusually rapid surfactant adsorption onto the substrate was necessary for the unique wetting properties of the trisiloxane surfactants. Based on results of Hardy [19], it was suggested that a thin precursor film spreads ahead of the main surfactant drop. Fast surfactant adsorption onto the substrate at the leading edge causes a reduction in the precursor film surfactant concentration, leading to a local liquid–vapor interfacial tension that is greater than the main drop. Because of this surface tension gradient, bulk flow is induced in the direction of the precursor film by the Marangoni effect. The droplet therefore spreads as it continually tries to catch up to the precursor film. While this argument is appealing, it is incomplete. The formation and maintenance of the precursor film are not explained, and the apparent uniqueness of the siloxane surfactants is related to unusual kinetics, which are somehow dependent on their peculiar molecular structure.

Subsequent to the work of Ananthapadmanabhan et al. [14], Zhu et al. [15] systematically investigated the wetting properties of aqueous dispersions and solutions of six siloxane surfactants, including $M(D'E_8OMe)M$. The spreading behavior on Parafilm was studied by measuring the area wet by a test droplet as a function of time. Experiments were performed to observe the spreading dependence on surfactant concentration, substrate surface roughness, relative humidity, and the size of the aggregates in turbid dispersions. The density and viscosity of these dilute dispersions and solutions were also obtained and found not to differ substantially from pure water: $\rho \approx 1.00$ g/cm^3, $\eta \approx 1$ cP.

Zhu et al. [15] reported that during the initial spreading period ($t < 2$ min), the area of Parafilm covered by fixed volume (≈ 8 μL) of aqueous $M(D'E_8OMe)M$ test mixture droplets increased linearly with both time and surfactant concentration for 0.007 wt% $< C <$ 0.16 wt%. The initial spreading rates decreased for $C >$ 0.16 wt%, but the final (limiting) covered areas were proportional to surfactant concentration up through at least 0.3 wt%. The limiting area was also directly proportional to the initial droplet volume. It is important to recall that the solubility limit of this compound is relatively low (0.007 wt%) and that the static surface tension was independent of concentration above 0.007 wt%. The apparent maximum in the initial spreading rate at 0.16 wt% is therefore highly intriguing and not yet well understood.

Zhu et al. [15] also discovered a strong spreading rate dependence on the size of the surfactant particles in aqueous $M(D'E_8OMe)M$ dispersions. By performing dynamic light scattering experiments, they determined that sonication narrows the particle size distribution compared to dispersions prepared by simply hand shaking and that the sonicated dispersions exhibited an average particle radius of approximately 260 Å, which is smaller than the 1370 Å radius characteristic of hand-shaken 0.16 wt% dispersions. These particle size distributions were found to depend only weakly on surfactant concentration. The sonicated dispersions spread considerably faster on Parafilm (by roughly a factor of 2.5) than the hand-shaken samples. Since the sonication did not appreciably alter either the viscosity or the surface tension of the dispersions, these results illustrate the importance of the nature of the dispersed second phase.

Stretched Parafilm has a considerably rougher surface than unstretched Parafilm. Interestingly, Zhu et al. [15] noted that spreading of aqueous M(D′E_8OMe)M dispersions occurred roughly four times faster on stretched Parafilm than on unstretched Parafilm but that the final covered area was reduced by about a factor of 2. M(D′E_8OMe)M aqueous dispersions did not exhibit rapid spreading on Parafilm in dry air, but the spreading rates increased with relative humidity above 20% with the fastest spreading observed at 100% RH. This suggested the involvement of a preexisting water film adsorbed on the Parafilm surface, which could relax the no-slip boundary condition associated with a dry interface. A preexisting water film would have a larger surface energy than a film of the aqueous-droplet-containing surfactant, and the resulting surface tension gradient would produce a strong Marangoni force that would facilitate the spreading of lower surface tension surfactant solution. However, it seems unlikely that such effects would be specific to the trisiloxane surfactants. Moreover, the existence of a water film on a hydrophobic surface like Parafilm had not been demonstrated.

By comparing the spreading behavior of aqueous dispersions and solutions of M(D′E_8OMe)M and five other silicone surfactants on Parafilm, Zhu et al. [15] were able to draw some intriguing conclusions. Two of the other surfactants investigated, M(D′E_{12}OH)M and M(D′E_{18}OH)M, form transparent micellar solutions. The remaining compounds [M(D′E_8OMe)M, M(D′E_5OH)M, M(D′E_8OAc)M, and M(D′E_8OH)M] form turbid dispersions at low concentration (<0.01 wt%). Rapid spreading on Parafilm, termed "superspreading" by Zhu et al. [15], was observed to occur only in the case of the turbid dispersions. The presence of a dispersed surfactant-rich phase alone was not sufficient to promote superspreading, however: aqueous dispersions of the somewhat analogous surfactants $C_{12}E_3$ and $C_{12}E_4$ did not effectively wet Parafilm. While this still suggests a special character for the trisiloxane hydrophobe, Hill et al. [16] demonstrated that its molecular geometry was not critical, since the linear siloxane surfactant MDM′E_8 also promoted superspreading. Based on these results, superspreading appeared to be unique to turbid dispersions of trisiloxane surfactants, and the spreading rate depended strongly on the size (and presumably the microstructure) of the dispersed aggregates, substrate roughness, and ambient relative humidity (see Table 3). Subsequent work performed in our laboratories, however, has demonstrated that some of these conclusions are not valid when spreading occurs on more hydrophilic substrates.

TABLE 3 Spreading Rate Results of 0.4 wt% Aqueous Surfactant Mixtures on Parafilm at 22°C and > 95% Relative Humidity, Obtained Via the Videobased Technique (see text)

Surfactant	Turbid?	Rate (mm^2/s^1)
$M(D'E_4OH)M$	Yes	5.85
$M(D'E_8OH)M$	Yes	4.95
$M(D'E_8OMe)M$	Yes	11.3
$M(D'E_8OAc)M$	Yes	4.1
$M(D'E_{12}OH)M$	No	<0.1
SDS	No	<0.1
DTAB	No	<0.1
$C_{12}E_3$	Yes	<0.1
$C_{12}E_4$	Yes	<0.1
$C_{12}E_5$	No	<0.1
C_8G_1	No	<0.1
$C_{10}E_3$	No	<0.1

III. SURFACE ENERGY MODIFICATION BY SELF-ASSEMBLED MONOLAYERS

The studies discussed above were restricted to Parafilm and polyethylene substrates. Substrate surface energy, which directly influences the spreading coefficient, should strongly influence the dynamics of wetting. Therefore, investigating the spreading rate dependence on substrate surface energy should shed important light on the seemingly unique behavior of aqueous dispersions of the trisiloxane surfactants.

Gold substrates can be readily modified by the deposition of organosulfur monolayers with different terminal functionalities [20]. Ideally, monolayers form a close-packed carpet that is strongly bound to the gold surface and exposes purposely introduced terminal groups (see Fig. 1b). Infrared spectroscopy and backscattering studies have revealed that alkane thiol monolayers terminated with methyl groups have crystalline order, forming a commensurate $\sqrt{3} \times \sqrt{3}$ structure on Au(111) surfaces, which is the predominant plane on a polycrystalline gold surface [21]. The chains tilt roughly 30° with respect to the surface normal to attain close packing. Mixed monolayers generally are less well ordered, which has hindered detailed characterization of these systems. However, it has been demonstrated amply that wetting of these surfaces by aqueous droplets is sensitive to the composition of the monolayers with respect to the terminal functional groups, with no obvious dependence on crystalline order or domain formation. Furthermore, recent investigations suggest that mixed monolayers have true molecular mixing [22]. Terminal groups such as methyl (CH_3) and hydroxyl (OH) groups produce hydrophobic and hydrophilic surfaces, respectively.

Mixed monolayers can be created by immersing gold substrates into 1 mM ethanol solutions containing different relative amounts of $HS(CH_2)_{14}CH_2OH$ and $HS(CH_2)_{15}CH_3$, which enables fabrication of a series of substrates ranging from very hydrophobic ($\theta > 90°$, $\cos\theta < 0$) to fairly hydrophilic (θ nearly 0°, $\cos\theta \approx 1$) (see Table 4). [23]. After rinsing excess organosulfur reagent from the gold surface and drying, it is possible to characterize the surface energy of the monolayer in terms of the apparent contact angle formed between the monolayer and pure water. The particular substrates discussed here were fairly rough, leading to substantial contact angle hysteresis and uncertainty. The substrate energy scale defined by these contact angles should be regarded as somewhat qualitative. Since differing adsorption kinetics for the two different organosulfur reagents. The contact angles of mixed monolayer must be measured, prevents their direct inference from the relative solution concentration.

IV. EXPERIMENTAL TECHNIQUES

Recent reports from our laboratories have described the spreading behavior on gold surfaces modified with the aforementioned mixed monolayers. The spreading rates were acquired by means of two different techniques, which are described here briefly. The first method exploits the radial sensitivity of a standard AT-cut quartz-crystal microbalance (QCM) [24,25]. AT-cut quartz is a low loss, shear mode piezoelectric material that can be made to resonate at a specific frequency by an alternating electric field applied by gold electrodes on opposite sides of a

TABLE 4 Contact Angles Measured Between 3.0 μL Water Drops and Mixed Monolayer Modified Substrates with Varying Concentrations of Hydrophilic (A = OH) and Hydrophobic (B = CH_3) Terminal Groups

Solution composition $HS(CH_2)_{15}OH{:}HS(CH_2)_{15}CH_3$	θ	$\cos\theta$
0:100	113	–0.39
30:70	92	–0.035
50:50	62	0.47
60:40	35	0.82
70:30	16	0.96
100:0	0	1.0
Parafilm[a]	99	–0.15

[a]Included for comparison.

quartz disk. The gold electrodes also can be modified with the organosulfur monolayers to allow systematic adjustment of the surface energy. The resonant frequency of the quartz crystal is related primarily to the combined thickness of the crystal and its electrodes, but it also depends on any external mass loading at the crystal surface. The frequency dependence on the mass loading is not uniform over the active area of the oscillator, which is defined by the overlapping electrodes. Mass deposited near the center of the active area has a significantly larger effect than the same mass deposited near the periphery. In the case of liquid loading, the shear wave is strongly attenuated by viscous damping, and the resonant frequency is influenced by the density and viscosity of the liquid. An AT-cut crystal oscillating at 5 MHz effectively responds to the mass of the fluid contained within approximately one damping decay length from the crystal surface ($\approx$ 250 nm for water). This, and the known radial mass sensitivity, provide a means by which the time-dependent areal extent of a spreading droplet can be determined by recording the oscillator frequency as a function of time after depositing a droplet near the center of an electrode. Because the droplet thickness is presumably always greater than the characteristic decay length, the effective mass loading increases as the droplet spreads over the monolayer modified electrode. The time dependence of the oscillator frequency is therefore related, through the radial mass sensitivity, to the area covered by the droplet as a function of time. Since the mass sensitivity can be independently determined by a number of methods, the spreading rate can be obtained easily from the slope of the covered area–time curve.

Lin et al. [24] used the technique just described to investigate the spreading rate dependence on substrate surface energy of aqueous dispersions of the siloxane surfactant M(D′E_8OH)M. Interestingly, the spreading rate did not simply increase with increasing substrate surface energy. In fact, a pronounced maximum in spreading rate was observed on surfaces of moderate hydrophobicity. This unanticipated result remains to be adequately explained and is discussed further below. The spreading rate also was found to be strongly dependent on surfactant concentration, exhibiting a maximum in spreading rate versus concentration, similar to the data reported by Zhu et al. [15]. This technique has been refined recently to yield more accurate spreading rates [25], demonstrating that the QCM is an effective tool for studying wetting dynamics, particularly for opaque substrates or surfaces that do not have sufficient contrast with test droplets for visual observation.

A second method for investigating the substrate surface energy dependence of the spreading rate is somewhat more straightforward [17]. The test droplet is deposited onto a mixed monolayer modified gold substrate under controlled temperature/humidity conditions, and the spreading event is simply videotaped using standard equipment. Recent advances in computer technology allow the videotape record to be digitized in real time for analysis. A digitization rate of 10 frames/s at 120×160 pixel resolution is sufficient to observe even rapid spreading, and the frame rate can be reduced for slower events. The area covered by the droplet on successive frames can be determined by image analysis methods, and change of the covered area with time can be obtained in a straightforward manner. This method requires sufficient contrast between the bare and droplet-covered portions of the substrate but, unlike the QCM method described above, the spreading need not be assumed to be radially uniform. Image analysis software such as NIH Image enable automated analysis of the digitized images.* Data typically were acquired under controlled (usually saturated) humidity conditions and at room temperature using 2.0 μL test droplets on 12 mm diameter gold films, prepared by electron beam evaporation onto quartz disks, modified with organosulfur monolayers. The area covered by the droplet generally increased linearly with time, and the spreading rate (mm^2/s) typically was determined from the slope of this dependence.

V. SURFACTANT-ENHANCED SPREADING BY TRISILOXANES

A. $M(D'E_4OH)M/H_2O$ Dispersions

The data obtained for $M(D'E_4OH)M$ dispersions revealed a pronounced maximum in spreading rate as a function of substrate surface energy, this maximum occurring at moderate surface energy ($0 < \cos\theta < 0.6$) for all concentrations investigated (Fig. 2a), similar to the behavior observed by Lin et al. [24]. This suggests that interactions at the substrate–liquid interface play an important role in the spreading behavior (see Sec. IX). Aqueous dispersions of $M(D'E_4OH)M$ were very turbid, and the concentrations examined were well above the break in the surface tension–concentration curve that marks the onset of aggregation. Nevertheless, the spreading rate increased in a roughly linear manner with increasing surfactant concentration. No maximum in spreading rate versus surfactant concentration was observed (Fig. 3). This suggests the importance of dynamic effects, since equilibrium interfacial tensions at these different concentrations are very similar. The linear behavior is also distinct from maxima in data for spreading rate versus concentration previously reported for $M(D'E_8OH)M$ [24] and $M(D'E_8OMe)M$ [65] aqueous dispersions.

B. Spreading of $M(D'E_8OH)M/H_2O$ Dispersions

The longer hydrophilic poly(oxyethylene) tail makes this compound more soluble than $M(D'E_4OH)M$. While aqueous dispersions of $M(D'E_8OH)M$ remained visi-

---

*NIH Image is a public domain image analysis program written by Wayne Rasband at the U.S. National Institutes of Health and available from the Internet by anonymous ftp from zippy.nimh.nih.gov/pub/nih-image or on floppy disk from the National Technical Information Service, 5285 Port Royal Rd., Springfield, VA 22161, part number PB93-504868. A macro for automated acquisition of spreading rate data and for analysis is available from the authors upon request.

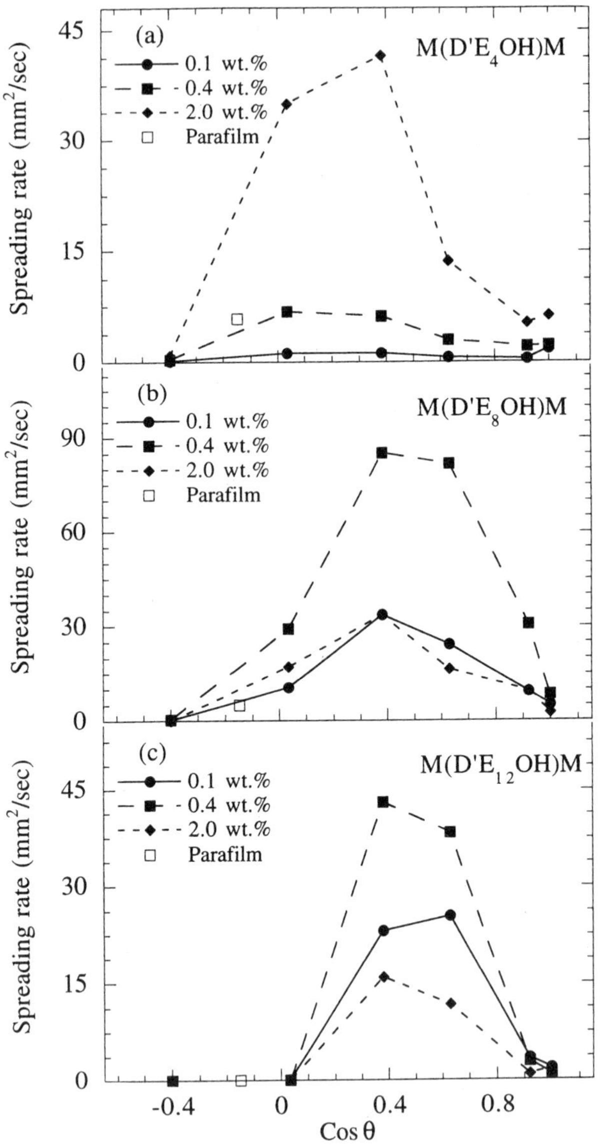

FIG. 2 Spreading rate dependence on substrate surface energy (at 22°C and > 95% relative humidity) for aqueous dispersions and solutions of (a) M(D′E$_4$OH)M, (b) M(D′E$_8$OH)M, and (c) M(D′E$_{12}$OH)M. The substrate surface energy was modified by the deposition of mixed organosulfur monolayers on rough gold-coated quartz crystals (rms roughness = 0.3 μm), and was characterized in terms of the cosine of the contact angle formed by a droplet of 18 MΩ deionized water on the monolayer surface. The open squares correspond to the spreading rates measured for a 0.4 wt% mixture on Parafilm, for which cos θ = –0.15. The error in the spreading rates is estimated to be ±10%.

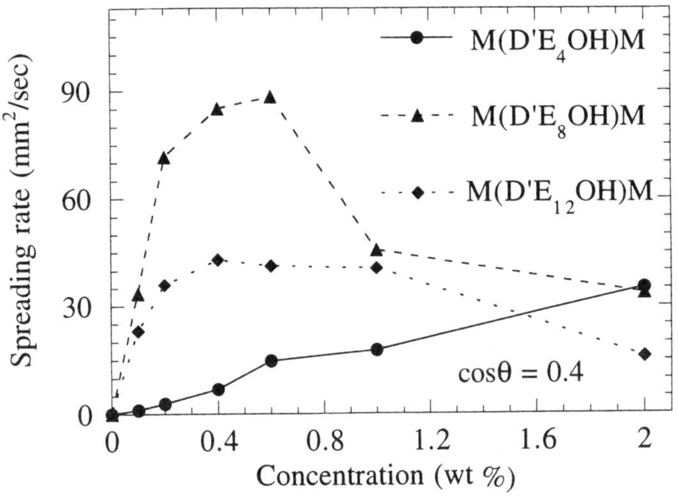

FIG. 3 Spreading rate dependence on the concentration of aqueous dispersions and solutions of M(D′E_4OH)M, M(D′E_8OH)M, and M(D′E_{12}OH)M. For clarity, only the results on the cos θ = 0.4 substrates are presented.

bly turbid at the concentrations examined, they were less so than the M(D′E_4OH)M/H_2O dispersions. The dependence of the spreading rate on surface energy and concentration for M(D′E_8OH)M is presented in Figs. 2b and 3. As with the M(D′E_4OH)M dispersions, the spreading rates achieved pronounced maxima on substrates having moderate surface energy. These maxima appear somewhat sharper and shifted to slightly greater surface energies than those exhibited by M(D′E_4OH)M. The shift may be related to the increased length of the hydrophilic poly(oxyethylene) chain. The spreading rates increased with concentration up to about 0.7 wt% and then progressively decreased as the concentration was increased (Fig. 3). This behavior is consistent with that reported for M(D′E_8OH)M by Zhu et al. and Lin et al. (although the maxima were somewhat broader and occurred at higher concentrations).

Studies were also performed to determine the effect of substrate roughness on the spreading rate obtained when M(D′E_8OH)M was used. Spreading on rough substrates was significantly faster (by as much as 50%) at all concentrations, surface energies, and humidities. This increase in spreading rate was systematic so that the major features (the maxima in rate vs. concentration and surface energy) were similar for smooth (rms roughness ≈ 80 Å by atomic force microscopy) and rough (rms roughness ≈ 0.30 μm by stylus profilometry) substrates. The enhanced spreading rate can be attributed to the increased capillarity associated with the rough surface, which would serve to enhance surface flow of the leading edge of the aqueous film [26].

Because the maximal spreading rates exhibited by M(D′E_8OH)M dispersions are larger than those of M(D′E_4OH)M, and since both spread on even the most hydrophobic surfaces, M(D′E_8OH)M would appear to be the more effective wetting agent and of more practical interest. The maximal spreading rates are also an order of magnitude larger than the rates observed on Parafilm ($\approx 5\ mm^2/s$). However, Parafilm is quite hydrophobic, exhibiting a contact angle for water of approximately 97° ($\cos\theta \approx -0.13$). The spreading rate data on Parafilm (indicated in Fig. 2 by the open square symbols) are consistent with observations on the mixed monolayer modified surfaces that have the same contact angle. Therefore, while these results demonstrate the importance of substrate surface energy, they generally support the conclusions revealed in earlier work with Parafilm and other strongly hydrophobic surfaces.

C. M(D′E_{12}OH)M/H_2O Solutions

Whereas M(D′E_4OH)M and M(D′E_8OH)M were shown to enhance spreading on Parafilm, spreading of M(D′E_{12}OH)M solutions on Parafilm was negligible. Therefore, this compound was thought to lack the qualities necessary to facilitate significant surfactant-enhanced spreading and was not classified as a superspreader. It was suggested that this distinction was related to the single-phase nature of M(D′E_{12}OH)M/H_2O solutions, which is a consequence of the increased solubility associated with the longer hydrophilic poly(oxyethylene) tail. Spreading rates measured on the mixed monolayer modified surfaces indicated that aqueous solutions of M(D′E_{12}OH)M did not spread well on substrates having surface energies similar to that of Parafilm (Fig. 2c). However, these solutions spread rapidly on more hydrophilic surfaces. In fact, while the maximal spreading rates are lower than those observed for M(D′E_8OH)M, they are very similar to those exhibited by M(D′E_4OH)M. *This indicates that mixtures need not be turbid (i.e., biphasic) to promote rapid spreading on relatively hydrophobic surfaces and, moreover that conclusions based on the limited data obtained on Parafilm are not general.* Based on the similarities between these data sets, it is reasonable to assert that similar mechanisms drive the enhanced spreading for the three compounds. These results demonstrate that the original superspreading experiments were somewhat limited in that they examined spreading only on substrates of hydrophobicity comparable to Parafilm. Use of the term "superspreading" therefore was abandoned and the term "surfactant-enhanced spreading" was adopted to describe cases for which this phenomenon occurs more generally over a wider range of substrates.

As in the case of the M(D′E_8OH)M data, the spreading rates for M(D′E_{12}OH)M exhibit maxima with respect to concentration and surface energy. However the transition between poor spreading and very rapid spreading with increasing substrate surface energy occurs much more abruptly for M(D′E_{12}OH)M than for either M(D′E_4OH)M or M(D′E_8OH)M. This difference in surface energy dependence is somewhat surprising in light of the similarity in molecular structures and the equilibrium surface tensions of M(D′E_{12}OH)M, M(D′E_8OH)M, and M(D′E_4OH)M aqueous dispersions and solutions (21.3, 21.0, and 21.0 mN/m, respectively). The narrower peak observed for M(D′E_{12}OH)M suggests increased spreading specificity to substrate surface energy (see Sec. IX). Furthermore, the shift in the maxima with increasing hydrophilic poly(oxyethylene) chain length (and solubility) toward higher surface energy continues the trend mentioned in the preceding section. Despite this shift, M(D′E_{12}OH)M solutions spread more poorly than either M(D′E_8OH)M or M(D′E_4OH)M dispersions on the most hydrophilic surfaces. A maximum in spreading rate as a function of concentration again is observed but is broader than that exhibited by the M(D′E_8OH)M dispersions (Fig. 3).

D. Spreading of M(D′E_4OH)M/M(D′E_{12}OH)M/H_2O Mixture

The preceding sections demonstrate intriguing behavior based on the length of the hydrophilic poly(oxyethylene) group. This length is expected to affect not only the turbidity but also the nature and microstructure of the dispersed aggregates in the test mixtures. The surfactants already studied were somewhat polydisperse owing to the nature of their synthesis. To investigate more thoroughly the effect of turbidity and the average length of the hydrophilic group, studies also were performed as a function of average tail length, which was varied by preparing 0.2 wt% aqueous mixtures containing different proportions of M(D′E_4OH)M and M(D′E_{12}OH)M. These mixtures can be described in terms of their average ethoxylation (AE) length, so that a mixture containing equal proportions of M(D′E_4OH)M and M(D′E_{12}OH)M has an AE of 8. Every sample, other than the M(D′E_{12}OH)M solution, was visibly turbid. However, a clear progression was apparent. The AE = 10.6 mixture was only slightly turbid, while AE = 5.3 and M(D′E_4OH)M mixtures were very turbid. Figure 4 summarizes the data for spreading rate versus surface energy data of these mixtures. As expected from the results described earlier, the most rapid spreading occurred on surfaces of moderate surface energy ($0 < \cos\theta < 0.6$) for each data set. The peaks in spreading rate became broader and less pronounced, shifting to slightly lower energy surfaces as AE is reduced. The shift is evident from comparison of the rates on the $\cos\theta \approx 0.0$ and $\cos\theta \approx 0.6$ surfaces. It is interesting to note that while M(D′E_{12}OH)M solutions spread very poorly on the more hydrophobic surfaces, the slightly turbid AE = 10.6 mixture spreads quite well on such substrates. In fact, the AE = 10.6 mixture spreads on all but one of the surfaces more readily than the other mixtures. These data therefore suggest a maximum in the spreading rate associated with the onset of turbidity, perhaps due to the size and/or nature of the dispersed aggregates. It is important to observe, however, that the AE = 8 mixture spreads considerably more slowly than the aqueous dispersions of M(D′E_8OH)M (Fig. 2a); hence these bimodal mixtures do not precisely mimic solutions or dispersions prepared from less polydisperse or monomodal compounds. The data presented in Fig. 4 clearly indicate a strong dependence of the spreading rate on average hydrophile length and surface energy, supporting the results presented in Fig. 2.

E. Spreading of M(D′E_8OMe)M/H_2O and M(D′E_8OAc)M/H_2O Dispersions

These trisiloxane surfactants differ from M(D′E_8OH)M with respect to the nature of the terminal group capping the poly(oxyethylene) chain. The end-cap group has been shown to strongly influence the phase behavior of aqueous systems [16]. These compounds also are of commercial interest and therefore were studied to provide comparison with the other siloxane surfactants. Both compounds form turbid dispersions in water at room temperature at the concentrations investigated, and the dependence of spreading rate on surface energy and concentration is qualitatively similar to the data presented on M(D′E_8OH)M (Fig. 5). Peaks in the spreading rate with respect to surfactant concentration and surface energy were

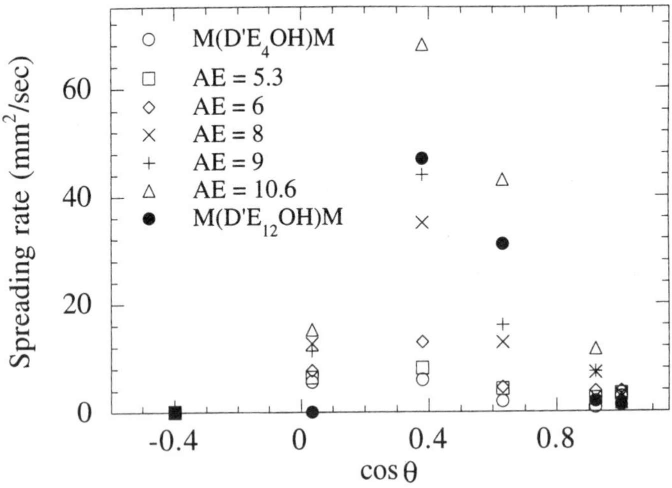

FIG. 4 Spreading rate dependence on substrate surface energy for 0.2 wt% aqueous surfactant mixtures (at 22°C and > 95% relative humidity). The average ethoxylation (AE) of the mixtures is controlled by systematic variation of the proportions of pure M(D′E_4OH)M and M(D′E_{12}OH)M in 18 MΩ water. Note that the AE = 8 mixture spreads more slowly than M(D′E_8OH)M dispersions (Fig. 2b).

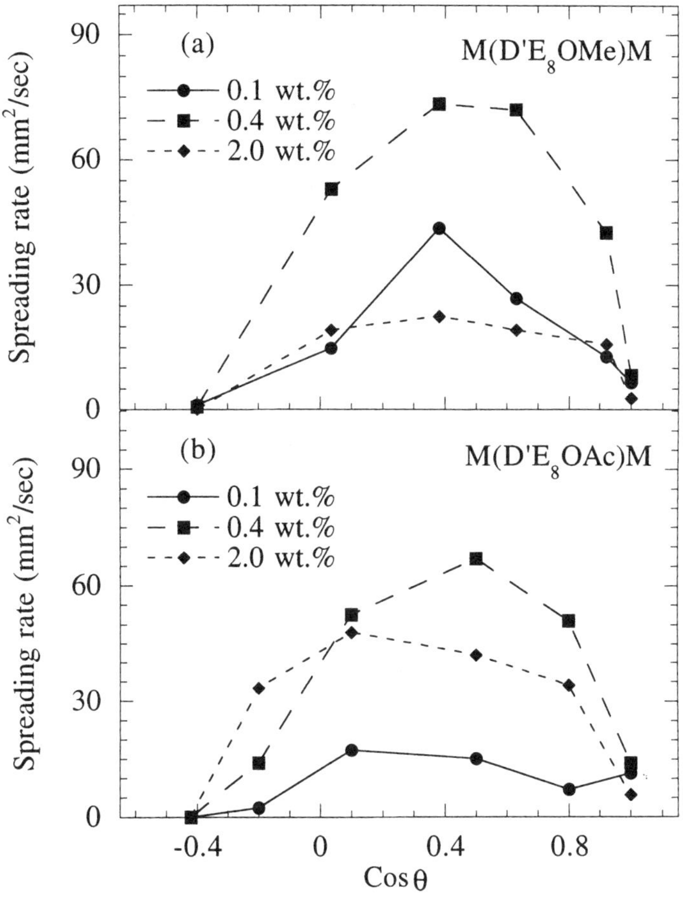

FIG. 5 Spreading rate dependence on substrate surface energy (at 22°C and > 95% relative humidity) for aqueous dispersions of (a) $M(D'E_8OMe)M$ and (b) $M(D'E_8OAc)M$.

clearly apparent. In the case of the $M(D'E_8OMe)M$ dispersions, the maximal rate shifts slightly to lower energy surfaces as the concentration is increased. Both $M(D'E_8OMe)M$ and $M(D'E_8OAc)M$ dispersions spread more rapidly on lower energy surfaces over wider concentration ranges than their $M(D'E_8OH)M$ counterparts, demonstrating the importance of the end-cap group to their wetting properties. Even though the maximal spreading rates are not quite as large as those exhibited by $M(D'E_8OH)M$, the increased ability of these dispersions to rapidly wet the most hydrophobic surfaces is of practical interest [27].

F. Dependence of Surfactant-Enhanced Spreading on Humidity

The spreading rate of aqueous M(D′E_8OMe)M dispersions on Parafilm was reported to be strongly dependent on relative humidity [15]. This was confirmed by the aforementioned videomicroscopy method (see below: Sec. VII.A Fig. 7). The spreading was greatly inhibited under dry conditions (< 20% relative humidity). Spreading occurs at higher humidities but does not increase appreciably between 40 and > 95% relative humidity. This behavior was tentatively attributed to the necessity for a preexisting water film on the substrate, which enables a surface tension gradient to be established so that spreading can be driven by Marangoni flow. When the spreading of M(D′E_8OH)M aqueous dispersions on the mixed monolayer modified substrates was investigated as a function of relative humidity, very different results were obtained (Fig. 6). Repeated experiments indicate that the spreading rates on these substrates remain basically constant from < 5 to > 95% relative humidity, and rapid spreading is possible under even very dry conditions. The observation of "dry" spreading has important implications regarding the mechanism driving the wetting and is further discussed below. The dis-

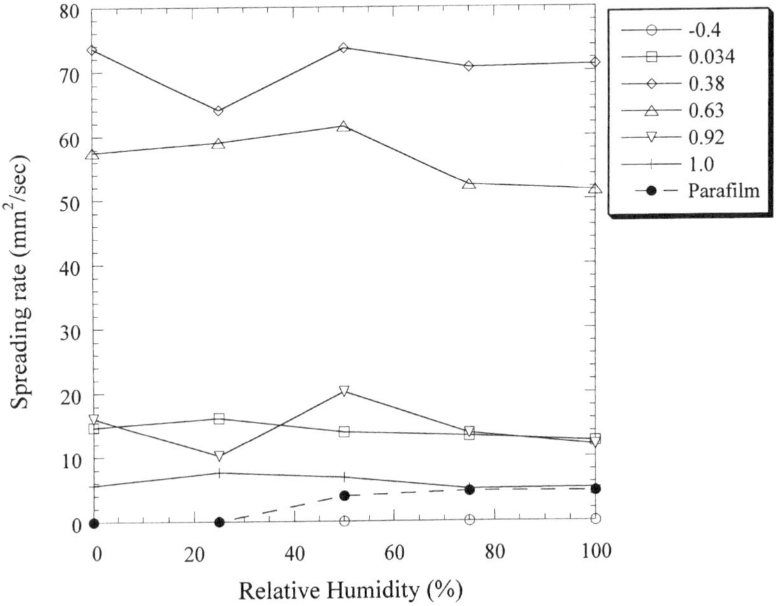

FIG. 6 Spreading rate dependence on relative humidity (at 22°C) of 0.2 wt% M(D′E_8OH)M aqueous dispersions on mixed monolayer modified gold substrates.

tinction in properties between the Parafilm and monolayer modified gold surfaces responsible for the different spreading behavior is not readily apparent. However, it is likely to be related to differences in surface chemistry or morphology. This unresolved behavior remains under active investigation.

VI. SPREADING BEHAVIOR OF ALKYL-ETHOXYLATED ALCOHOLS

Even though the aggregate microstructure and turbidity apparently were not critical for the observation of trisiloxane surfactant-enhanced spreading, the results above provide little insight into which, if any, other (nonsiloxane) surfactants would enhance spreading. As Ananthapadmanabhan et al. [14] noted, the trisiloxane compounds are atypical amphiphilic molecules. The trisiloxane head group is unusually hydrophobic for its size (its hydrophobicity is roughly comparable to that of a linear dodecyl group) with an effective "length" of 9.7 Å, vs. 15 Å for a linear dodecyl group, and a molecular volume of approximately 530 $Å^3$, vs. 350 $Å^3$ for a linear dodecyl group). It is conceivable that this unconventional structure is the source of the unique wetting behavior. This prompted us to examine whether more conventional surfactants were capable of promoting spreading of aqueous solutions.

The trisiloxane group has been shown to exhibit hydrophobicity similar to that of a linear dodecyl hydrophobe [28]. The aqueous phase behavior of nonionic ethoxylated alcohol surfactants $CH_3(CH_2)_{m-1}(OCH_2CH_2)_n$—$OH(C_mE_n)$ has been extensively investigated [29]. These studies have demonstrated significant similarities between the trisiloxane–water and ethoxylated dodecyl alcohol–water phase diagrams. Since the alkyl head group presents a considerably smaller cross section, shorter poly(oxyethylene) tails, relative to the trisiloxanes, are required to form phases of similar curvature. The ethoxylated dodecyl alcohol compounds $C_{12}E_n$, with $n = 3, 4, 5$, and $C_{10}E_3$ might therefore be expected to mimic the aggregation behavior of trisiloxane-based surfactants with longer poly(oxyethlene) tails. Consequently, their spreading behavior was investigated on the mixed monolayer modified substrates.

A. $C_{10}E_3/H_2O$ Dispersions

Aqueous dispersions of $C_{10}E_3$ were visibly turbid at all concentrations tested. These dispersions did not spread well over Parafilm or the low energy mixed monolayer surfaces ($\cos\theta < 0.25$) (Fig. 7). However, very rapid wetting was apparent on somewhat higher energy surfaces. In fact, the maximal rates are comparable to those exhibited by $M(D'E_8OH)M$ dispersions. As in the case of the $M(D'E_{12}OH)M$ micellar solutions, there appears to be a surface energy below which $C_{10}E_3$ dispersions cease to spread rapidly. The peak in spreading rate with

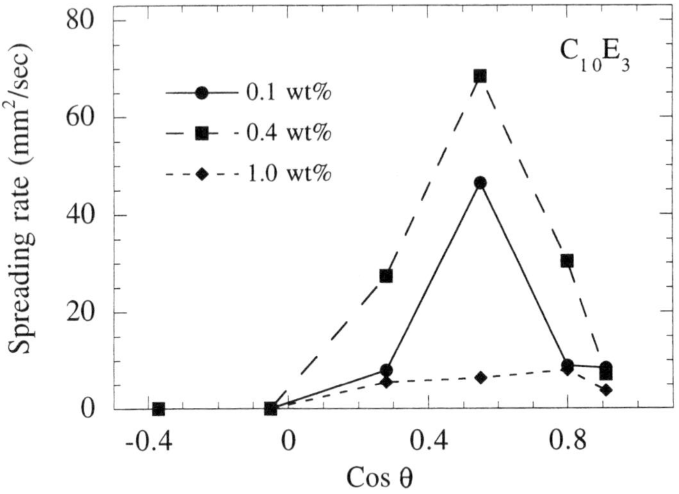

FIG. 7 Spreading rate dependence on substrate surface energy (at 22°C and > 95% relative humidity) for aqueous solutions of $C_{10}E_3$.

respect to substrate surface energy was shifted toward more hydrophilic surfaces compared with the peaks observed for the trisiloxane based compounds, largely because of the depressed spreading on the lower surface energy substrates. A maximum in spreading rate with respect to surfactant concentration was also observed, resembling the peak exhibited by the M(D′E_8OH)M dispersions with the maximal rates occurring near 0.5 wt% (see below: Sec. VII.B,) (Fig. 9). The observation of rates and wetting characteristics similar to those exhibited by the trisiloxane compounds (including highly linear area–time curves) suggests a similar spreading mechanism. *More important, this clearly indicates that the characteristics of surfactant-enhanced spreading are not unique to siloxane surfactants and suggests that the phenomenon may be fairly general.*

B. $C_{12}E_3/H_2O$ Dispersions

Like the $C_{10}E_3$ mixtures, aqueous mixtures of $C_{12}E_3$ were visibly turbid at all concentrations tested as a result of the presence of a dispersed phase (lamellar liquid crystal). Again, no spreading on the lower energy substrates was observed. The maximal rates were considerably smaller than those exhibited by the $C_{10}E_3$ dispersions, but the peak in spreading rate with respect to substrate surface energy qualitatively was similar, occurring over roughly the same region of substrate surface energy (Fig. 8a). Unlike the $C_{10}E_3$ dispersions, no maximum in rate dependence on surfactant concentration was observed for the $C_{12}E_3$ dispersions; the

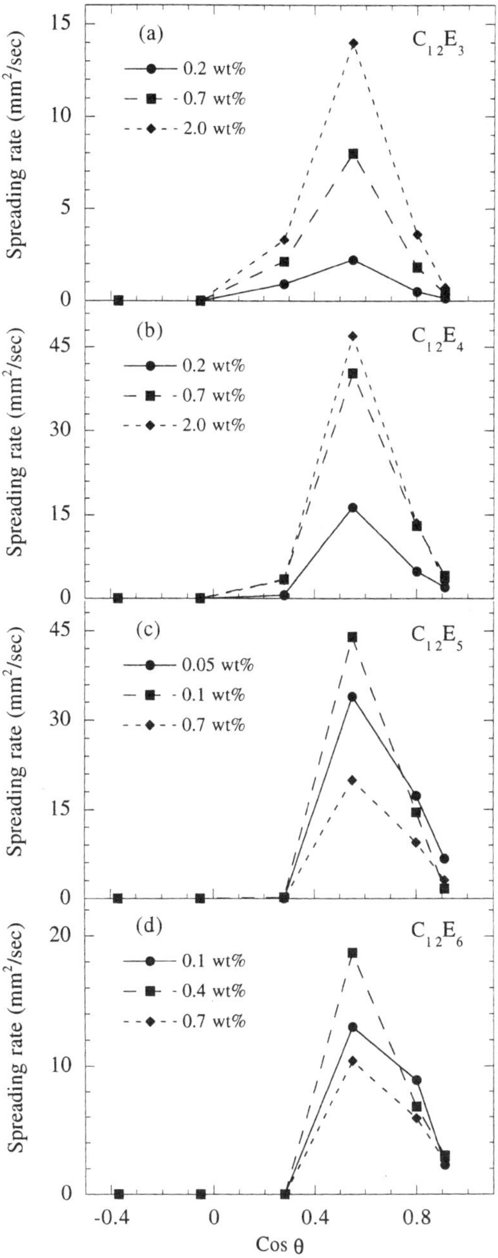

FIG. 8 Spreading rate dependence on substrate surface energy (at 22°C and > 95% relative humidity) for aqueous dispersions and solutions of (a) $C_{12}E_3$, (b) $C_{12}E_4$, (c) $C_{12}E_5$, and (d) $C_{12}E_6$. These plots are similar qualitatively to those presented in Fig. 2.

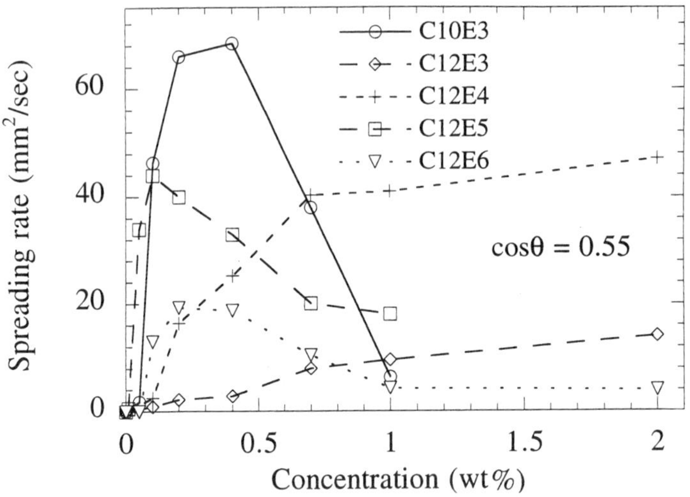

FIG. 9 Spreading rate dependence on the concentration of aqueous dispersions or solutions of $C_{10}E_3$, $C_{12}E_3$, $C_{12}E_4$, $C_{12}E_5$, and $C_{12}E_6$. For clarity, only the results on the cos θ = 0.55 substrates are presented. These data are similar qualitatively to the data on the $M(D'E_nOH)M$ (n = 4, 8, 12) dispersions and solutions presented in Fig. 3.

largest concentration investigated exhibited the highest rate (Fig. 9). Despite the lack of a spreading rate maximum with respect to surfactant concentration, the $C_{12}E_3$ aqueous dispersions clearly exhibit surfactant-enhanced spreading, indicating that this maximum is not a necessary feature of surfactant-enhanced spreading systems.

C. $C_{12}E_4/H_2O$ Dispersions

The longer hydrophilic chain of $C_{12}E_4$ increases the solubility compared to $C_{12}E_3$, leading to less turbid aqueous dispersions. The $C_{12}E_4$ dispersions (Fig. 8b) also exhibit enhanced spreading characteristics, with maximal rates substantially larger than those achieved with $C_{12}E_3$ (Fig. 8a). Again, appreciable spreading rates were observed only on substrates having significantly higher surface energies than Parafilm. The peak in spreading rate with respect to substrate surface energy is somewhat sharper than that exhibited by $C_{12}E_3$, largely because spreading is poorer on the lower surface energy side of the peak. As in the case of the trisiloxanes, increasing the hydrophilic tail length results in an increased sensitivity to surface energy, particularly on the lower surface energy side. That is, the peak in spreading rate does not shift simply to higher energy substrates with increasing surfactant hydrophilicity. As in the case of the $C_{12}E_3$, no maximum in the spreading rate dependence on concentration was observed, although the rates did not increase greatly above 0.7 wt% (Fig. 9). The $C_{12}E_4$ dispersions were distinct, however, in that macroscopic phase segregation was observed to occur in the micropipet tip at high surfactant concentration (> 1 wt%), and the corresponding spreading rates may not be as accurate as the other data. Such rapid phase segregation was not observed in any of the other systems examined.

D. Spreading of $C_{12}E_5/H_2O$ Solutions

$C_{12}E_5$ forms clear micellar solutions at the concentrations investigated. The maximal rates exhibited by these $C_{12}E_5$ solutions are comparable to those exhibited by the $C_{12}E_4/H_2O$ dispersions, and it is apparent that these solutions also exhibit surfactant-enhanced spreading (Fig. 8c). As in the case of the $C_{10}E_3$ dispersions (but not the $C_{12}E_3$ or $C_{12}E_4$ dispersions), the $C_{12}E_5$ solutions exhibited a pronounced maximum in spreading rate with respect to surfactant concentration (Fig. 9). The spreading of these solutions on lower energy substrates appears to be disfavored further relative to the $C_{12}E_3$ or $C_{12}E_4$ systems, and the maximum in spreading rate with respect to surface energy is shifted further toward higher energy substrates. These trends are similar qualitatively to the behavior exhibited by the dispersions and solutions of M(D′E_nOH)M, $n = 4, 8, 12$. For a given hydrophobic group, increasing the length of the poly(oxyethylene) hydrophilic group inhibited spreading on low energy surfaces. The maxima in rate versus surface energy became correspondingly sharper as the spreading became more surface selective with increasing n. These results demonstrate that the observation of surfactant-enhanced spreading and the accompanying characteristics for aqueous alkyl-ethoxylated alcohol surfactant systems is not critically dependent on turbidity, resembling the behavior observed for the trisiloxane–water systems.

E. $C_{12}E_6/H_2O$ Solutions

$C_{12}E_6$ also formed nonturbid micellar solutions at the concentrations investigated. The maximal rates exhibited by the $C_{12}E_6$ solutions (Fig. 8d) were substantially lower than those exhibited by the $C_{12}E_5$ solutions or $C_{12}E_4$ dispersions, but remained higher than those exhibited by the $C_{12}E_3$dispersions. These data suggest that maximal spreading rates are associated with turbidity, consistent with the data obtained for M(D′E_nOH)M, $n = 4, 8, 12$, aqueous systems and for mixtures with various average ethoxylation (AE) lengths. The spreading rate peaks with respect to substrate surface energy were not shifted appreciably relative to the peaks exhibited by the $C_{12}E_5$ solutions. Like the $C_{12}E_5$ solutions, the $C_{12}E_6$ solutions exhibited a peak in spreading rate versus surfactant concentration, with the maximal rate shifted to somewhat higher concentrations (Fig. 9). It is interesting to note that the two alkyl ethoxylate surfactants forming nonturbid aqueous solutions ($C_{12}E_5$, C_{12} E_6) exhibit peaks in spreading rate versus surfactant concentration, while the two forming turbid aqueous dispersions ($C_{12}E_3$, $C_{12}E_4$) do not. This would suggest another general trend; however, the data on the turbid $C_{10}E_3$ dispersions serve as a clear counterexample, with pronounced spreading rate maxima versus concentration occurring near 0.5 wt%. Moreover, the turbid two-phase M(D′E_8R)M aqueous systems also exhibited spreading rate maxima with respect to surfactant concentration. These results again indicate that while microstructure influences spreading rates, it is not responsible for the general characteristics of surfactant-enhanced wetting.

VII. GLUCOSIDE (C_8G_1) SURFACTANTS

Observation of the rapid spreading exhibited by the C_mE_n systems eliminated the uniqueness of the trisiloxane hydrophobic moeity as the source of surfactant-enhanced spreading, indicating that this phenomenon can be extended to other surfactants. To investigate the spreading behavior dependence on the hydrophilic portion of the surfactant molecules, the spreading behavior of aqueous solutions of nonionic octyl-β-D-glucopyranoside (C_8G_1) was examined (Fig. 10). The mixtures investigated were clear micellar solutions at surfactant concentrations well above the critical micelle concentration (cmc ≈ 1.6×10^{-3} M for C_8G_1). As in the case of the aqueous C_mE_n dispersions and solutions, no spreading was observed on very low energy substrates, but reasonably large spreading rates were observed on sufficiently high energy substrates. The data shown in Fig. 10 indicate that glucoside surfactants can also enhance spreading. The spreading rates are highest on surfaces of moderate surface energy, and a maximum in spreading rate versus concentration again is readily apparent. The peak in spreading rate versus substrate

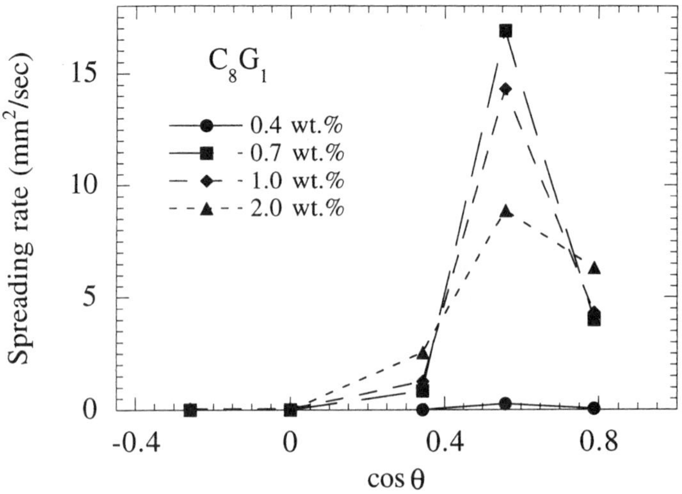

FIG. 10 Spreading rate dependence on substrate surface energy (at 22°C and > 95% relative humidity) for aqueous solutions of C_8G_1.

surface energy occurs in roughly the same position as the peak exhibited by the $C_{12}E_3$ dispersions and solutions. The observation of surfactant-enhanced spreading in three distinct systems with different combinations of hydrophobic and hydrophilic moieties illustrates further that the phenomenon is a general one.

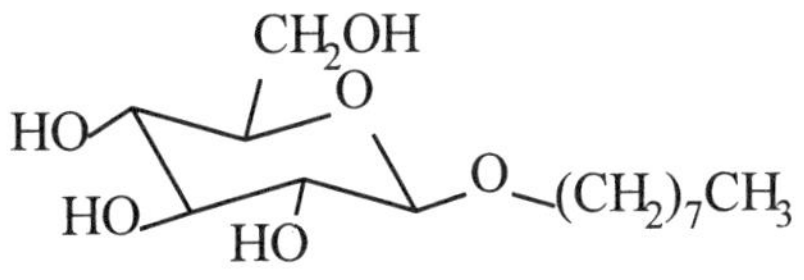

C8G1

VIII. IONIC SURFACTANTS

The spreading of two well-known ionic surfactants on mixed monolayer modified surfaces and on Parafilm was also studied Concentrations well above, near, and well below the respective critical micelle concentrations were investigated ($cmc_{SDS} = 8.4 \times 10^{-3}$ mol/L, $cmc_{DTAB} = 0.065$ mol/L). These aqueous mixtures were clear solutions at the concentrations studied. Neither cationic DTAB nor anionic SDS solutions spread on Parafilm. In fact, limited spreading of these solutions (< 2 mm^2/s) was observed to occur only on the most hydrophilic surfaces ($\theta \approx 32°$, $\cos\theta \approx 0.85$). No maxima in rate versus surface energy could be resolved, and the most rapid spreading occurred on the highest surface energy substrate. Both compounds appeared to display weak maxima in rate versus concentration at concentrations near their respective cmc. The rates were somewhat difficult to determine, however, since the area–time curves increased less than linearly and exhibited significant negative curvature because these solutions often did not wet the entire surface. Consequently, the rates were determined from the portion of the event during which spreading was apparent, and the values obtained represented considerable overestimations. The poor spreading rates, lack of a maximum in rate versus surface energy, and nonlinear area–time curves suggest that these compounds do not promote significant surfactant-enhanced spreading of the type exhibited by the trisiloxane, ethoxylated alcohol, and alkyl glucoside dispersions and solutions.

In contrast to SDS and DTAB, we have identified two other ionic surfactants (cationic DDAB and anionic AOT) (Table 1) that promote surfactant-enhanced spreading of aqueous mixtures on hydrophobic substrates [30]. The DDAB/H_2O and AOT/H_2O systems exhibit pronounced spreading rate peaks as a function of both substrate surface energy and surfactant concentration, similar to the nonionic systems. The maximal spreading rates are somewhat lower (≈ 15 and 25 mm^2/s, for the DDAB/H_2O and AOT/H_2O systems, respectively) but remain similar to rates exhibited by the nonionic systems.

Because both anionic AOT and cationic DDAB promote rapid spreading of aqueous mixtures on fairly hydrophobic substrates, surfactant-enhanced spreading appears to be a very general phenomenon exhibited by a wide variety of surfactant systems and is clearly not specific to nonionic surfactant systems. The distinction between SDS and AOT and between DTAB and DDAB suggests that molecular geometry (hence microstructure) and the relative strength of the hydrophobic and hydrophilic groups play important roles in the ability of a particular surfactant to promote rapid spreading of aqueous systems on hydrophobic substrates. The anionic hydrophilic groups of AOT and SDS are similar, as are the cationic hydrophilic groups of DDAB and DTAB. However, the dual-chain hydrophobic groups of AOT and DDAB are significantly bulkier than the linear hydrocarbon chains of SDS and DTAB. Branched or bulkier hydrophobes tend to favor lamellar microstructures (turbid dispersions) in aqueous solutions, whereas linear or smaller hydrophobes tend to favor spherical micellar microstructures (clear solutions). Surfactant adsorption at the liquid–solid interface near the leading edge of a droplet might be favored by a tendency to form a lamellar microstructure, this adsorption creating interfacial gradients that result in Marangoni flow (flow in the direction of an increasing interfacial tension gradient) and drive the spreading. Conversely, poor surfactant adsorption at the liquid–solid interface could result in slow or negligible spreading. The tendency of AOT and DDAB toward flatter microstructures, relative to SDS and DTAB, may increase surfactant adsorption efficiency and enhance spreading rates at the liquid–solid interface. However, the enhancement of spreading due, as well, to micelle-forming surfactants makes it clear that microstructure is not critical.

Spreading of the ionic surfactant systems was further investigated by addition of salt to increase electrostatic screening. The SDS/H_2O/NaCl and DTAB/H_2O/NaCl systems did not exhibit enhanced spreading characteristics at the concentrations investigated. However, unexpected results were obtained upon addition of salt to the AOT/H_2O and DDAB/H_2O systems. Spreading rates of the DDAB/H_2O/NaCl system decrease dramatically with increasing salt concentration, and spreading is effectively halted at 1 wt% NaCl. Addition of salt to the AOT/H_2O system has the opposite effect, and spreading rates increase dramatically with increased salt concentration. In fact, spreading rates are more than three times larger at 1 wt% NaCl than without added salt, and the maximal rates exhibited by the DDAB/H_2O/NaCl mixtures are comparable to the highest rates observed in the nonionic systems.

The increased electrostatic screening due to added electrolyte should affect the anionic AOT and cationic DDAB aqueous systems somewhat similarly. It is not yet clear why added salt can greatly increase spreading rates for AOT/H_2O mixtures while it greatly reduces the spreading rates of DDAB/H_2O mixtures. A possible explanation involves coordination of the salt cation with the hydroxyl functional groups present at the surface of the mixed thiol monolayer modified substrates. Such coordination would produce a positively charged interface that would strongly enhance the adsorption of anionic AOT while retarding adsorption of cationic DDAB. Again assuming that the spreading is driven by Marangoni flow induced by surfactant adsorption at the solid–liquid interface, enhanced AOT adsorption could result in higher spreading rates for this system, while diminished DDAB adsorption could explain the reduced spreading rates of the DDAB/H_2O/NaCl mixtures. The effect of added salt on the spreading behavior of ionic surfactant aqueous systems remains highly intriguing and is a topic of active investigation.

IX. POSSIBLE MECHANISMS

Despite some theoretical progress [31–35], adequate models of the mechanics and dynamics of the spreading of surfactant solutions remain to be developed, largely because the phenomenon is so complex. Spreading of liquids on solids is generally thought to be driven by the decrease in interfacial free energy (which is equivalent to a positive spreading coefficient) associated with replacing the solid–vapor interface with liquid–solid and liquid–vapor interfaces. The equilibrium interfacial tensions clearly are important. However, spreading is a dynamic process, and instantaneous values of dynamic interfacial tensions should play a more important role [36]. Dynamic interfacial tensions depend on factors such as the rate of surface deformation, adsorption kinetics and adhesion energy, and surfactant transport, solubility, and solution microstructure [35]. Because of this complex dependence, dynamic interfacial tensions are difficult to characterize and interpret. The absence of well-developed models, makes it impossible to discuss the spreading rate results presented above in other than rather qualitative terms.

Four characteristics common to all the systems we investigated are difficult to explain by means of existing theoretical models: (1) maximal rates of spreading that are too high for surface diffusion or movement of a three-phase contact line under the constraint of the no-slip boundary condition, (2) a maximum versus surfactant concentration, (3) a maximum versus substrate surface energy, and (4) turbid dispersions that seem to spread faster than clear micellar solutions. Each of these points is discussed below.

The maximal spreading rates (80–90 mm^2/s) are quite high, too high to be explained in terms of surface diffusion of the liquid over a dry substrate [30]. Tiberg and Cazabat [37] recently determined the surface diffusion coefficient of the trisiloxane surfactants M(D′E_8OH)M, M(D′E_{12}OH)M, and MDM′E_8 to be < 10^{-6} cm^2/s by ellipsometric investigations of the spreading of a thin precursor film associated with the deposition of a bulk droplet on both hydrophilic and hydrophobic substrates. Interestingly, the area covered by this precursor film increased linearly with time. However, with such a low diffusion constant, days instead of seconds are necessary for the precursor film to cover 1 cm^2 of substrate. Therefore, dry surface diffusion cannot explain the rapid spreading exhibited by the aqueous surfactant dispersions and solutions. Tiberg and Cazabat also reported that the spreading rate of precursor films increased dramatically with increasing relative humidity above 65%. In fact, the spreading became too fast to be followed by their ellipsometric

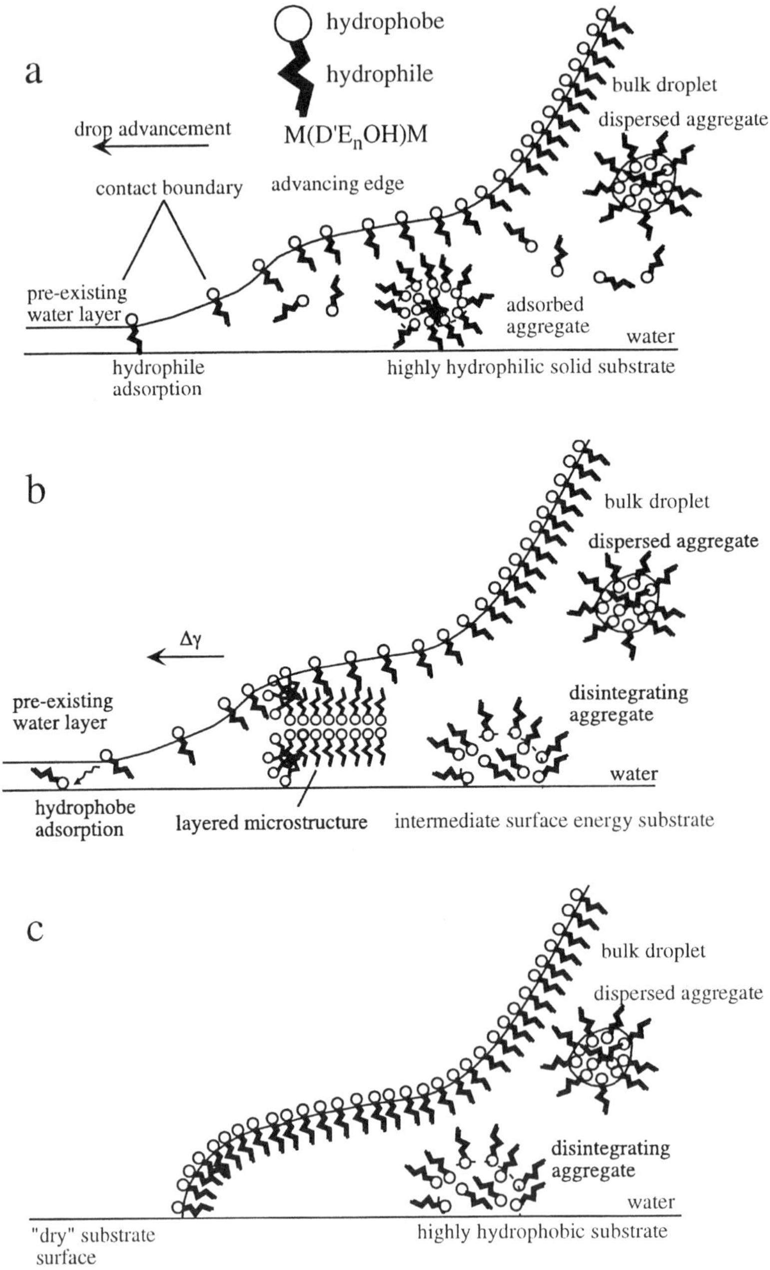
a
hydrophobe
hydrophile
M(D'E$_n$OH)M
drop advancement
contact boundary
advancing edge
bulk droplet
dispersed aggregate
pre-existing
water layer
adsorbed
aggregate
water
hydrophile
adsorption
highly hydrophilic solid substrate
b
bulk droplet
dispersed aggregate
Δγ
pre-existing
water layer
disintegrating
aggregate
water
hydrophobe
adsorption
layered microstructure
intermediate surface energy substrate
c
bulk droplet
dispersed aggregate
disintegrating
aggregate
water
"dry" substrate
surface
highly hydrophobic substrate

FIG. 11 Schematic representation of the processes that may be responsible for surfactant enhanced spreading. In (a) and (b) the drop introduced to the surface actually is in contact with a preexisting water layer. The contact line defining the boundary of the leading edge of the droplet and the preexisting water layer is continuous and may not be very distinct. However, the surfactant concentrations at the air–water interfaces of the droplet and the region just beyond this contact line will differ throughout the spreading event because the contact line across the preexisting water layer, which initially is devoid of surfactant, is continuously advancing. A surface tension gradient ($\Delta\gamma$) between a preexisting water layer and the surfactant–water droplet favors Marangoni flow and rapid spreading of the aqueous droplet.

(a) On highly hydrophilic substrates the preexisting water layer will be present. However, spreading on these surfaces is observed to be slow. This effect can be attributed to adsorption of the surfactant hydrophiles on the substrate, which forces the hydrophobe to point away from the substrate surface and presents a hydrophobic barrier to drop advancement. The surfactant molecule, with its hydrophile bound to the substrate, may pin the drop at the contact line joining the leading edge of the drop and preexisting water film. The kinetics of desorption of the surfactant hydrophile from the substrate may govern the spreading rate. Surfactant aggregates may adsorb on hydrophilic substrates intact, since there is no driving force to adsorb the hydrophobes on these surfaces. Consequently, aggregate disintegration may be slower than on hydrophobic substrates.

(b) On surfaces of intermediate surface energy, a preexisting water layer necessary for Marangoni flow is still likely, but the tendency for the hydrophobe of the surfactant to adsorb to the substrate will increase. Such adsorption can instigate aggregate disintegration, resulting in the efficient delivery of surfactant to the air–water interface when disintegration occurs in the vicinity of the leading edge of the advancing drop. This is illustrated here for a flat-layered microstructure unraveling at the leading edge, although the specific microstructure of the aggregates responsible for enhanced spreading has not been identified. Direct surfactant adsorption on the substrate in the region just beyond the contact line between the advancing drop and preexisting water layer, depicted here by the arrow, will also reduce the concentration of surfactant at the opposing air–water interface and augment the surface tension gradient between these regions. The difference in thickness of the preexisting water layer in (a) and (b) is only for clarity.

(c) On a highly hydrophobic substrate, a preexisting water layer is considerably less likely, prohibiting the creation of the surface tension gradient necessary for Marangonic flow. Very hydrophobic surfaces do not favor formation of a continuous water film, which prevents the creation of the surface tension gradient necessary for Marangoni flow and accounts for the negligible spreading on these surfaces. Spreading on these surfaces also is less thermodynamically favorable because the spreading coefficient is less positive.

technique under supersaturated conditions. These results indicate that an adsorbed water film on the surface plays an important role in the spreading.

Rapid spreading of a liquid droplet is inhibited by viscous dissipation, and so fluid viscosity is a critical parameter opposing the driving force which is due to the reduction in interfacial energy. In fact, in a continuum analysis a singularity results when the viscous dissipation of a classical fluid is calculated at very short scales near the wetting line associated with a no-slip boundary condition between the spreading liquid and the solid substrate [38,39]. Such a singularity would prevent any wetting of solid surfaces. Obviously, spreading does occur, and this boundary condition must be relaxed in some way to allow slip at the liquid–solid interface. Although the physical basis and molecular processes involved in this slip need to be elucidated, it is conceivable that slip can be achieved by a preexisting water layer on the surface. Contact of the liquid droplet with a preexisting water layer would relax the no-slip boundary condition at the junction of the advancing edge of the droplet, permitting Marangoni flow and high spreading rates. This is illustrated in Fig. 11, which depicts the leading edge region of an advancing droplet in contact with a preexisting water layer present on hydrophilic and intermediate energy surfaces. The contact boundary of the leading edge of the droplet and the preexisting water layer is continuous and may not be very distinct. However, the surfactant concentrations at the air–water interfaces of the droplet and the region just beyond this contact boundary will differ throughout the spreading event because the contact boundary continues to advance across the preexisting water layer, which initially is devoid of surfactant.

The Marangoni effect [19,40] describes flow in the direction of increasing surface tension induced by a surface tension gradient at the air–liquid interface, which creates a tractive force that results in fluid motion in the adjacent bulk liquid. The gradient, produced by contact between the surfactant-containing droplet and a preexisting water layer, can be augmented by dynamic processes described below. Marangoni flow is known to be much more rapid than spreading by surface diffusion over a dry substrate [41]. Such flow could produce the high spreading rates we observed for aqueous surfactant dispersions and solutions and is consistent with the covered area initially increasing linearly with time. However, the lack of an observed increase in spreading rate with relative humidity (and presumably the thickness of any preexisting water film) on the mixed monolayer modified substrates argues against the presence of an adsorbed film of water on the surface, and therefore against explaining the rapid spreading by such a Marangoni flow. Nevertheless, we cannot eliminate the possibility that even under nominally dry conditions, a water layer rapidly forms on the surface as a result of evaporation and adsorption of water vapor from the water droplet (even before contact with the surface).

The most surprising observation in our studies has been that the maxima in spreading rates are a function of surfactant concentration and substrate surface energy. Maxima in physical processes usually signal competitive effects. The driving force for spreading must be related to the reduction in surface energy accompanying the process, probably augmented by Marangoni effects. Our results indicate that spreading of surfactant solutions (as opposed to pure liquids, where such maxima are not seen) also involves competitive inhibitory processes whose rate varies with surfactant concentration and substrate surface energy. We offer the following speculation as to the nature of those processes.

The case of substrate surface energy can be considered first. We assume that a surface can be made sufficiently hydrophobic that a given liquid will no longer spread over it. From the definition of the spreading coefficient in Eq. (1) above, it is clear that for spreading to occur, the surface tension of the liquid must always be less than that of the solid. Therefore the spreading rate must go to zero with decreasing substrate surface energy. At higher surface energies, the spreading rate should increase to some finite value. The precise dependence of the rate on substrate surface energy is not indicated by these elementary considerations. Surprisingly, below an intermediate value of substrate surface energy, the spreading rate decreases. A possible source of this behavior is the difference in the orientation or adsorption of surfactants on (uncharged) hydrophilic versus hydrophobic surfaces [42]. Different adsorption orientations were found by Tiberg and Cazabat [37] in their work on spreading of pure liquid trisiloxane surfactants.

Surfactants adsorb weakly on high energy surfaces, and at sufficiently high concentrations they may adsorb as a bilayer. A dilute surfactant solution allowed to dry on a high energy surface such as glass can form a monolayer with the hydrophobic groups oriented toward the air. This renders the surface hydrophobic, and a subsequent drop of the surfactant solution usually will not spread over its own monolayer—this is called the autophobic effect. Surfactant that adsorbed in this orientation, at or just beyond the advancing edge of the drop (in the region defined by the preexisting water layer) would pin the spreading drop because it would represent a hydrophobic barrier to drop advancement. We suggest that this process could be one of the inhibitive processes responsible for the maximum in spreading rate versus substrate surface energy. An increasing tendency of the hydrophile to adsorb with increasing substrate surface energy could explain the decrease in spreading rate to the high energy side of the maximum versus substrate surface energy. It can also be argued that the dynamic interfacial tension between the surfactant solution and the substrate is increasing with increasing substrate surface energy because the ability of the surfactant to lower the surface energy of the interface by adsorbing there is decreasing.

In contrast, surfactant adsorption on hydrophobic surfaces will tend to direct the hydrophobe toward the surface and the hydrophile toward the liquid phase. Adsorption near the advancing edge of the aqueous film will decrease hydrophobic character of the surface in this region, which will favor spreading of the aqueous film. Furthermore, substrate adsorption of surfactant molecules in front of the advancing edge will tend to deplete the surfactant at the opposing air–water interface, thereby augmenting the surface tension gradient (indicated by the arrow in Fig. 11a, b). Differences in adsorption interactions may also explain the observed shift of the spreading rate peak toward higher energy hydrophilic surfaces with increasing poly(oxyethylene) chain length for both the trisiloxane surfactants and the ethoxylated alcohol surfactants.

Only the fluorocarbon surfactants exceed the trisiloxane surfactants in ability to lower aqueous surface tension. However, fluorocarbon surfactants do not promote rapid spreading on paraffinic surfaces [13]. Even though γ_L is smaller for fluorocarbon surfactant solutions, γ_{LS} against hydrocarbon substrates is large, making S less positive and inhibiting spreading. Values of γ_{LS} for the trisiloxane surfactants against hydrocarbon substrates are relatively small [36] and comparable to typical values for nonionic hydrocarbon surfactants. Thus, it is the combination of low γ_L and γ_{LS} for the trisiloxane surfactants on hydrocarbon substrates that is responsible for their spreading.

Most of the systems we investigated also exhibit a maximum in spreading rate as a function of surfactant concentration. However, unlike the maxima versus substrate surface energy, this maximum varied from quite narrow to extremely broad, and depended on surfactant type and substrate. Interestingly, in all cases it occurred at concentrations much higher (10–100 $\times$) than the critical aggregation concentration (cac). During spreading of a drop of surfactant solution to a thin wetting film, the areas of the air–liquid and solid–liquid interfaces increase substantially. To sustain rapid spreading, surfactant must adsorb quickly at the new interfacial area. Obviously, rates and mechanisms of transport of surfactant to the interfaces are a vital factor in controlling this process. Dynamic surface tensions of surfactant solutions fall from values near pure water at very short times to equilibrium values at long times. The surface tension fall rate continues to increase to concentrations well above the cac. Thus, it is reasonable to expect that spreading rates would increase with increasing surfactant concentration. However, the data indicate that at some point a competitive process assumes control and lowers the spreading rate. One possible candidate is an increasing fluid viscosity with increasing surfactant concentration. However, Zhu et al, found that the viscosities of $M(D'E_8OMe)M$ dispersions [14] do not show a significant increase in the appropriate concentration range. Still, these were low shear viscosities. High shear values, or large amplitude extensional viscosity values, might be more appropriate to the actual flow conditions near the edge of the spreading droplet, and could be much larger. The autophobic effect mentioned above is another possibility.

Another aspect of our spreading results requiring explanation is the observation that within a homologous series of surfactants such as $M(D'E_n)M$, $n = 6$–12, and $C_{12}E_j$, $j = 3$–5, the spreading is always fastest for the homologs that form dispersions of a surfactant-rich phase. We believe that this effect can be attributed to two factors: particles of a surfactant-rich phase that (1) are transported rapidly to the interfaces under conditions of convective flow and (2) favor rapid delivery of surfactant to the interfaces just behind the advancing edge of the spreading aqueous film. These points are illustrated in Fig. 11.

Surfactant transport from the bulk to an interface is thought to involve at least five processes:

1. Diffusion of unassociated surfactant molecules to the region immediately adjacent to the interface
2. Aggregate (dispersed particle, micelle or vesicle) diffusion
3. Aggregate dissociation to individual surfactant molecules
4. Surfactant molecules reorienting and entering the interface
5. Convection of aggregates by mass average liquid motion

Of these processes, the second is usually too slow to be significant. Recent work [36] indicates that surfactant transport in vesicle dispersions can be faster than in micellar solutions. Since vesicles are larger than micelles, this cannot be because process 2 is substantially faster.

Aggregate dissociation (process 3) could be faster, but convective flow during spreading might lead to capture of particles of surfactant-rich phase (including vesicles) by the interface, followed by disintegration of the particle and efficient incorporation of surfactant molecules into the interface. Note that the method used by Svitova [36] to measure dynamic surface and interfacial tensions (the drop volume method) involves flow in and around the interface.

Recent observations in our laboratory have revealed aggregate disintegration during spreading of turbid aqueous dispersions trisiloxane surfactants on mineral oil surfaces [43]. Disintegration events were observed to correlate with stepwise motion of the leading edge of the aqueous film. We believe that similar events also are occurring in turbid surfactant dispersions during spreading on solid substrates. If aggregate disintegration occurs just behind the contact boundary of the advancing aqueous film and the preexisting water layer, this process can deliver a large amount of surfactant to both the expanding air–liquid and solid–liquid interfaces. Surfactant delivered to the air–liquid interface behind the advance drop will produce high interfacial concentrations of surfactant in this region, thereby maintaining the surface tension gradient between this region and the preexisting water layer region as the drop advances. Marangoni flow can thus be sustained under this condition. Different kinetics on different substrates can explain, at least partially, the influence of substrate surface energy on spreading rates. A surfactant aggregate can adsorb on a hydrophilic surface while remaining intact. In contrast, for surfactant hydrophobes to bind to the surface, adsorption on a hydrophobic surface must be accompanied by substantial reorganization, leading to disintegration of the aggregate. Consequently, the decreasing spreading rate with increasing substrate surface energy may reflect diminished aggregate disintegration, which in turn diminishes surfactant delivery to the air–liquid interface. In the systems we have investigated, the surfactant-rich phase is a bilayer microstructure, although in a few cases vesicles or micelles were likely. The flat geometry of bilayer microstructures may facilitate movement of surfactant into the two-dimensional air–liquid interface, and adsorption onto the substrate surface.

While these arguments are consistent with the existing data, they remain speculative in that we have no actual measurements of surfactant molecule orientation and adsorption on the surfaces of interest. Limited dynamic surface and interfacial tension data are available, but a better understanding of their relationship to adsorption and surfactant microstructures is essential. Further study clearly is necessary to elucidate the critical events responsible for rapid spreading of surfactant solutions and dispersions on surfaces, and how these events depend on substrate energy and surfactant concentration.

X. SUMMARY

Recent studies of the spreading dynamics of aqueous surfactant dispersions and solutions have demonstrated the importance, to the rapid wetting of fairly hydrophobic surfaces, of substrate surface energy, surfactant concentration, substrate roughness, and relative humidity. It is now apparent that the observed "superspreading" of trisiloxane surfactant dispersions on Parafilm is a particular example of a more universal phenomenon, which we have termed "surfactant-enhanced spreading." The characteristics of surfactant-enhanced spreading include rapid wetting of fairly hydrophilic substrates, a linear increase in area with time in the initial stages of spreading, a spreading rate maximum on surfaces of intermediate surface energy and, in some cases, a maximum in spreading rate with respect to surfactant concentration.

Surfactant-enhanced spreading does not appear to be directly related to any identifiable surfactant microstructure, nor is it unique to a particular hydrophobic or hydrophilic moiety. In fact, aside from the poor spreading exhibited by the ionic surfactant solutions SDS and DTAB, the observation of surfactant-enhanced spreading appears to be general for nonionic surfactants.

Trisiloxane surfactants remain remarkable for their ability to promote wetting of even very hydrophobic substrates. However, aqueous dispersions of other surfactants, notably $C_{10}E_3$, exhibit maximal rates similar to those of the trisiloxanes, albeit on substrates of slightly higher energy. This result does not support the suggestion by Ananthapadmanabhan et al. [13] that the rapid wetting properties of the trisiloxane surfactants are most likely related to unusually fast kinetics associated with their atypical molecular structure. Similarities have also been observed in the spreading rate dependence on hydrophilic chain length for the trisiloxanes and ethoxylated alcohol surfactants. Increased hydrophilic moiety length results in poorer spreading on hydrophobic surfaces, shifting the peak in spreading rate versus substrate surface energy to correspondingly higher surface energies. Spreading rates were found to be higher on rough substrates and, unlike the results on Parafilm, rapid spreading was observed on mixed monolayer modified gold substrates under very low humidity conditions. The nature of the end-cap group was found to be important in as much as M(D′E_8OMe)M and M(D′E_8OAc)M wet very hydrophobic substrates more effectively than M(D′E_8OH)M. Even with these new observations, however, the mechanism driving the rapid wetting remains elusive.

ACKNOWLEDGMENTS

This work was supported by the Center for Interfacial Engineering (CIE), an NSF Engineering Research Center. The authors are also grateful to L. M. Frostman and C. M. Yip (University of Minnesota) for providing the organosulfur reagents and to T. Knutsen (University of Minnesota) and Rochelle Carleton (University of North Carolina) for technical assistance.

REFERENCES

1. Marmur, A. *Adv. Colloid Interface Sci. 19*, 75 (1983).
2. Zabkiewicz, J. A., and Gaskin, R. E., in N. P. Chow, C. A. Grant, A. M. Hinshalwood, and E. Simundsson, Eds., *Adjuvants and Agrochemicals*, Vol. 1, *Mode of Action and Physiological Activity*, CRC Press, Boca Raton, FL, 1989, p. 142.
3. Gaskin, R. E., and Kirkwood, R. C., in *Adjuvants and Agrochemicals*, Vol. 1, *Mode of Action and Physiological Activity*, (N. P. Chow, C. A. Grant, A. M. Hinshalwood, and E. Simundsson, eds.), CRC Press, Boca Raton, FL, 1989, p. 129.
4. Knoche, M., Tamura, H, and Bukovac, M. J., *J. Agric. Food Chem. 39*, 202 (1991).
5. Knoche, M., *Weed Rese. 34*, 221 (1994).
6. Roggenbuck, F. C., Rowe, L., Penner, D., Petroff, L., and Burow, R., *Weed Technol. 4*, 576 (1990).
7. Stevens, P. J. G., *Pestic. Sci. 38*, 103 (1993).
8. Murphy, G. J., Policello, G. A., and Ruckle, R. E., *Brighton Crop Protection Conference, Weeds, 1*, 355 (1991).
9. Petroff, L. J., U.S. Patent 4,784,799 (1988).
10. Bailey, D. L., U.S. Patent 3,562,786 (1971).
11. Policello, G. A., U.S. Patent 5,104,647, (1992).
12. Hill, R. M., in *Mixed Surfactant Systems, ACS Symposium Ser. Vol. 501* (P. M. Holland and D. N. Rubingh, eds.), American Chemical Society, Washington, DC, 1992, p. 278.
13. Murphy, D. S., Policello, G. A., Goddard, E. D., and Stevens, P. J. G., in *Pesticide Formulations and Application Systems, ASTM STP 1146*, Vol. 12 (B. N. Devisetty, D. G. Chasin, and P. D. Berger, eds.), American Society for Testing and Materials, Philadelphia, 1993, p. 45.
14. Ananthapadmanabhan, K. P., Goddard, E. D., and Chandar, P., *Colloids Surf. 44*, 281 (1990).
15. (a) Zhu, X., Ph.D. thesis, University of Minnesota 1992. (b) Zhu, X., Miller, W. G., Scriven, L. E., and Davis, H. T., *Colloids Surf. 90*, 63 (1994).

16. Hill, R. M., He, M., Davis, H. T., and Scriven, L. E. *Langmuir 10*, 1724 (1994).
17. Stoebe, T, Lin, Z., Hill, R. M., Ward, M. D., and Davis, H. T., *Langmuir*, *12*, 337 (1996).
18. Nilsson, F., and Soderman, O., *Langmuir 12*, 902 (1996).
19. Hardy, W. B., *Collected Works*, Cambridge University Press, Cambridge, 1936.
20. (a) Blackman, L. C. F., and Dewar, M. J. S., *J. Chem. Soc.* 1957, 162. (b) Blackman, L. C. F., and Dewar, M. J. S., *J. Chem. Soc.* 1957, 171. (c) Bain, C. D., Evall, J., and Whitesides, G. M., *J. Am. Chem. Soc. 111*, 7155 (1989). (d) Sanassy, P., and Evans, S. D., *Langmuir 9*, 1024 (1993). (e) Whitesides, G. M., and Laibinis, P. E., *Langmuir 9*, 1024 (1990). (f) Holmes-Farley, S. R., Bain, C. D., and Whitesides, G. M., *Langmuir 4*, 921 (1988).
21. Porter, M. D., Bright, T. A., Allara, D. L., and Chidsey, C. D., *J. Am. Chem. Soc. 109*, 3559 (1987).
22. Bertilsson, L., and Liedberg, B *Langmuir 9*, 141 (1993).
23. (a) Laibinis, P. E., Fox, M. A., Folkers, J. P., and Whitesides, G. M., *Langmuir 7*, 3167 (1991). (b) Bain, C. D., and Whitesides, G. M., *J. Am. Chem. Soc. 111*, 7164 (1989).
24. Lin, Z., Hill, R. M., Davis, H. T., and Ward, M. D., *Langmuir 10*, 4060 (1994).
25. Lin, Z., Stoebe, T., Hill, R. M., Davis, H. T., and Ward, M. D., *Langmuir 12*, 345 (1996).
26. Zisman, W. A., in *Contact Angle, Wettability and Adhesion* (F.M. Fowkes, ed.), ACS Advances in Chemistry Ser., No. 43, American Chemical Society, Washington, DC, 1964, p. 1.
27. Petroff, L. J., Romenesko, D. J., and Bahr, B. C., U.S. Patent 4,933,002 (1990).
28. Gradzielski, M., Hoffmann, H., Robisch, P., and Ulbricht, W., *Tenside Surf. Detergents 27*, 366 (1990).
29. (a) Mitchell, J. D., Tiddy, G. J. T., Waring, L., Bostock, T., and McDonald, M. P., *J. Chem. Soc., Faraday Trans. 1 79*, 975 (1983). (b) Bleasdale, T. A., and Tiddy, G. J. T., *NATO ASI Ser., Ser. C 324*, 397 (1990). (c) Strey, R., Schomaker, R., Roux, D., Nallet, F., and Olsson, U., *J. Chem. Soc., Faraday Trans. 86*, 2253 (1990). (d) Lang, J. C., and Morgan, R. D., *J. Chem. Phys. 73*, 5849 (1980).
30. Stoebe, T, Hill, R. M., Ward, M. D., and Davis, H. T., submitted for publication.
31. de Gennes, P. G., *Rev. Mod. Phys. 57*, 827 (1985).
32. Tetetzke, G. F., Davis, H. T., and Scriven, L. E., *Chem. Eng. Commun. 55*, 41 (1987).
33. Blake, T. D., De Coninck, J., and D'Ortona, U., *Langmuir 11*, 4588 (1995).
34. Blake, T. D., *Wettability, Surfactant Sci. Ser. 49*, 251 (1993).
35. Dukhin, S. S., Kretzschmar, G, and Miller, R., *Dynamics of Adsorption at Liquid Interfaces*, Elsevier, Amsterdam, 1995.
36. Svitova, T., Hoffmann, H., and Hill, R. M., *Langmuir 12*, 1712 (1996).
37. Tiberg, F., and Cazabat, A.-M., *Europhys. Lett. 25*, 205 (1994).
38. Huh, C., and Scriven, L. E., *J. Colloid Interface Sci. 35*, 85 (1971).
39. Blake, T. D., *Colloids Surf. 47*, 135 (1990).
40. Thomson, J., *Philos. Mag. 10* (4th ser.), 330 (1855).
41. Troian, S. M., Herbolzheimer, E., and Safran, S. A., *Phys. Rev. Lett. 65*, 333 (1990).
42. Rosen, M. J., *Surfactants and Interfacial Phenomena*, Wiley-Interscience, New York 1989, p. 57.
43. Stoebe, T., Hill, R. M., Ward, M. D., and Davis, H. T., submitted for publication.

12

Ternary Phase Behavior of Mixtures of Siloxane Surfactants, Silicone Oils, and Water

RANDAL M. HILL Central Research and Development, Dow Corning Corporation, Midland, Michigan

X. LI* Department of Chemical Engineering and Materials Science, University of Minnesota, Minneapolis, Minnesota

H. TED DAVIS Dean, Institute of Technology and Department of Chemical Engineering and Materials Science, University of Minnesota, Minneapolis, Minnesota

**Current affiliation*: Applied Materials Inc., Santa Clara, California.

I. INTRODUCTION

Surfactants self-assemble in aqueous solution to form microstructures such as globular and cylindrical micelles. At higher concentrations liquid crystalline phases are formed. Most of these structures are able to solubilize significant amounts of nonpolar oils. Ternary mixtures of water, surfactant, and oil, which spontaneously self-assemble to form isotropic, low viscosity liquid phases, are called microemulsions.* The ternary phase behavior of mixtures of hydrocarbon surfactants, water, and organic oils has been extensively studied, and the principles governing such systems are well understood [1–3].

Nonionic hydrocarbon surfactants containing polyoxyethylene polar groups (the C_iE_j homologous series) exhibit a lower consolute temperature (LCT) boundary, or cloud point, transforming from one phase at lower temperatures to two phases above the boundary. Mixtures of alkyl ethoxylate surfactants with nonpolar organic oils exhibit an upper consolute temperature (UCT) boundary transforming from two phases at lower temperatures to one phase above the boundary. The superposition of these opposite behaviors in ternary mixtures creates a three-phase body in the temperature range in which the boundaries overlap [1]. Isotropic liquid phases spring from the binary sides of the temperature-composition phase prism (Fig. 1 [p. 318]), which are analogous to the one-phase regions in the binary systems. These include droplet microemulsions analogous to micellar solutions (L_1), bicontinuous microemulsions analogous to the sponge phase (L_3), and inverted droplet microemulsions analogous to the surfactant-rich isotropic phase (L_2). The fundamental principles governing the phase behavior has been described in terms of the hydrophile–lipophile balance (HLB), the phase inversion temperature (PIT), and the classical Winsor sequence of $\underline{2}\Phi$ to 3Φ to $\overline{2}\Phi$, or Winsor type I to III to II, with increasing temperature, or, more generally, with any thermodynamic field variable [2].

The trisiloxane surfactants shown in the accompanying diagram are used in many commercial applications such as wetting agents, agricultural adjuvants, and paint additives [5]. Despite their wide usage, study and understanding of these siloxane surfactants have been mostly limited to their wetting behavior and surface

*Since it is common, especially in the patent literature, to describe a variety of transparent mixtures of water, oil, and surfactants, as microemulsions, it is important that we make some careful distinctions with regard to the nature of these mixtures. Many patents refer to systems that are transparent by virtue of index of refraction matching as microemulsions. Such systems may also be described as gels having quite high viscosity. Other patents and articles describe systems that are transparent by virtue of having a particle size less than half the wavelength of light as microemulsions. When specific and critical processing pathways are required to prepare such mixtures, the mixtures are obviously not spontaneously self-assembling. Much of the literature referred to in Sec. I of this chapter refers to systems of these types. Our discussion in Secs. II–IV of our recent work with the trisiloxane surfactants refers to true microemulsions, which are equilibrium isotropic phases containing oil, water, and surfactant [1].

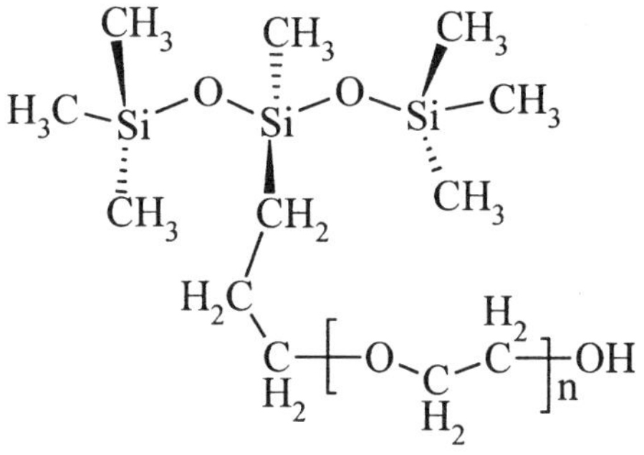

Molecular structure of the trisiloxane polyoxyethylene surfactants

activity. Recently the binary phase behavior and microstructure of trisiloxane and polymeric siloxane surfactants in water have been investigated. The self-assembly of siloxane surfactants in water is discussed elsewhere in this volume [5,6].

Emulsification of silicone oils, especially polydimethylsiloxane (PDMS), by hydrocarbon surfactants is extensively described in the patent art and in the literature [7–9]. Methods and compositions to prepare microemulsions of amine functional silicone polymers are also well known [10–13]. Katayama et al. [14] and Gee [15] have presented studies of the phase behavior of the amine functional silicones. In a frequently cited patent, Gee [16] described a method of preparing emulsions of a polyorganosiloxane (silicone polymer) yielding particle sizes sufficiently small to render the product transparent, or at least translucent. Anderson et al. [17] prepared monodisperse emulsions of silicone oil by hydrolysis of dimethyldiethoxysilane. Dumoulin [18] also disclosed compositions that could be used to form transparent microemulsions or gels containing diorganopolysiloxanes. A number of patents have described procedures for preparing microemulsions of silicone polymers by a route involving polymerization of lower molecular weight starting materials [19–22].

Messier et al. [23] present a isothermal ternary phase diagram for the water/surfactant/PDMS ($MD_{46}M$) system, where the surfactant is a blend of alkyl phenol ethoxylate surfactants that shows a very narrow one-phase region along the water–surfactant axis; the rest of the diagram is two phases. These workers used small-angle x-ray (SAXS) scattering techniques to show that hexamethyldisiloxane (MM) swells the lamellar phase formed by $C_{10}E_5$ while the higher molecular weight silicone oils do not.

John et al. [24] assert that "polydimethyl siloxane, as an oil phase, has difficulty forming single-phase microemulsions [and that] phase behavioral studies using this oil are also difficult as clean phase separation is seldom obtained." To investigate the phase behavior of systems containing silicone oils, these authors dissolved a relatively low molecular weight polymeric silicone oil in a polar organic oil with known phase behavior and determined how the phase diagram evolved with increasing levels of polymeric oil. The starting system consisted of water, $C_{12}E_6$, and isopropyl myristate. At 16% silicone oil (as a fraction of the total oil present), the three-phase body no longer closed off at high surfactant concentrations to form one phase microemulsion. The authors conceded that this might be due to the silicone oil preferentially partitioning into the excess oil phase.

Binks and Dong [25] investigated emulsions and equilibrium phase behavior of PDMS in mixtures with the nonionic C_iE_j surfactants. They found little solubilization of this polymeric oil into either micelles or surfactant-rich third phases. Hoffmann and Stürmer [26] investigated the solubilization of several low molecular weight linear and cyclic silicone oils in solutions of a cationic hydrocarbon surfactant containing rodlike micelles. They found that these solutions solubilized larger quantities of the silicone oils than alkanes of the same molecular weight. The silicone oils influenced the microstructure of the surfactant solution in a very similar way to the alkanes. Steytler et al. [27] report the formation of water-in-oil microemulsions stabilized by AOT in low molecular weight silicone oils. All the work described above involves conventional hydrocarbon surfactants.

Blehm and White [28] patented the preparation of microemulsions of cyclic and short linear dimethylsiloxanes by means of mixtures of several cosurfactants with an unusual cationic hydrocarbon surfactant containing a reactive alkoxy silane group. Several phase diagrams with hexane as the oil were presented, but no actual examples involving silicone oils were given. This patent, like much of the patent art, uses the term "microemulsion" to mean a transparent mixture of water, surfactant, and oil. It is stated that "because of the small average particle size microemulsions are clear." The details given regarding mixing procedures and the claims for freeze–thaw stability suggest that these are not spontaneously self-assembling systems.

Gee and Keil [29] and Keil [30] disclosed transparent mixtures of water and low molecular weight silicone oils using a silicone–polyoxyalkylene copolymer as the surfactant, including potential uses of such mixtures for personal care applications. Other workers have also described the use of polymeric silicone surfactants to prepare transparent mixtures of silicone oils and aqueous phases [31–33]. These mixtures are sometimes called microemulsions in the literature in the sense of being transparent mixtures of water, surfactant, and oil. They may be transparent because of small particle size, or because of index of refraction matching. They are usually sensitive to mixing procedures and processes and therefore are not spontaneously self-assembling microemulsions. Dahms and Zombeck [34] discussed the stabilization mechanisms of these unusual emulsifiers. Gasperlin et al. [35,36] and Baquerizo et al. [37] investigated the microstructure and rheology of this type of transparent mixture. A few studies have discussed emulsification of hydrocarbon oils by siloxane surfactants [38].

Other than these systems based on polymeric siloxane surfactants, there have been few studies of the ternary phase behavior and microstructure of ternary systems consisting of water, siloxane surfactants, and silicone oil. Kunieda et al. [39] reported the phase behavior of several polyoxyethylene trisiloxane surfactants, including some results for ternary mixtures with decane and a cyclic silicone oil, octamethylcyclosiloxane (D_4). They found three phases, including middle phase microemulsion, at 25°C for intermediate polyoxyethylene chain lengths for both oils. Iwanaga et al. [40] report an isothermal phase diagram for a rake-type silicone polyoxyethylene copolymer with D_4 showing L_α and H_2 liquid crystal phases, but no microemulsion.

Mayer [7] discusses the use of mixtures of low molecular weight silicon-based surfactants and cosurfactants to prepare a self-dispersing microemulsion of silicone agents used to impart water repellency to building materials. The specific surfactants used are not disclosed, but they are described as being themselves reactive so that they bind to the surfaces of the building materials and become part of the water repellency treatment.

Von Berlepsch and Wagner [41] investigated the ternary phase behavior of several small silicon-containing surfactants with hexamethyldisiloxane (MM) in water. The surfactants included a silane and a trisiloxane polyether surfactant, and a silane, a trisiloxane, and a carbosilane amine surfactant (the molecular structures denoted by these names are given in Chapter 1 of this volume). The polyether surfactants exhibited behavior similar to the C_iE_j surfactants, while the amines required the addition of cosurfactant to form microemulsions. All the surfactants were relatively inefficient, requiring about 15% surfactant to form one-phase microemulsion of a 1:1 water/oil mixture.

Since the trisiloxane surfactants are similar in many respects to the C_iE_j series, it would be very interesting to systematically compare the ternary phase behaviors of mixtures of a homologous series of the trisiloxane surfactants with water and low molecular weight silicone oils. We have recently determined the ternary phase behavior of a series of homologous trisiloxane surfactants with low molecular weight cyclic and linear silicone oils [42,43]. The remainder of this chapter reviews and summarizes those results.

II. TERNARY PHASE BEHAVIOR

Studies of the ternary phase behavior of nonionic surfactants often make use of two types of cut through the ternary phase prism (Fig. 1) [44,45]. A cut varying the temperature and surfactant concentration at constant oil-to-water ratio is called a "fish cut" because the three-phase body and its contiguous V-shaped region of one-phase microemulsion can be said to form the body and tail of a fish. A cut varying the temperature and oil concentration at constant surfactant concentration is called a channel cut because of the presence of a continuous channel of one-phase microemulsion (connecting to the L_3 region on the water–surfactant binary) from the water-rich to the oil-rich side of the diagram (usually found at low to

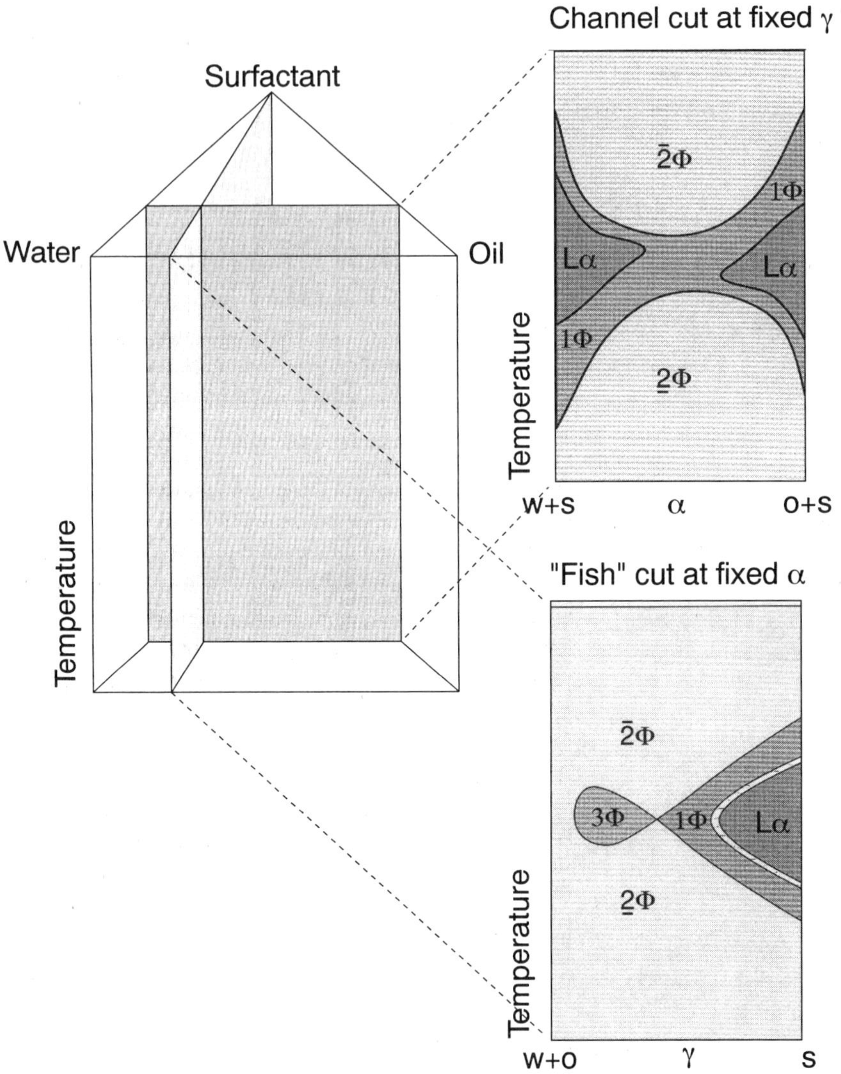

FIG. 1 Diagram of the Gibbs phase prism illustrating the fish and channel cuts.

moderate surfactant concentrations). We have used studies of both types to investigate the ternary phase behavior of mixtures of siloxane surfactants with silicone oils and water. The independent composition variables are defined as α and γ, where α is the weight ratio of oil to oil plus water:

$$\alpha = \text{weight oil/(weight water + weight oil)}$$

and γ is the weight ratio of surfactant in the total system:

$$\gamma = \text{weight surfactant/(weight water + weight surfactant + weight oil)}$$

In the phase diagrams that follow, the notation 1Φ denotes a one-phase isotropic liquid phase or microemulsion; 3Φ denotes a region of three liquid phases with the surfactant present primarily in the middle phase; $\underline{2}\Phi$ indicates two liquid phases with the surfactant present primarily in the lower, water-rich phase; $\overline{2}\Phi$ indicates two liquid phases with the surfactant present primarily in the upper, oil-rich phase. In addition, L_α indicates lamellar liquid crystal phase; H_1 denotes the normal hexagonal phase liquid crystal, I_1 denotes the normal cubic phase; and I_3 denotes the inverted (oil-continuous) cubic phase.

III. MATERIALS AND METHODS

A shorthand notation is used for the trisiloxane surfactants that is derived from the organosilicon literature [46,47], in which these surfactants are denoted $M(D'E_n)M$, where the symbols are defined as follows:

M = the trimethyl siloxy group, $(CH_3)_3SiO_{1/2}$—

D′ = —$O_{1/2}Si(CH_3)(R)O_{1/2}$—, where R is a polyoxyethylene group attached to the silicon by way of a propyl spacer

E_n = polyoxyethylene, —$(CH_2CH_2O)_nH$

The methods and materials used in this investigation were described in detail by Li [48] and are reviewed only briefly here.

The molecular structure of the trisiloxane surfactants in this study was shown above. Binary and ternary phase behavior was characterized for trisiloxane surfactants with polyoxyethylene (EO_n) groups with $n = 6$, 8, 10, 12, and 18. The trisiloxane surfactants were prepared by hydrosilylation of 1,1,1,3,5,5,5-heptamethyltrisiloxane with the appropriate allyl polyoxyethylene derivatives and chloroplatinic acid catalyst. The 1,1,1,3,5,5,5-heptamethyltrisiloxane was distilled to better than 95 purity prior to hydrosilylation, thus, the trisiloxane hydrophobe was essentially monodisperse, while the polyoxyethylene groups were polydisperse ($M_w/M_n \approx 1.1$ by gel permeation chromatography).

The molecular structure of the cyclic silicone oil, octamethylcyclotetrasiloxane ($C_8H_{24}O_4Si_4$ or D_4) is:

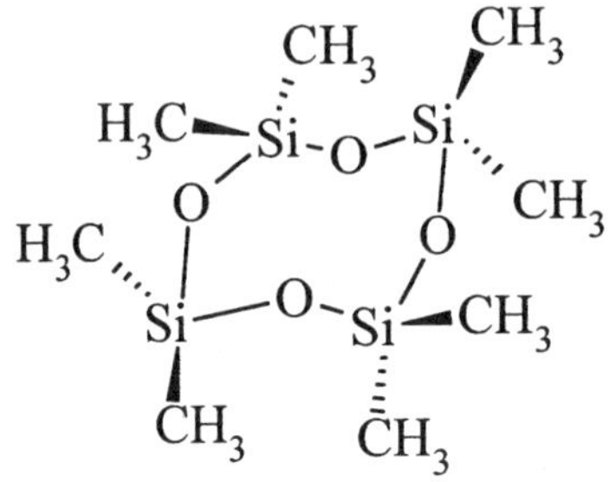

The molecular structure of decamethylcyclopentasiloxane ($C_{10}H_{30}O_5Si_5$ or D_5) is:

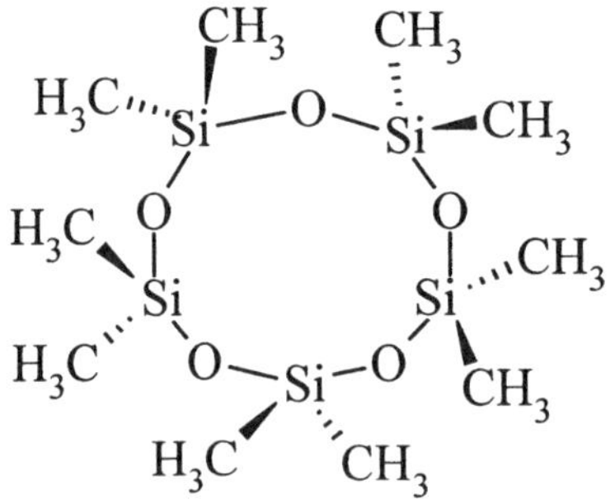

D_4 and D_5 were purchased from Fluka. The molecular structure of the short linear silicone oil, decamethyltetrasiloxane (MD_2M) is:

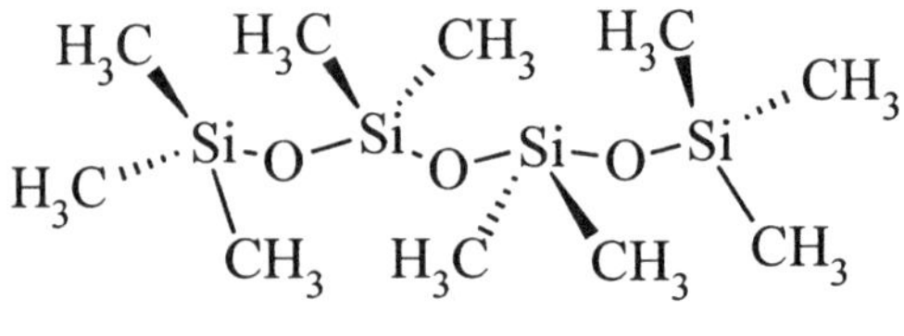

MD_2M was purchased from United Chemical Technologies, Inc. (Bristol, PA). All chemicals were used as received.

The molecular weights, densities, and molar volumes of the four silicone oils used in this study are listed in Table 1. In spite of the presence of highly polar Si–O–Si bonds in these molecules, they are nonpolar hydrophobic oils, more similar to the alkanes than to (e.g.) polyethers. This is demonstrated directly by the strong immiscibility of water and silicone oils. [49].

Fresh deionized water was used to prepare mixtures of water, surfactant, and oil by weight. Along with a small magnetic stir bar, each solution of known composition was contained in a 7 mL, 1 cm diameter, sealed glass tube with 0.1 mL volumetric tick marks, which were submerged in a glass water bath with precise temperature control (0.1°C). The sample in the water bath was viewed between crossed polarizers for the examination of birefringence. A low power laser beam was also used to detect faint levels of turbidity. At temperatures 2°C apart, and then 0.2°C apart near transition temperatures, samples were well mixed with stirring or gentle hand shaking and allowed to reach equilibrium. After sufficient time to reach equilibrium had passed, samples were inspected for evidence of turbidity and birefringence. The onset, or disappearance, of turbidity or birefringence was taken to mark a phase transition.

The transition from two phases to three phases with rising temperature requires careful observation to detect a middle, third layer. Lamellar and hexagonal liquid crystal phases are easily identified by their characteristic texture or appearance when viewed using a polarized light microscope. Further details of the microstructure of the liquid crystal phases, such as interaggregate spacing, were obtained from small-angle x-ray scattering and cryotransmission electron microscopy (cryo-TEM). Details of SAXS and cryo-TEM methods are given by Li [48]. The cubic liquid crystal phases are a special problem because they are not birefringent. We identified the cubic phases from their extremely high viscosity and from SAXS data.

TABLE 1 Physical Properties of the Silicone Oils Used in This Study

	MW	Density at 20°C (g/cm^3)	Molar volume (cm^3/mol)
D_4	296.61	0.956	310.26
D_5	370.77	0.959	386.62
MM	162.38	0.764	212.54
MD_2M	310.69	0.854	363.81

IV. RESULTS AND DISCUSSION

A. The Binary Water/M(D′E_6)M System

The binary phase diagram for mixtures of M(D′E_6)M and water is shown in Fig. 2. A water-rich isotropic phase, W, is found at very dilute M(D′E_6)M concentrations ($< \approx 0.1$ wt %, not shown on the diagram). An extensive region of lamellar phase liquid crystal occupies the central portion of the phase diagram. SAXS spectra of samples in this region show maxima at wave vectors, q, in the ratio of 1:2, which is characteristic of lamellar phases [50]. We mark a dashed vertical line on the diagram at 15 wt % surfactant because the precise location of the phase boundary in this region is uncertain. Other workers have also noted difficulties with other surfactants in the dilute region, due in part to sensitivity to shear when a highly swollen lamellar phase is present [51]. The thickness of surfactant bilayer d_s was estimated from SAXS results to be about 34.4 Å. In the region denoted W + L_α in Fig. 2, the mixture was cloudy and weakly birefringent. A dispersion of mostly globular unilamellar vesicles of 50–500 nm, along with some small vesicles trapped in larger ones, appears in cryo-TEM micrographs of samples in this region [48].

Figure 2 also shows a narrow region above and extending to the left of the L_α region, which we have labeled L_3, and which we show as contiguous with a region we have labeled L_2. We found 3–5°C wide corridor of transparent and isotropic phase, above the L_α phase at all concentrations between 1 and 80 wt % surfactant. We have labeled the water-rich end of this band L_3, and the surfactant-rich end L_2. Li et al. [42] observed that samples near the low concentration end of this region were flow birefringent, consistent with the behavior of the L_3 phase i.e., the sponge phase having a disordered bilayer microstructure [52]. At higher concentrations, the flow birefringence disappears, consistent with the behavior of the L_2 phase. We do not know how or where in the one-phase corridor the L_3 phase becomes the L_2 phase. We simply report the results as we observed them. We have, however, previously pointed out [53] that there is no fundamental reason why the microstructures associated with L_3 and L_2 *must be* separated by a *first-order* phase transition. Moreover, He et al. [54] earlier demonstrated that the E_{12} trisiloxane surfactant M(D′E_{12})M evolves continuously (in a broad temperature band) from a microstructure analogous to L_1 to a spongelike microstructure analogous to L_3 to a microstructure analogous to L_2, without undergoing any phase transitions. M(D′E_{10})M shows a similar behavior (see Sec. IV.D). We speculate that a similar progression may be occurring within a narrower temperature band for M(D′E_6)M. Finally, we should point out that far from being forbidden, or even atypical, continuous evolution with no phase boundaries from bicontinuous microstructures analogous to L_3 to microstructures analogous to L_2 is, in fact, the rule in microemulsions, as is illustrated by the schematic of the microstructures in a fish-cut phase diagram shown by Strey et al. [55].

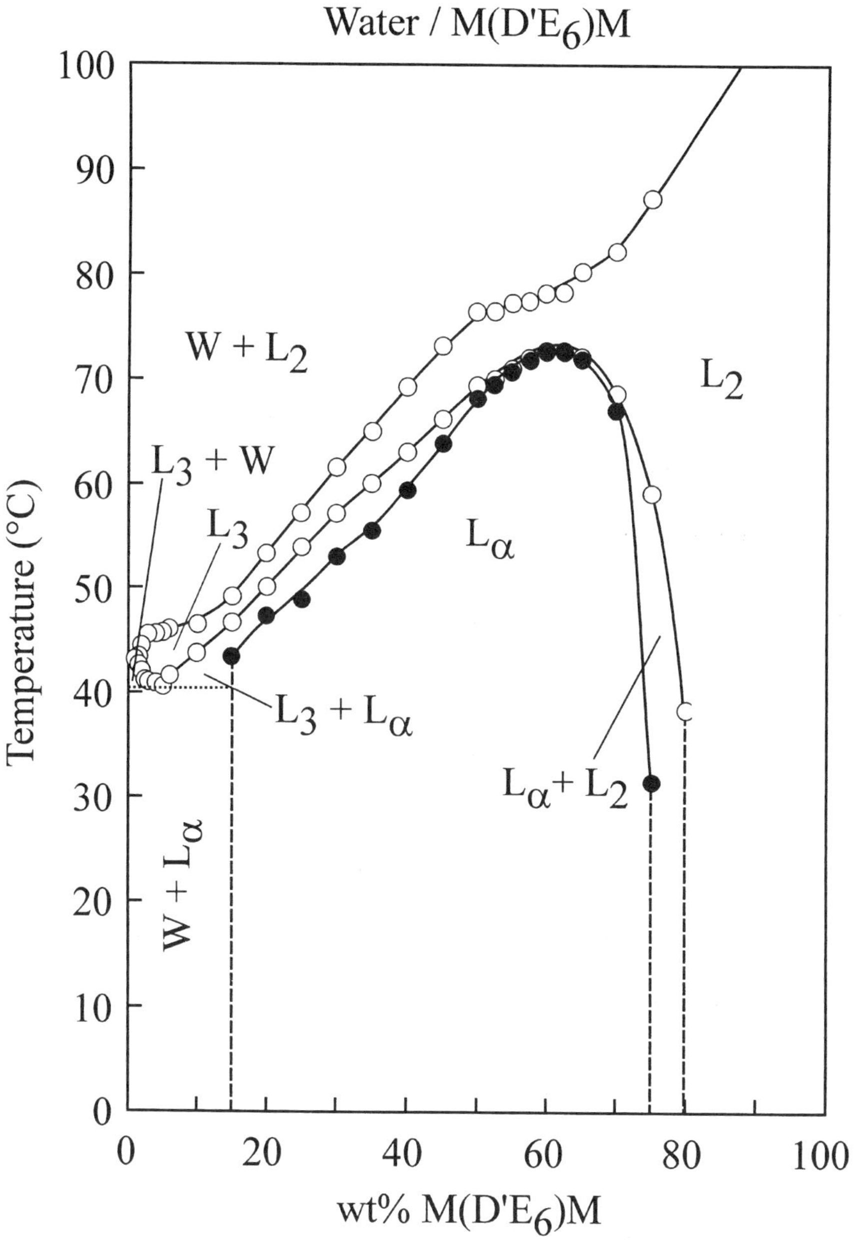

FIG. 2 Binary phase diagram of $M(D'E_6)M$/water. (From Ref. 42, used with permission.)

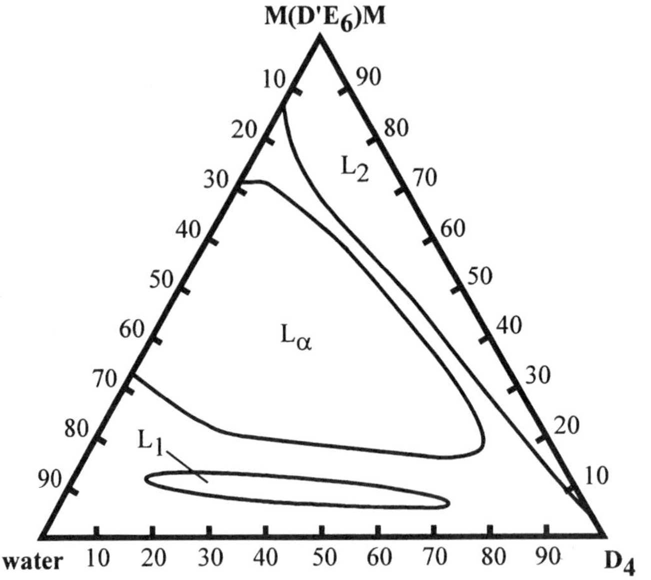

FIG. 3 Isothermal ternary phase diagram of the water/M(D′E_6)M/D_4 system at 20°C. (From Ref. 42, used with permission.)

B. The Ternary Water/M(D′E_6)M/D_4 and D_5 Systems

Isothermal phase diagrams for the water/M(D′E_6)M/D_4 and water/M(D′E_6)M/D_5 ternary systems at 20°C are shown in Figs. 3 and 4, respectively. The two diagrams are quite similar, and both have a large region of lamellar phase, L_α. The L_α phase in the ternary and binary systems share the same origin—the added oil is solubilized into the hydrophobic part of surfactant bilayer, causing it to swell. In the regions denoted as L_α phase, polarized light microscopy reveals a mosaic texture in samples of the water/M(D′E_6)M/D_4 system and oily-streaks texture [56,57] in samples of the water/M(D′E_6)M/D_5 system. Both textures are fingerprints of the lamellar phase. There are two narrow regions of microemulsion in each ternary phase diagram. One is near the M(D′E_6)M/silicone oil axis (labeled L_2). This is a water-lean microemulsion and exists almost continuously from the surfactant corner to the oil corner. The other microemulsion region (labeled L_1) is roughly parallel to the water–oil axis and contains about 10 wt % surfactant.

Figure 5 is a fish-cut phase diagram for the water/M(D′E_6)M/D_4 system at α = 50 wt %. This diagram shows that $\underline{2}\Phi$, a water-continuous, two-phase mixture with the surfactant predominantly in the aqueous phase, is formed in the low temperature, low surfactant concentration region. At higher temperatures, $\overline{2}\Phi$, an oil-

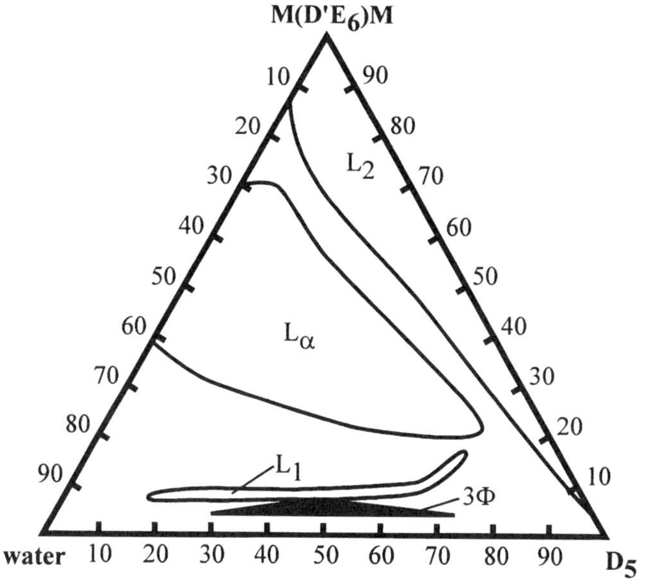

FIG. 4 Isothermal ternary phase diagram of the water/M(D′E_6)M/D_5 system at 20°C. (From Ref. 42, used with permission.)

continuous, two-phase mixture with the surfactant predominantly in the oil phase, is formed. In between is a region of one-phase microemulsion that is narrow at low surfactant concentrations and widens rapidly above 50% surfactant. A region of lamellar phase, L_α, is found below the microemulsion region. This diagram is similar in appearance to Fig. 2, the phase diagram for the binary water/M(D′E_6)M system. No three-phase body is observed—the oil is completely incorporated into the sponge phase at the lowest surfactant concentrations we investigated, even at relatively high weight fractions of oil. In effect, Fig. 5 shows only the upper half of the fish "tail."

Figure 6 is a fish-cut phase diagram for the water/M(D′E_6)M/D_5 system at α = 48.93 wt %. The fish has been shifted upward and to the right compared with D_4 (which is the usual trend with increasing molecular weight), and now the three-phase body and some of the lower branch of the tail have become visible.

C. The Ternary Water/M(D′E_8)M/D_4, D_5, and MD_2M Systems

The binary phase diagram for M(D′E_8)M in water is given by Hill et al. [53]. Like M(D′E_6)M, there is a large region of L_α and some uncertainty regarding the location of phase boundaries in the dilute region. Figure 7 is an isothermal phase dia-

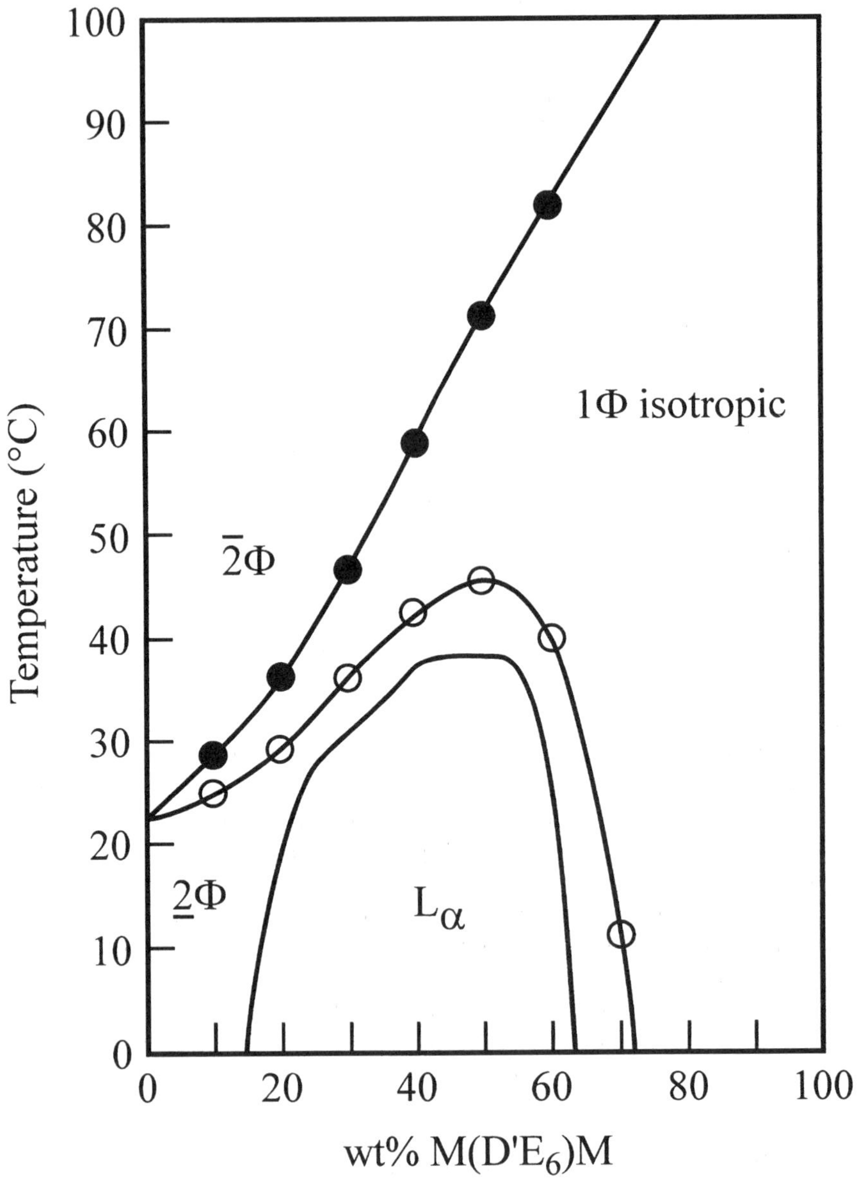

FIG. 5 Fish-cut phase diagram of the water/M(D′E_6)M/D_4 system at $\alpha = 50$ wt %, where α is weight oil/(weight oil + weight water). (From Ref. 42, used with permission.)

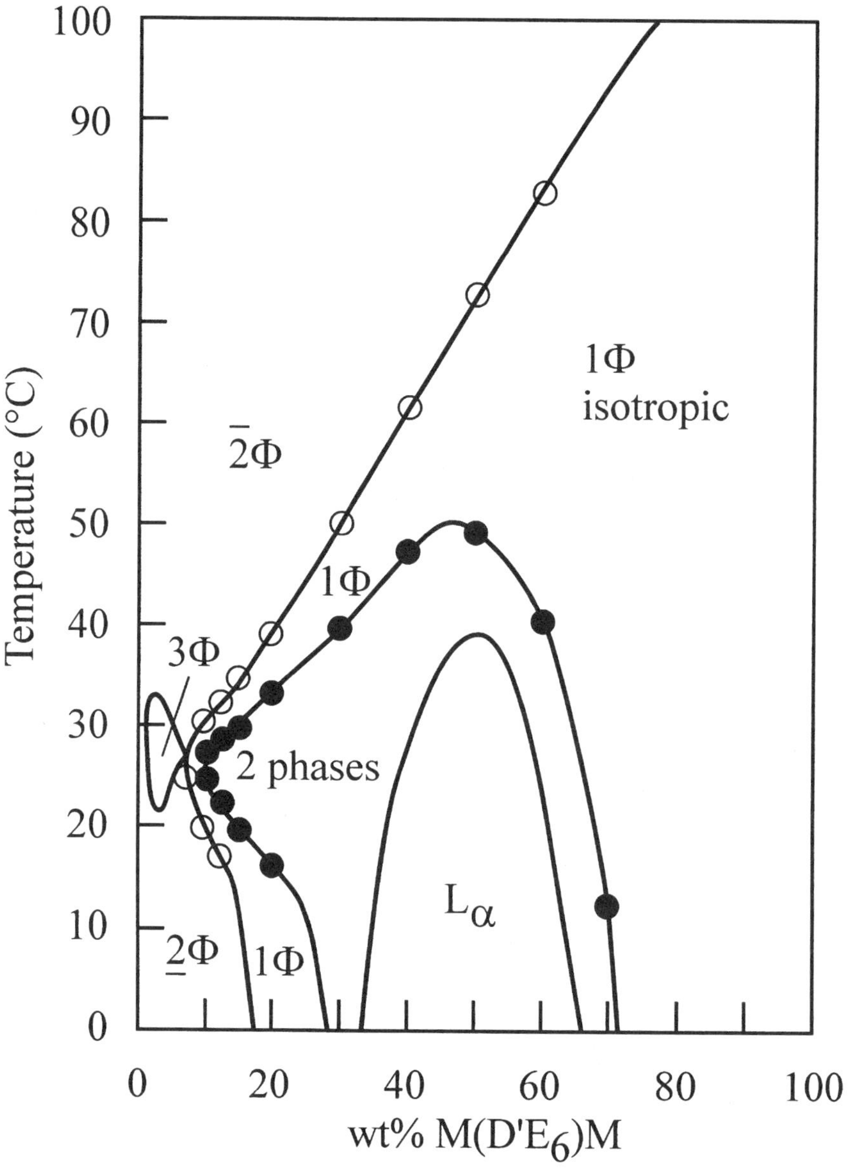

FIG. 6 Fish-cut phase diagrams of the water/M(D′E_6)M/D_5 system at α = 48.93 wt %. (From Ref. 42, used with permission.)

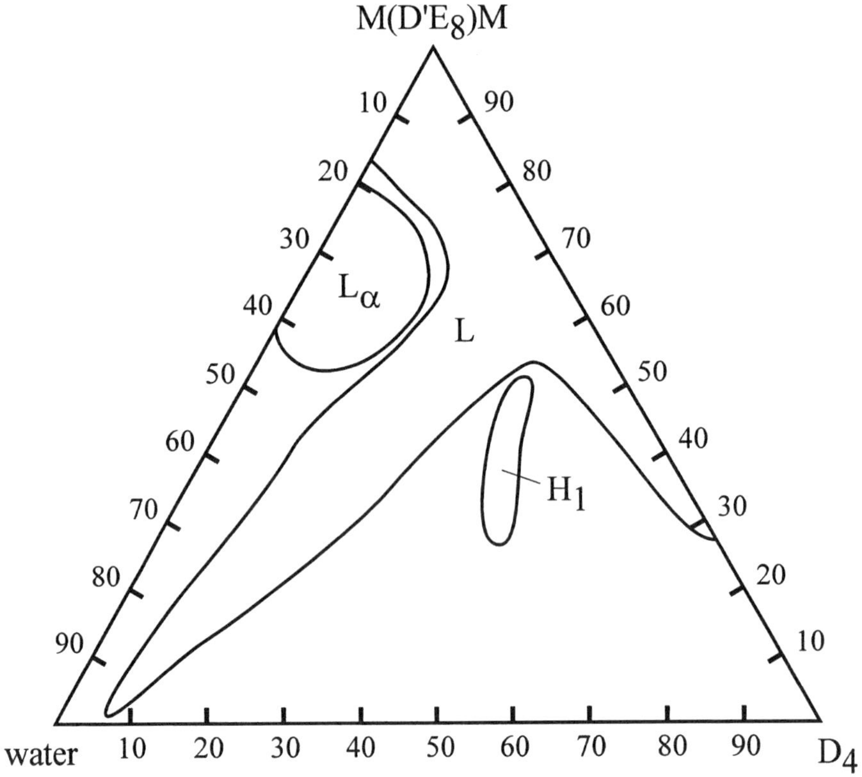

FIG. 7 Isothermal ternary phase diagram of the water/M(D′E_8)M/D_4 system at 25°C.

gram for the water/M(D′E_8)M/D_4 system at 25°C. Isotropic surfactant-rich microemulsion is formed in the surfactant corner, and along the surfactant–oil axis. A relatively wide band of microemulsion extends from the center of this region almost all the way to the water corner. The L_α phase formed by the M(D′E_8)M/water binary swells with added oil to about 15%. Higher levels of oil lead to formation of microemulsion. At even higher proportions of oil a small region of H_1, or hexagonal, phase is found (there is no H_1 phase in the binary diagram). This sequence shows that increasing amounts of oil lead to progressively higher curvature aggregates, from L_α to what is probably a bicontinuous microemulsion, to H_1. Compared to M(D′E_6)M, the L_α region has contracted toward the water–surfactant axis, and there is no single phase isotropic region (L_1 in Fig. 4) in the lower part of the diagram.

In a channel-cut phase diagram for the water/M(D′E_8)M/D_4 system at $\gamma = 0.2$ (Fig. 8), a channel of microemulsion is formed with added oil, which connects to

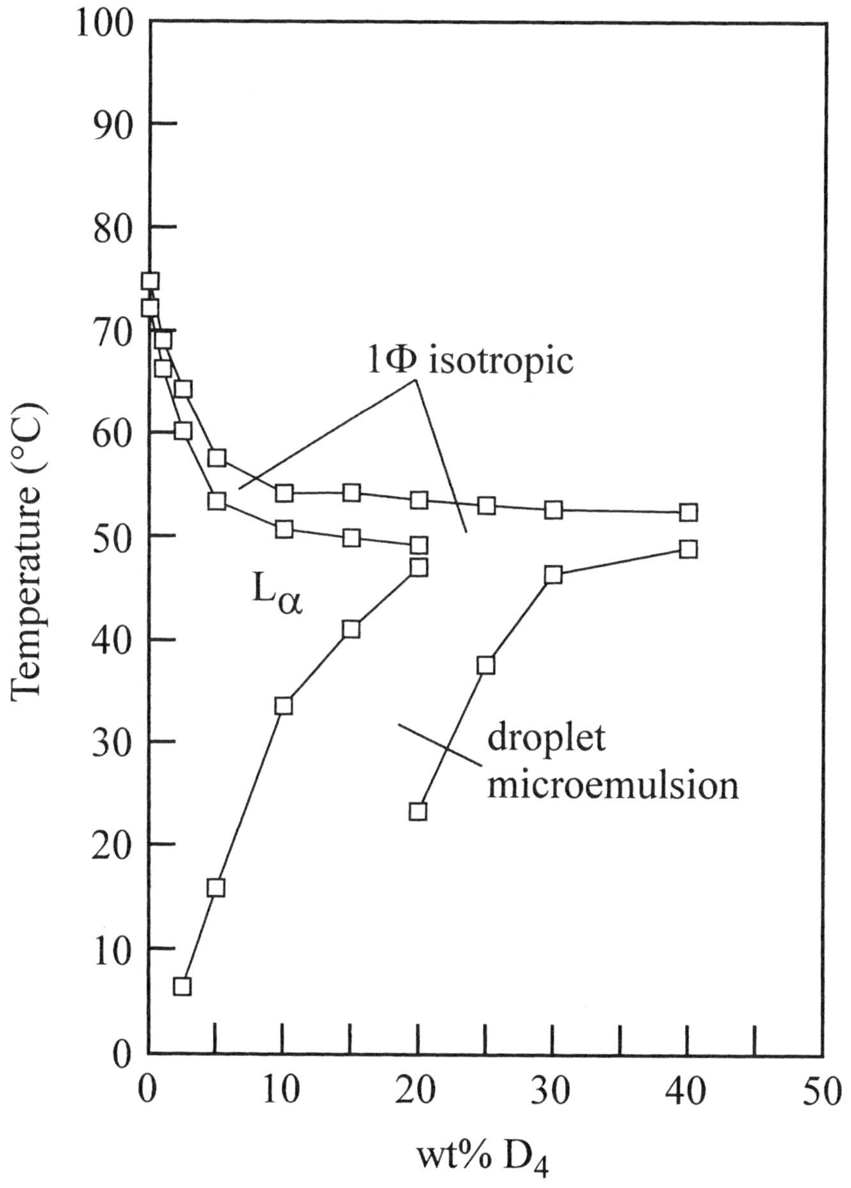

FIG. 8 Channel-cut phase diagram of the water/M(D′E$_8$)M/D$_4$ system at γ = 20 wt %,

the L_3 region of the binary system. Adding D_4 up to about 5% shifts the L_3 region progressively toward lower temperatures. Above 5% D_4 the L_3 channel runs nearly horizontal until it merges with another single isotropic phase channel. This channel begins at lower temperature and low D_4 concentrations in the diagram, where the phase is a water continuous microemulsion. The channel merges with the L_3 channel at about 50°C and the channel formed by the merger continues nearly horizontally to what is presumably bicontinuous microemulsion in the 50/50 water/oil region. At about 20% D_4 the water continuous microemulsion appears bluish hazy in a strong light, indicating that the particle size is relatively large, between 200–500 nm. The water continuous microemulsion channel and the L_3 channel evolve continuously to the bicontinuous microemulsion region without undergoing a first-order phase transition.

Figure 9 shows a series of fish-cut phase diagrams for the water/M(D′E_8)M/D_4 system, varying α from 0.4 to 0.8. The body of the fish slants somewhat (especially at $\alpha = 0.4$), probably as a result of the polydispersity of the surfactant. These diagrams show the classical $\underline{2}\Phi$ to 3Φ to $\bar{2}\Phi$ progression with increasing temperature. As the proportion of oil in the system increases, the fish shifts upward and to higher surfactant concentrations. At $\alpha = 0.5$, C_{min}, the minimum amount of surfactant required to form a one-phase microemulsion from a 1:1 mixture of water and oil, is about 12–13%. Thus, this surfactant is only moderately efficient for making microemulsions of this oil. For clarity, a region of L_α in the middle of each fish tail has been omitted from the diagram.

The midpoint of the temperature range of the 3Φ and the 1Φ regions is called the phase inversion temperature, or PIT. It is also called the HLB temperature because it is the temperature at which the hydrophilicity and the lipophilicity of the surfactant are exactly balanced for a given oil. For ethoxylate surfactants the PIT is useful as a guide to surfactant selection for emulsion stabilization. Oil-in-water emulsions are said to be relatively stable when stored at temperatures 25–60°C lower than their PIT, while water-in-oil emulsions are relatively stable when stored at temperatures 10–40°C higher than their PIT [58]. PIT of M(D′E_8)M in for D_4 in deionized water at $\alpha = 0.5$ is about 56°C and increases by about 2–3°C for every 0.1 unit increase in oil-to-water ratio.

The influence of NaCl on the phase behavior of the water/M(D′E_8)M/D_4 system at $\alpha = 0.5$ is shown in Fig. 10 for 5, 10, and 15% NaCl. Addition of NaCl shifts the phase boundaries to lower temperatures and smaller surfactant concentrations (exactly the opposite of the effect of oil-to-water ratio). As the 3Φ body is shifted toward the axis, it becomes smaller and is difficult to detect at 15% NaCl. At 15% NaCl, C_{min} is about 7%, indicating that with the added salt, this surfactant is very efficient at stabilizing microemulsions of D_4. Channel-cut phase diagrams show that the lower, droplet microemulsion branch becomes broader with added salt. These two effects—shifting the microemulsion region to lower temperatures and stabilizing the microemulsion phases over a wider composition range—are gen-

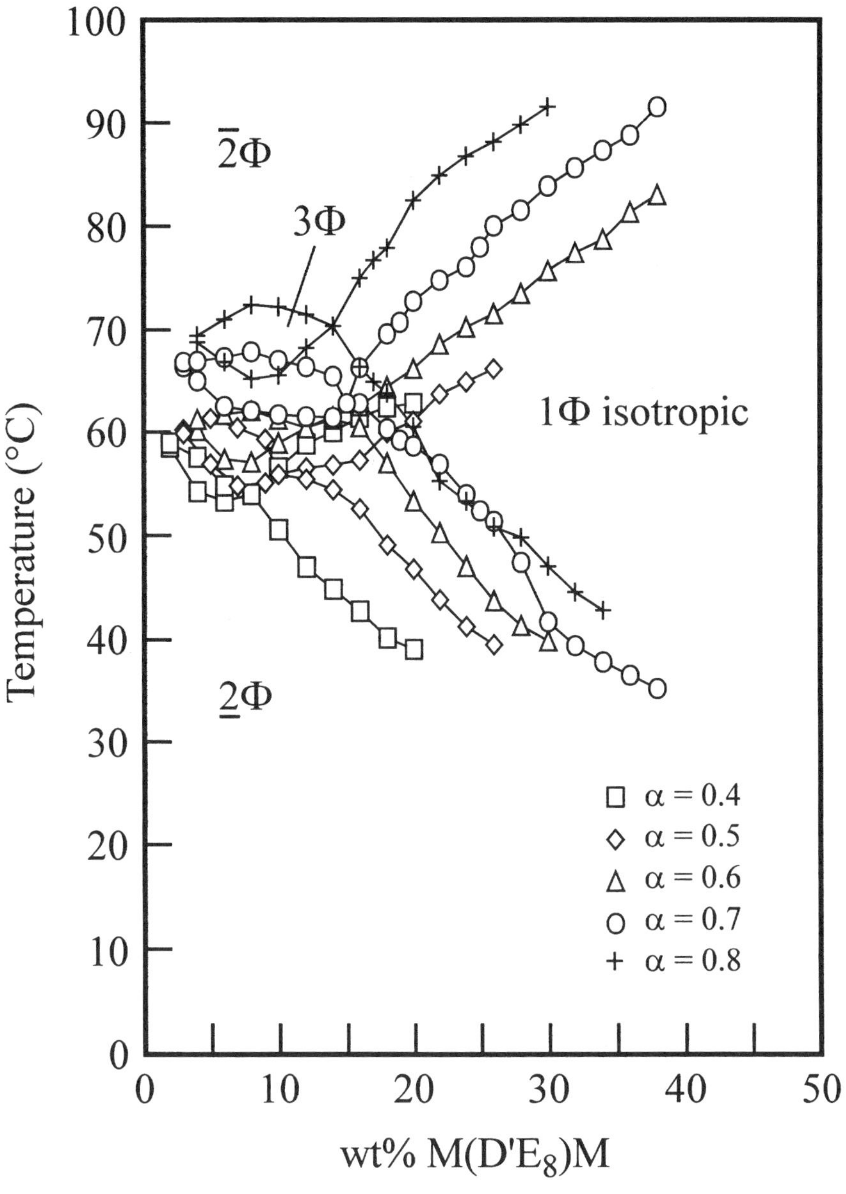

FIG. 9 Fish-cut phase diagrams of the water/M(D′E_8)M/D_4 system at α = 0.4–0.8.

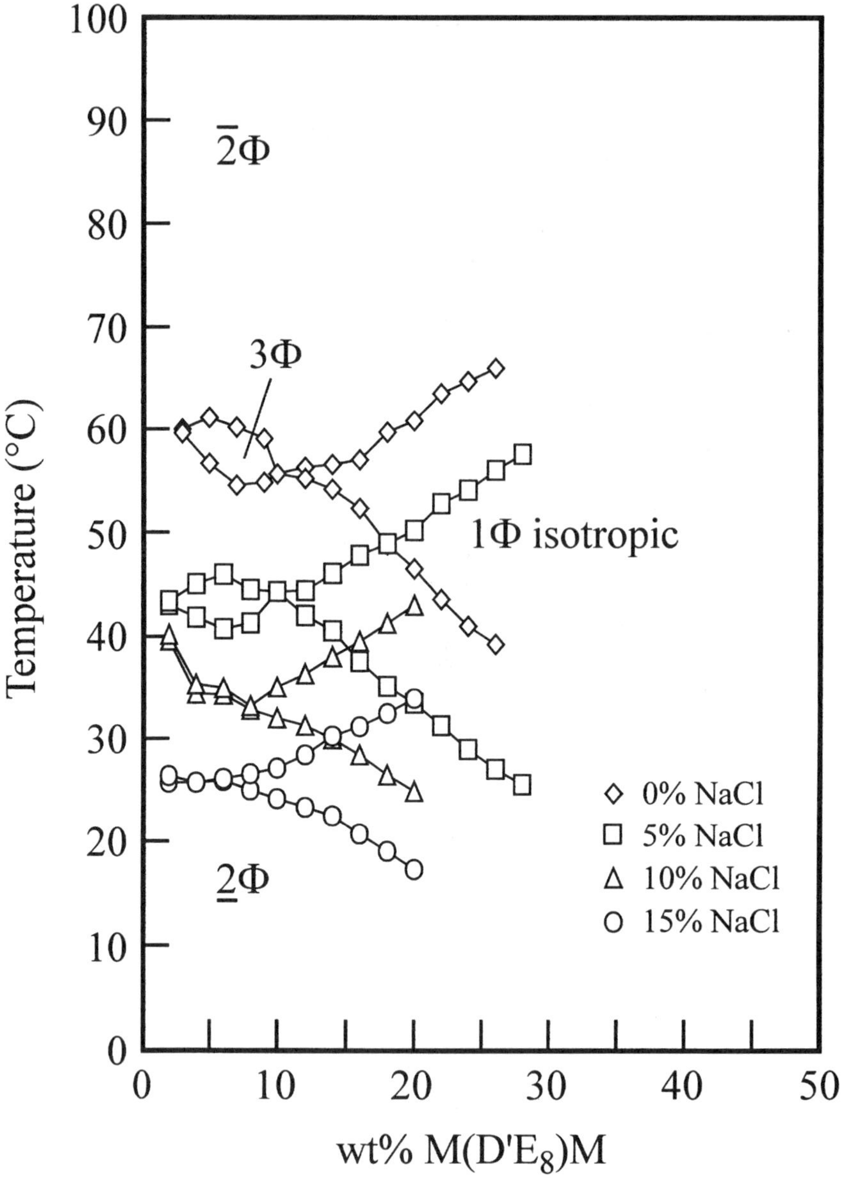

FIG. 10 Fish-cut phase diagrams of the water/M(D′E$_8$)M/D$_4$ system at $\alpha = 0.5$ versus salt concentration.

eral and are due to the effect of NaCl on the solubility of the surfactant in the water [29]. NaCl salts out the polyoxyethylene groups of the surfactant, shifting the partitioning of the surfactant between water and oil toward the oil. In other words, NaCl has an effect on the system similar to those of temperature and shortening the polyoxyethylene chain. NaCl also compresses the R_g of the polyoxyethylene groups, making the polar groups smaller and changing the packing of the surfactant molecules in the interface and probably also influences the flexibility of the surfactant monolayer.

Figure 11 shows a series of fish-cut phase diagrams for $M(D'E_8)M$ with a larger cyclic silicone oil D_5, and with the two short linear silicone oils, MM, and MD_2M. Compared to D_4, the fish for D_5 is shifted to higher temperature and surfactant concentrations, while MM is shifted to lower temperatures and surfactant concentrations. MD_2M gives a fish very close to the one for D_5. This is consistent with the relative molecular weights and molar volumes of these oils (see Table 1). Increasing the molecular weight of the oil decreases the solubility of the surfactant in the oil and shifts the UCT boundary for the surfactant–oil pair to higher temperatures. This causes the fish, which is the superposition of the binary UCT and LCT, to shift upward and to the right. It is likely that the cyclic and linear oils do not shift with the same coefficient.

D. The Binary Water/$M(D'E_{10})M$ System

The binary phase diagram of mixtures of $M(D'E_{10})M$ and water is shown in Fig. 12. Between 30 and 44°C, mixtures of $M(D'E_{10})M$ and water form a single isotropic liquid phase, L, at all concentrations. A similar broad band having a single phase was observed for $M(D'E_{12})M$, and the microstructural evolution within that region analyzed [54]. Below 30°C a region of lamellar phase, L_α, is found near 72 wt % surfactant, and a region of normal hexagonal phase, H_1, is found below 20°C near 50 wt % surfactant. H_1 phase does not appear in the binary phase diagrams for the E_6 or E_8 homologs. The larger E_{10} hydrophilic head group in $M(D'E_{10})M$ assists in the formation of the H_1 phase, which requires greater curvature of the sheet-like surfactant structures toward the oil side. In ternary systems, relative sizes of the hydrophobic and hydrophilic moieties of a surfactant are not the only factors controlling curvature. Interactions between the oil molecules and the hydrophobic parts of the surfactant can play a role. For instance, H_1 phase occurs in the water/$M(D'E_8)M/D_4$ ternary phase diagram, even though H_1 does not appear in the water/$M(D'E_8)M$ binary diagram. Above 44°C an LCT boundary is found. cryo-TEM micrographs of 5 wt % $M(D'E_{10})M$ in water at 20°C show spherical and elongated micelles [48]. Above the LCT boundary, two additional phases, L_α, and L_3, were identified. Samples from the lower region are statically birefringent, which identifies them as L_α, while samples in the upper region exhibit flow birefringence, which identifies that region as L_3.

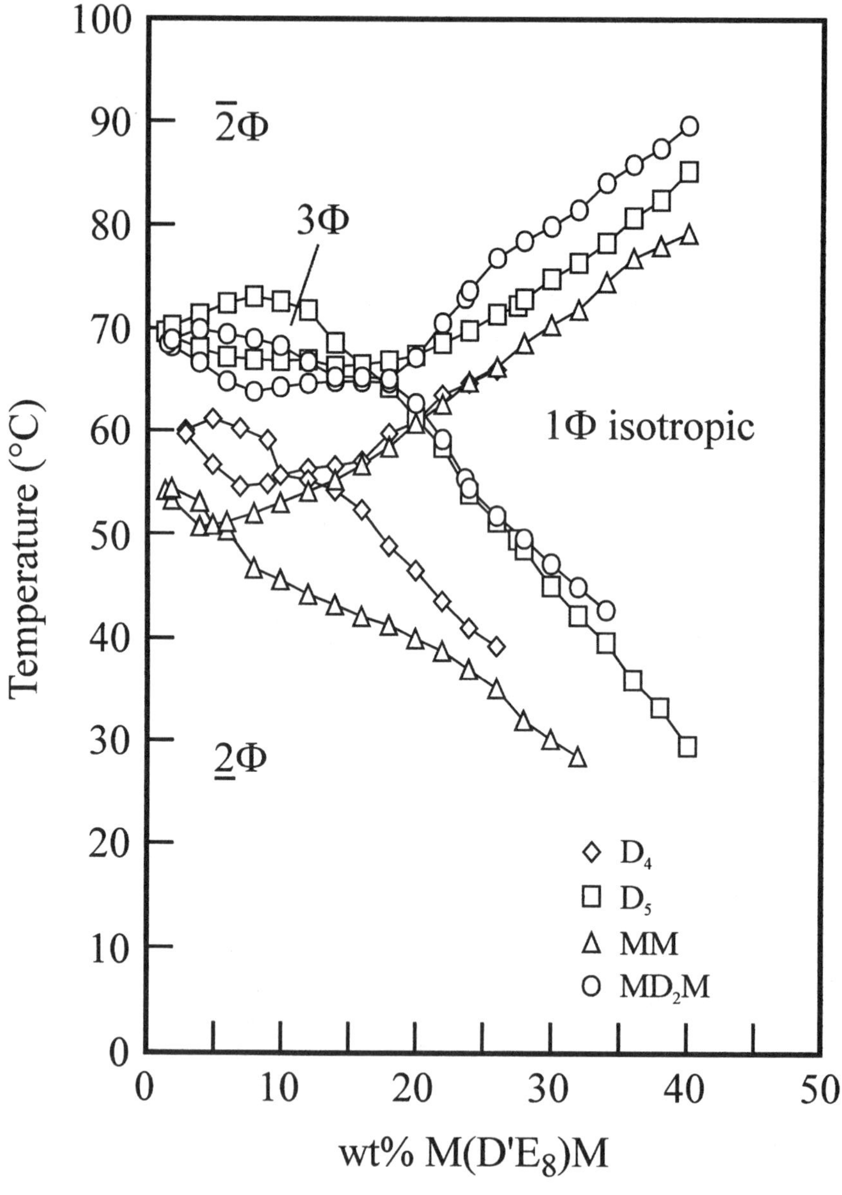

FIG. 11 Fish-cut phase diagrams of the water/M(D′E_8)M/silicone oil system at $\alpha = 0.5$ for different silicone oils: D_4, D_5, MM, and MD_2M.

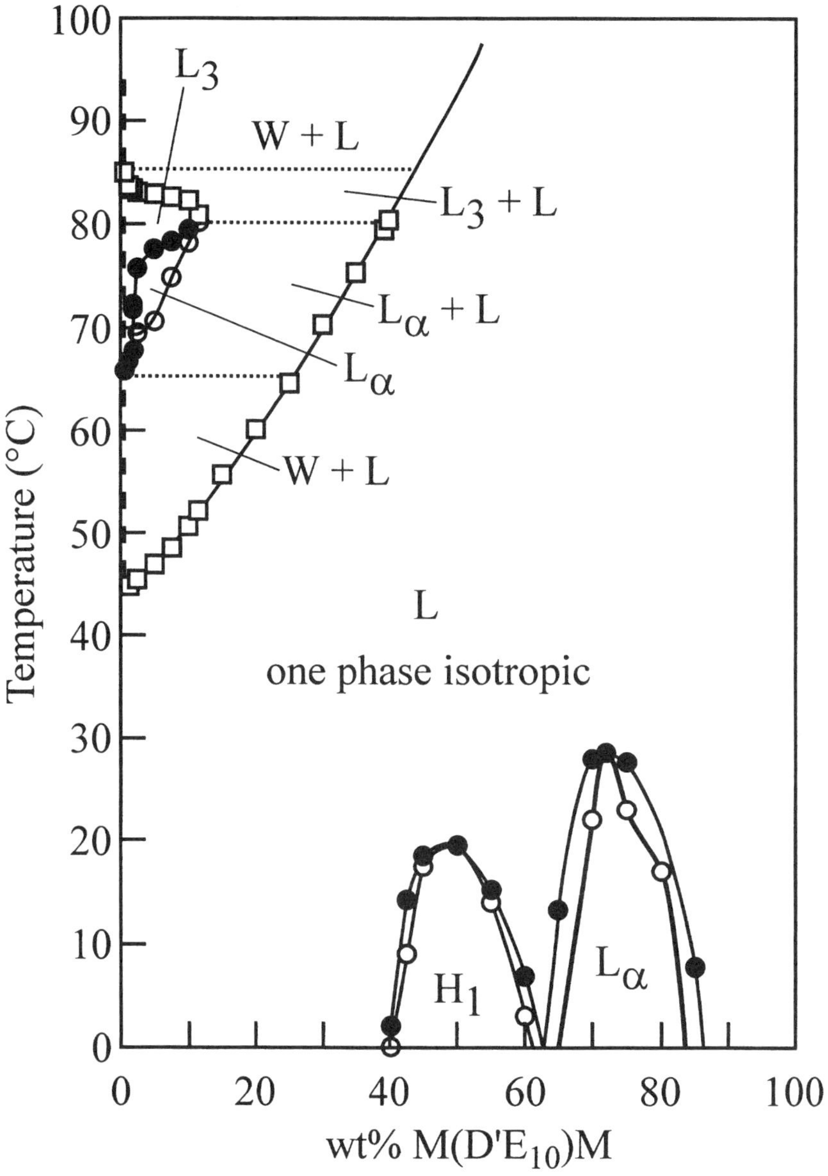

FIG. 12 Binary phase diagram of M(D′E_{10})M/water. (From Ref. 42, used with permission.)

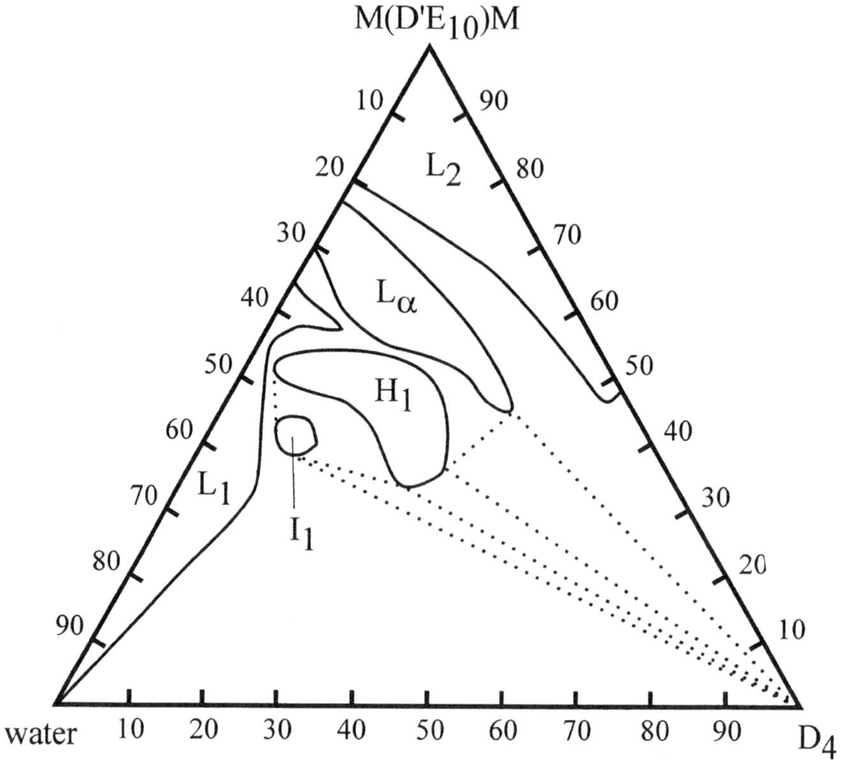

FIG. 13 Isothermal phase diagram of the water/M(D′E_{10})M/D_4 system at 20°C. (From Ref. 42, used with permission.)

E. The Ternary Water/M(D′E_{10})M/D_4 System

Figure 13 shows an isothermal phase diagram of the ternary water/M(D′E_{10})M/D_4 system at 20°C. Two microemulsion regions are present, one in the surfactant-rich corner and the other along the water–surfactant axis. In addition to the lamellar phase, there is a hexagonal phase that originates from the H_1 phase in the binary system. In the binary system, the H_1 phase is only seen below 20°C, whereas addition of 7 wt % D_4 raises the melting point of the phase, causing it to appear at room temperature. The Lα phase shows the mosaic texture between crossed polarizers, and SAXS peaks with wave vectors in the ratio of 1:2. The H_1 phase shows a fan-like texture and SAXS peaks with wave vectors in the ratio of $1:\sqrt{3}:2$, representing hexagonally packed, semi-infinite rodlike micelles. At lower surfactant concentrations we found a cubic liquid crystal phase. This phase is isotropic and extremely viscous. Three peaks were observed in the SAXS spectrum, in the ratio of $\sqrt{3}:\sqrt{4}:\sqrt{14}$, which is consistent with identification of the phase as I*a*3*d*, body-centered cubic phase,* but positive identification cannot be made from these three peaks alone. The presence of this highly ordered phase in the ternary system alone, and the higher melting point of the H_1 phase, demonstrate that addition of oil promotes increased ordering of the surfactant microstructures. Again a large curvature microstructure, I_1, is seen in the ternary phase diagram for $M(D'E_{10})M$ but was not observed for the E_6 or E_8 homologs.

F. The Ternary Water/$M(D'E_{12})M/D_4$ System

The binary phase diagram for $M(D'E_{12})M/D_4$ was given by He et al. [54]. Because of its long hydrophilic polyoxyethylene head group, $M(D'E_{12})M$ has little solubility in D_4, and a very large lower miscibility gap (UCT) exists between $M(D'E_{12})M$ and D_4 that extends well beyond the upper miscibility gap (LCT) of the $M(D'E_{12})M$/water system.

An isothermal ternary phase diagram for the water/$M(D'E_{12})M/D_4$ system at 25°C is shown in Fig. 14. There is a one-phase isotropic microemulsion region extending along the water/$M(D'E_{12})M$ axis and around the surfactant corner. This ternary phase diagram exhibits a rich liquid crystal phase behavior, containing regions of hexagonal H_1, lamellar L_α, and cubic I_1 and I_3 phases. There is also a small region of low viscosity, isotropic liquid phase on the low surfactant concentration side of the I_3 cubic phase. Comparing this diagram with Fig. 15 shows that this region is connected to the region nearer to the surfactant corner at higher temperatures, while at 25°C the two regions are separated by multiphase regions.

All the diagrams for $M(D'E_{12})M$ show complex phase behavior. In all six of the diagrams shown here (Figs. 14–19), two two-phase regions sometimes appear as contiguous because the narrow three-phase region, which must be present between them, was not detected. For example, the low concentration, low temperature region of Fig. 15, and the corresponding region of Fig. 14, show a $\underline{2}\Phi$ region contiguous with a two-phase, $I_1 + D_4$ region with no intervening three-phase region because we did not detect such a region even though it must be present. Samples in the regions labeled I_1 and I_3 in the ternary diagram are transparent, optically isotropic, and extremely viscous. There are three types of lyotropic cubic liquid crystal phase: I_1, I_2, and I_3. The first, I_1, is usually identified as close-packed spherical micelles or oil-swollen micelles with water as the continuous phase. I_3 is close-packed inverse micelles or water-swollen micelles with either surfactant or oil as the continuous phase. I_2 is a bicontinuous intermediate between the two extremes. We identify the cubic phase region closest to the water corner as I_1 because of its location (especially relative to the H_1 region), and because pulsed field gradient NMR self-diffusion measurements indicate that it is water continu-

*The spacing of peaks in powder spectra of the cubic phases obeys $d_{hkl} = a/(h^2 + k^2 + l^2)^{1/2}$ according to which the $\sqrt{3}:\sqrt{4}:\sqrt{14}$ peaks are reflections from the 111, 200, and 123 planes, respectively.

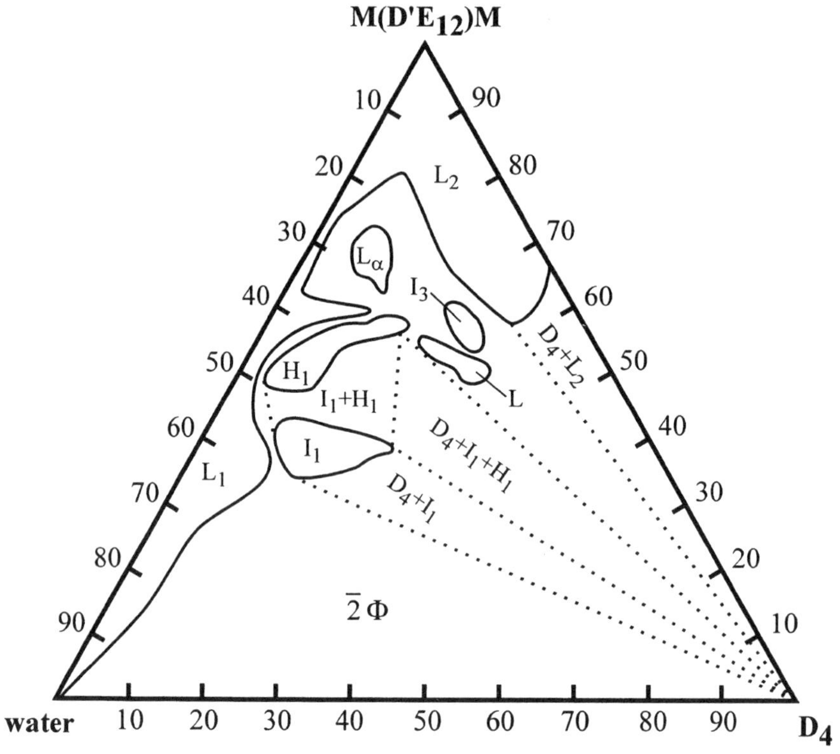

FIG. 14 Isothermal ternary phase diagram of water/M(D′E$_{12}$)M/D$_4$ at 25°C: L$_2$, a surfactant-rich isotropic phase; I$_1$, a water-continuous cubic liquid crystal phase; and I$_3$, an oil- (or surfactant-) continuous cubic liquid crystal phase. (From Ref. 42, used with permission.)

ous. Similarly, NMR self-diffusion measurements indicate that samples from the region we have labeled I$_3$ are oil continuous. Five peaks were detected in the SAXS spectrum of a sample from the I$_1$ region. They have wave vectors in the ratios $\sqrt{3}:\sqrt{4}:\sqrt{8}:\sqrt{11}:\sqrt{12}$. This is consistent with diffraction from the (111), (200), (220), (311) and (222) planes, respectively. The indices of the planes indicate that this particular cubic phase is probably a face-centered cubic (fcc) belonging to the F23 space group. Three peaks were detected in the spectrum of the I$_3$ sample in the ratios of $\sqrt{3}:\sqrt{4}:\sqrt{8}$. The similarity in the SAXS spectra of the two cubic phases indicates that the two are similar in that each has a cubic structure of close-packed spherical micelles.

Figure 15 shows a fish-cut phase diagram for the ternary water/M(D′E$_{12}$)M/D$_4$ system at α = 48.85% (equal volume of oil and water). This diagram shows a

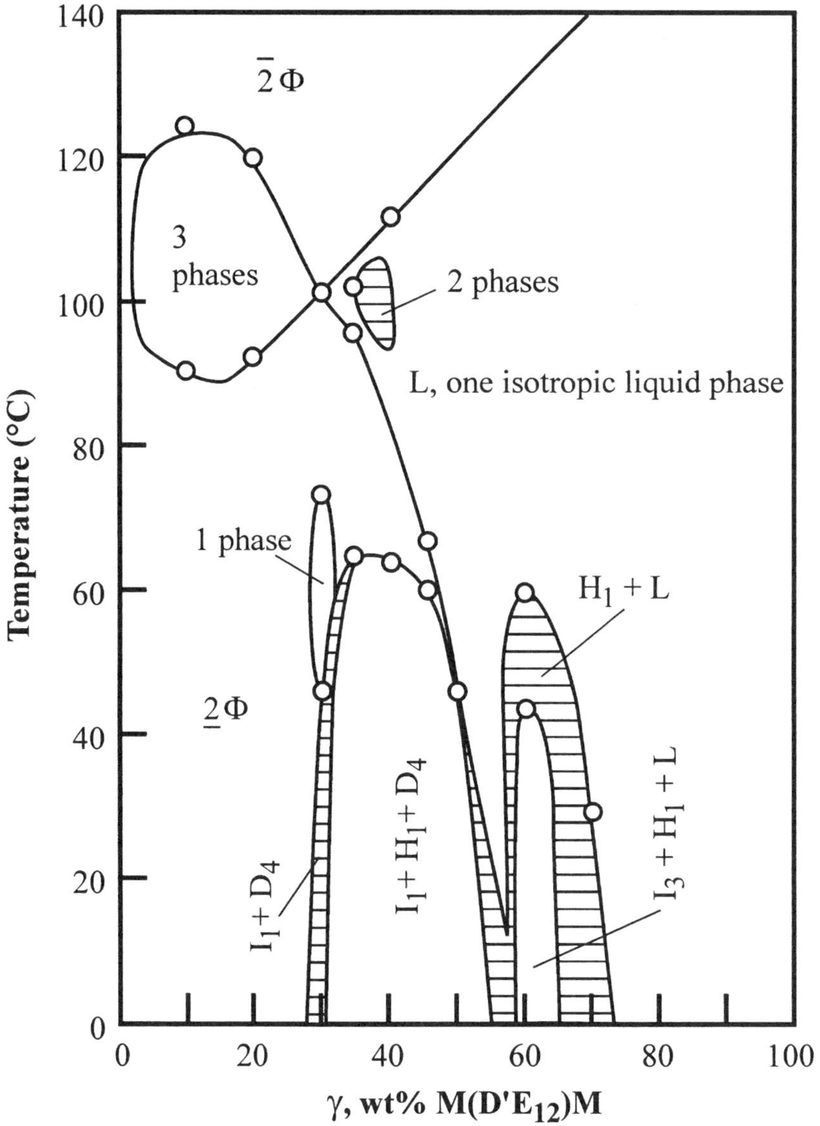

FIG. 15 Fish-cut phase diagram for the water/M(D′E_{12})M/D_4 system at α = 48.85%. (From Ref. 42, used with permission.)

three-phase body and a tail of one-phase microemulsion. However, it also contains complex multiphase regions along the lower boundary of the tail. In the $\bar{2}\Phi$ region below the three-phase body, M(D′E$_{12}$)M is soluble in D_4 but not in water, so the surfactant resides primarily in the oil-rich upper phase. In the $\underline{2}\Phi$ region, M(D′E$_{12}$)M is soluble in water but not in D_4 so the surfactant resides primarily in the water-rich lower phase. Embedded in the fish tail are liquid crystal phases and their mixtures. The thermal stability of these liquid crystals is higher than their counterparts in the binary water/M(D′E$_{12}$)M system. With increasing proportion of oil, the three-phase region expands and moves to higher temperatures, and C_{min} shifts to higher surfactant concentrations [48].

G. The Ternary Water/M(D′E$_{12}$)M/D$_5$ System

The isothermal ternary phase diagram of the water/M(D′E$_{12}$)M/D$_5$ system at 25°C is shown in Fig. 16. This diagram is similar to the D_4 ternary shown in Fig. 14.

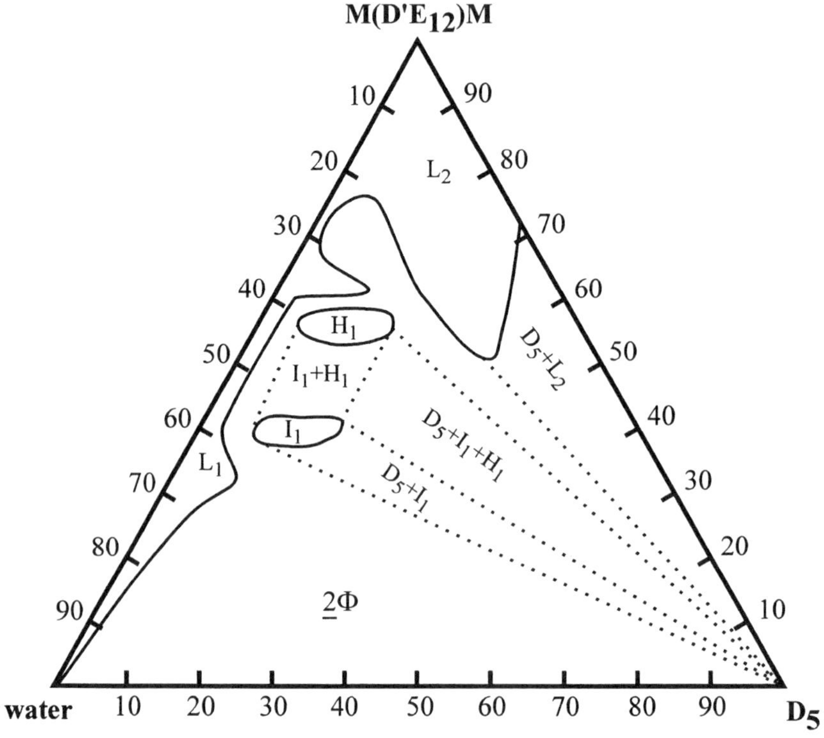

FIG. 16 Isothermal ternary phase diagram of the water/M(D′E$_{12}$)M/D$_5$ system at 25°C. (From Ref. 42, used with permission.)

Both form a narrow band of microemulsion along the water–surfactant axis, and a region of surfactant-rich microemulsion. The D_5 ternary also forms the cubic I_1 phase and the hexagonal H_1 phase, but the regions are smaller than for D_4. No lamellar L_α or inverse cubic I_3 phases were found in the D_5 ternary. The SAXS spectrum of a sample from the cubic I_1 region has peaks with wave vector ratios of $\sqrt{3}:\sqrt{4}:\sqrt{8}:\sqrt{11}:\sqrt{12}$., very similar to the I_1 in the D_4 ternary.

Figure 17 shows a fish-cut phase diagram at $\alpha = 48.93\%$ that contains equal volumes of water and D_5. Compared to the D_4 system in Fig. 15, the three-phase region is larger and occurs at higher temperatures. The I_3 phase inside the fish tail of the D_4 system is absent for D_5, consistent with the isothermal phase diagram in Fig. 16. C_{min} at the 1:1 water-to-oil (volume) ratio for the D_5 system has been shifted to higher surfactant concentration ($\gamma = 48\%$) and temperature ($T = 126°C$) compared with the D_4 system ($\gamma = 32\%$, $T = 108°C$). Thus, more surfactant and higher temperature are required to form microemulsion for D_5 than for D_4, consistent with the higher molecular weight of the D_5. Note that 32–48% surfactant is a rather high level of surfactant, indicating that $M(D'E_{12})M$ is not very efficient in forming microemulsions with these oils [59].

H. The Ternary Water/$M(D'E_{12})M/MD_2M$ System

MD_2M is a linear tetrasiloxane oil that contains two more methyl groups and one less oxygen atom than D_4. Its molecular weight lies between D_4 and D_5, but because of its lower density, its molar volume is very close to that of D_5. Figure 18 shows an isothermal ternary phase diagram for the water/$M(D'E_{12})M/MD_2M$ system at 25°C. This system also forms a narrow band of microemulsion along the water–surfactant axis, a region of surfactant-rich microemulsion, and cubic I_1, and hexagonal H_1 liquid crystal phases. The liquid crystal regions are substantially larger than either of the cyclic oils. MD_2M is a very flexible short chain molecule [60], whereas D_4 is a relatively stiff ring with the approximate proportions of a hockey puck [61]. This difference, and the closeness of the molar volumes of MD_2M and D_5, may explain why the MD_2M ternary generally resembles the D_5 system more than the D_4 system.

Figure 19 shows a fish-cut phase diagram for the water/$M(D'E_{12})M/MD_2M$ system at $\alpha = 46.06\%$. This diagram shows a large fish-tail region, almost all of which is isotropic phase except for small two-phase regions near C_{min}. The oil-continuous cubic I_3 phase is absent, as it is for D_5.

Both C_{min} and $T(C_{min})$ increase linearly with α for D_4 in the α range studied [48]. At $\alpha = 50\%$, where the solutions contain approximately equal volumes of water and oil, C_{min} for D_5 and MD_2M are about twice the value for D_4. $T(C_{min})$ for D_5 and MD_2M are also about 20°C higher than for D_4. Thus, at elevated temperature, D_4 needs less surfactant to form a microemulsion with water than either D_5 or MD_2M, although the three oils have similar ternary phase behavior at ambient temperature.

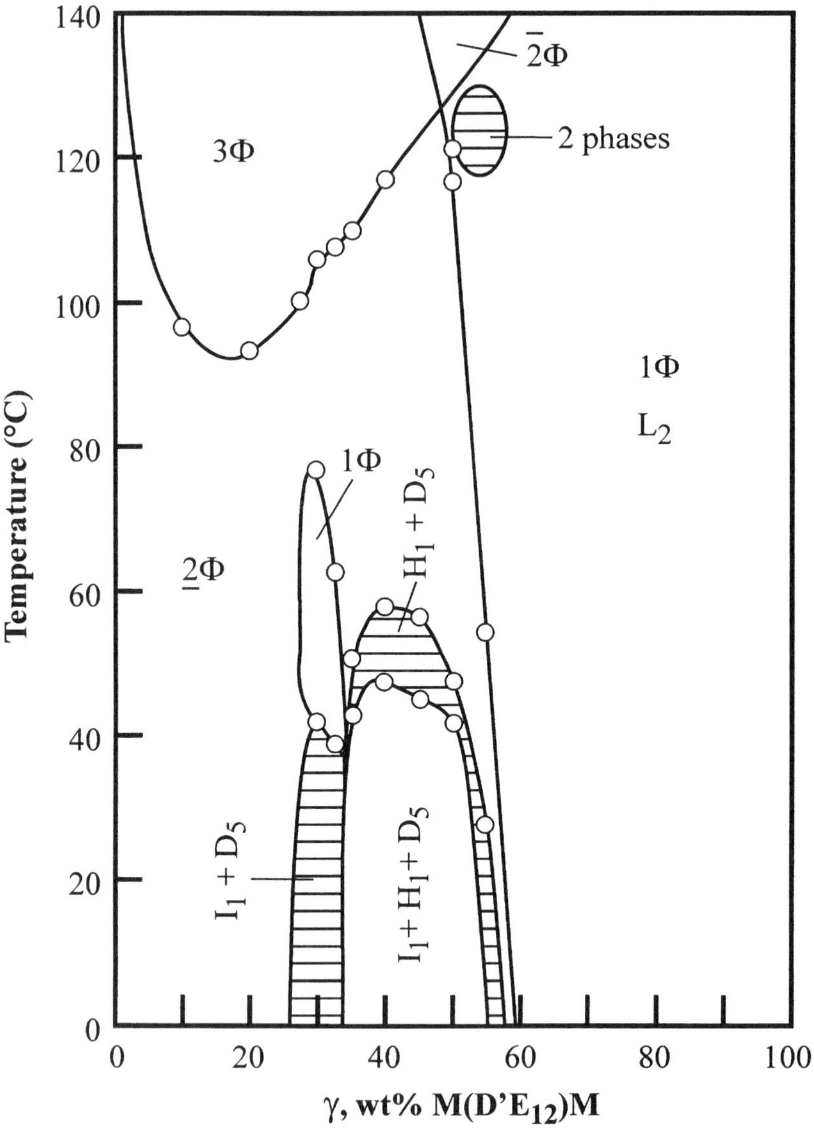

FIG. 17 Fish-cut phase diagram for water/$M(D'E_{12})M/D_5$ at $\alpha = 48.93\%$. (From Ref. 42, used with permission.)

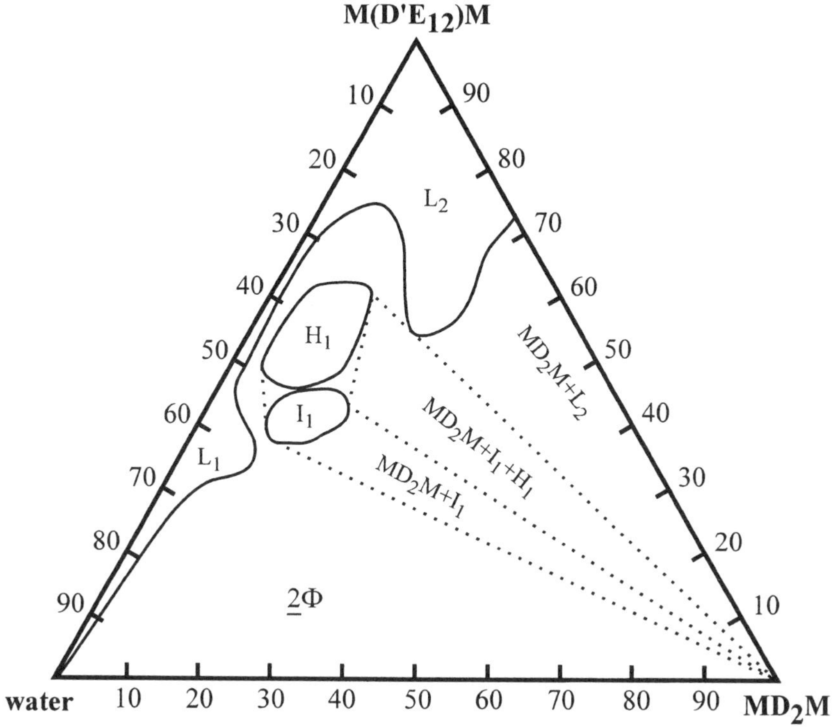

FIG. 18 Isothermal ternary phase diagram of the water/M(D′E_{12})M/MD_2M system at 25°C. (From Ref. 42, used with permission.)

I. The Ternary Water/M(D′E_{18})M/D_4 System

The binary phase diagram for M(D′E_{18})M in water is given by He et al. [62]. This surfactant forms a region of hexagonal, H_1, phase between 55 and 75 wt % surfactant, up to about 50°C. The huge E_{18} hydrophilic group prevents formation of the low curvature lamellar phase in the binary phase diagram. An isothermal ternary phase diagram for the water/M(D′E_{18})M/D_4 system is shown in Fig. 20 [63]. There is a very small region of isotropic liquid phase in the surfactant corner, and two narrow regions along the water–surfactant axis, corresponding to oil-swollen micellar solution and L_2. A region of oil-swollen H_1 extends upward from the M(D′E_{18})M/water axis to about 25% oil. In the center of the diagram is a large region of I_1, extending to near 50% D_4. We identify this phase as I_1, hence as water continuous, because it lies to the water-rich side of the H_1 region.

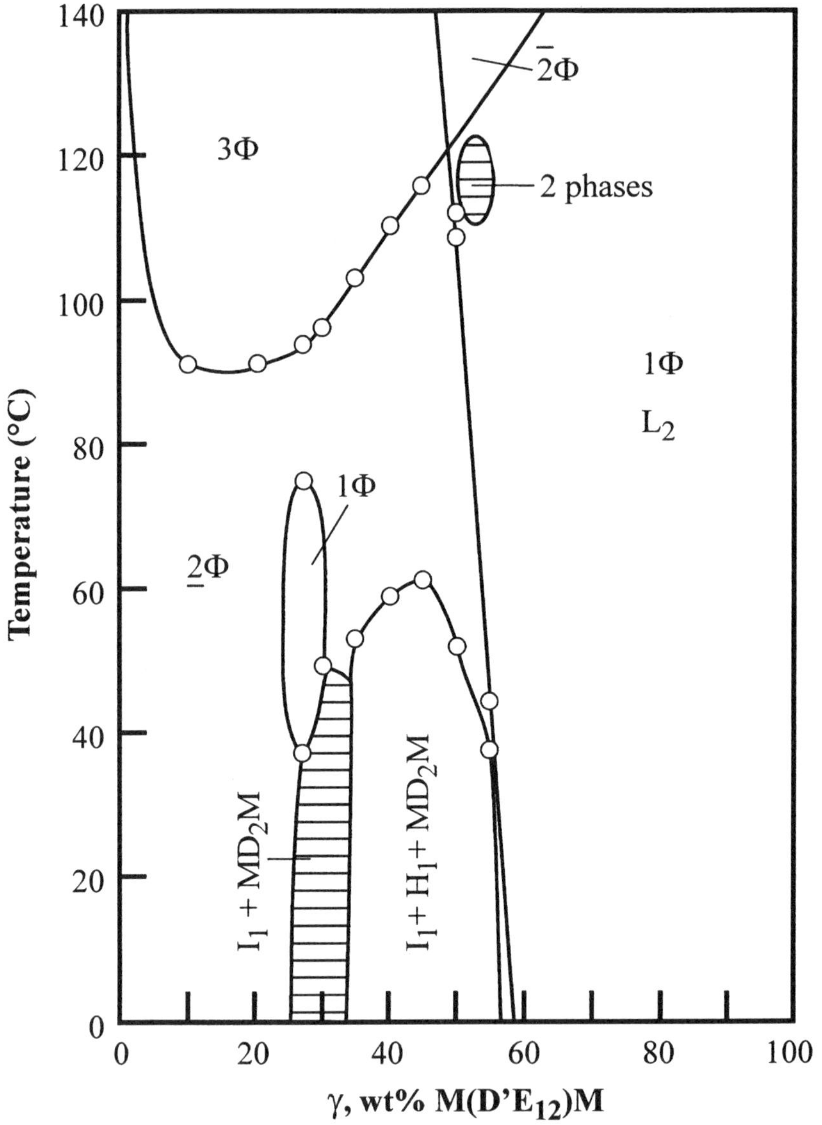

FIG. 19 Fish-cut phase diagram of the water/M(D′E_{12})M/MD_2M system at $\alpha = 46.06\%$. (From Ref. 42, used with permission.)

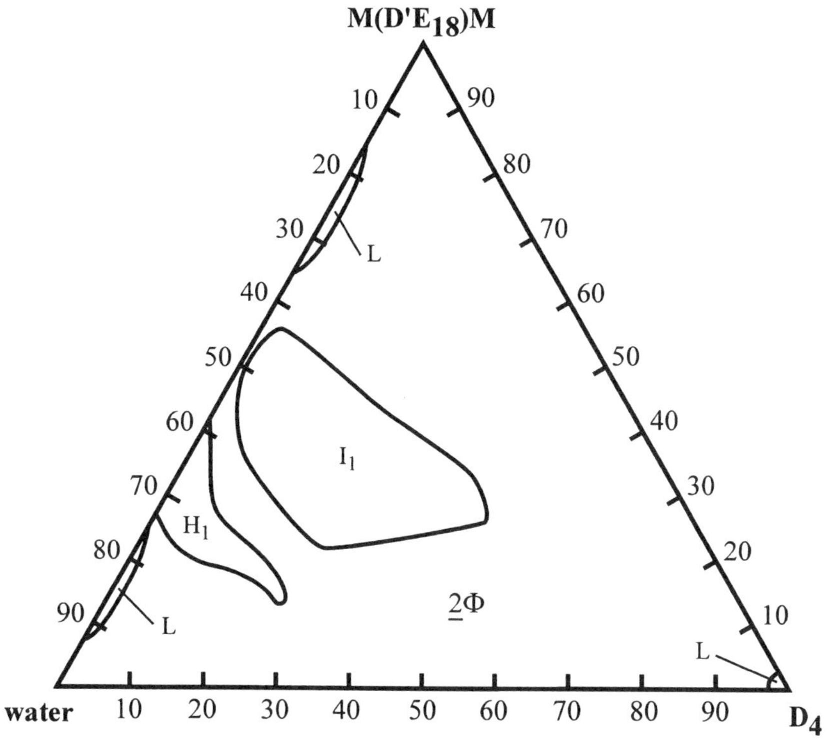

FIG. 20 Isothermal phase diagram of the water/M(D′E_{18})M/D_4 system at 25°C.

V. CONCLUSION

The technologies of preparing emulsions of polydimethlysiloxane and microemulsions of amine functional PDMS have been known for some time. Recently, procedures and compositions to prepare microemulsions of PDMS by a polymerization route have been reported. Other workers have shown how to use a particular class of silicone polyether copolymer to prepare transparent water-in-oil systems by means of low molecular weight silicone oils. Several recent studies report the aqueous phase behavior of the trisiloxane surfactants demonstrating their similarities and differences with the C_iE_j surfactants. We have extended this work to investigate the ternary phase behavior of mixtures a series of homologous trisiloxane polyoxyethylene surfactants with four low molecular weight silicone oils. Increasing the size of the surfactant's polyoxyethylene group makes the surfactant more hydrophilic, preferring higher curvature aggregates; in addition, the formation of microemulsions shifts to higher temperatures. Salt has the expected effect of making the surfactants behave like shorter polyoxyethylene homologs. Increasing the molecular weight of the oil decreases the lipophilicity of the surfactant, shifting the formation of microemulsions to higher temperatures and requiring higher concentrations of surfactant. Although the trends are similar to the behavior of C_iE_j surfactants and alkanes, there are significant differences. Because of the different shape of the trisiloxane surfactants compared with alkyl ethoxylates, the balance between packing considerations and the strength of interactions between the surfactant and the water or the oil is different, shifting the phase behavior into significantly different ranges of temperature and composition space.

ACKNOWLEDGMENTS

The project was supported by the National Science Foundation through the Center for Interfacial Engineering (CIE) at the University of Minnesota and Dow Corning Corporation. The phase diagrams for the $M(D'E_{18})M$ systems were done by Mengtao He at the University of Minnesota.

REFERENCES

1. Davis, H. T.; Bodet, J. F.; Scriven, L. F.; Miller W. G., in *Physics of Amphiphilic Layers* (Langevin, D.; Meunier, J., eds.), Springer-Verlag, 1987.
2. Davis, H. T.; *Colloids Surf.* A *91*:9 (1994).
3. Kahlweit, M.; Strey, R.; Firman, P., *J. Phys. Chem. 90*:671 (1986).
4. Shah, D. O., ed., *Micelles, Microemulsions, and Monolayers: Science and Technology*. Marcel Dekker, New York, 1998.
5. Hill, R. M., Chapter 1, this volume.
6. Hoffmann, H.; Ulbricht, W., Chapter 4, this volume.
7. Mayer, H., *Tenside Surf. Deterg. 30*:90 (1993).
8. Lang, J. C.; Morgan, R. D., *J. Chem. Phys. 73*:5849 (1980).
9. Laughlin, R. G., Aqueous phase science of cationic surfactant salts, in *Cationic Surfactants: Physical Chemistry* (Rubingh, D. N., Holland, P. M., eds.), Surfactant Science Series, Vol. 37. Marcel Dekker, New York, 1991.
10. Merrifield, J. H.; Thimineur, R. J.; Traver, F. J., U.S. Patent 5,244,598 (1993), to General Electric Co.
11. Berthiaume, M. D.; Merrifield, J. H., U.S. Patent 5,683,625 (1997), to General Electric Co.
12. Gee, R. P., U.S. Patent 5,852,110 (1996), to Dow Corning Corp.
13. Gee, R. P., Europatent 138192 B1 (1988), to Dow Corning Corp.

14. Katayama, H.; Tagawa, T.; Kunieda, H., *J. Colloid Interface Sci. 153*:429 (1992).
15. Gee, R. P., *Colloids Surf. A 137*:91 (1998).
16. Gee, R. P., U.S. Patent 4,620,878 (1986), to Dow Corning Corp.
17. Anderson, K. R.; Obey, T. M.; Vincent, B., *Langmuir 10*:2493 (1994).
18. Dumoulin, J., U.S. Patent 3,975,294 (1976), to Rhone-Poulenc S.A.
19. Ona, I.; Ozaki, M., U.S. Patent 4,935,464 (1990), to Toray Silicone Co. Ltd.
20. Glover, S. O.; Graiver, D., U.S. Patent 4,824,890 (1989), to Dow Corning Corp.
21. Graiver, D.; Tanaka, O., U.S. Patent 4,999,398 (1991), to Dow Corning Corp.
22. Cekada, Jr., J.; Weyenburg, D. R., U.S. Patent 3,433,780 (1969), to Dow Corning Corp.
23. Messier, A.; Schorsch, G.; Rouviere, J.; Tenebre, L., *Prog. Colloid Polym. Sci. 79*:249 (1989).
24. John, A. C.; Uchiyama, H.; Nakamura, K.; Kunieda, H., *J. Colloid Interface Sci. 186*:294 (1997).
25. Binks, B. P.; Dong, J., *Colloid Surf A 132*:289 (1998).
26. Hoffmann, H.; Stürmer, A., *Tenside, Surf., Deterg. 30*:5 (1993).
27. Steytler, D. C.; Dowding, P. J.; Robinson, B. H.; Hague, J. D.; Rennie, J. H. S.; Leng, C. A.; Eastoe, J.; Heenan, R. K., *Langmuir 14*:3517 (1998).
28. Blehm, L. M.; White, W. C., U.S. Patent 4,842,766 (1989), to Dow Corning Corp.
29. Gee, R. P.; Keil, J. W., U.S. Patent 4,122,029 (1978), to Dow Corning Corp.
30. Keil, J. W., U.S. Patent 4,268,499 (1981), Dow Corning Corp.
31. Gum, M. L., U.S. Patent 4,782,095 (1988), to Union Carbide Corp.
32. Gum, M. L., U.S. Patent 4,801,447 (1989), to Union Carbide Corp.
33. Starch, M. S., U.S. Patent 4,311,695 (1982), to Dow Corning Corp.
34. Dahms, G. H.; Zombeck, A., *Cosmet. Toiletries 110*:91 (1995).
35. Gasperlin, M.; Kristl, J.; Smid-Korbar, J., *Int. J. Pharm. 107*:51 (1994).
36. Gasperlin, M.; Kristl, J.; Smid-Korbar, J., *STP Pharma Sci. 7*:158 (1997).
37. Baquerizo, I.; Gallardo, V.; Parera, A.; Ruiz, M. A., *J. Cosmet. Sci. 49*:89 (1998).
38. Smid-Korbar, J.; Krist, J.; Stare, M., *Int. J. Cosmet. Sci. 12*:135 (1990).
39. Kunieda, H.; Taoka, H.; Iwanaga, T.; Harashima, A., *Langmuir 14*:5113 (1998).
40. Iwanaga, T.; Shiogai, Y.; Kunieda, H., *Prog. Colloid Polym. Sci. 110*:225 (1998).
41. Von Berlepsch, H.; Wagner, R., *Prog. Colloid Polym. Sci. 111*:107 (1998).
42. Li, X.; Hill, R. M.; Washenberger, R. M.; Scriven, L. E.; Davis, H. T., *Langmuir*, in press, 1999.
43. Hill, R. M., *Langmuir*, manuscript in preparation, 1999.
44. Kahlweit, M., and Strey, R., *Angew. Chem., Int. Ed. Engl. 24*:654 (1995).
45. Friberg, S., and Kunieda, H., *Bull Chem. Soc. Jpn. 54*:1010 (1981).
46. Noll, W., *The Chemistry and Technology of Silicones*, Academic Press, New York, 1968.
47. Bailey, D. L., U.S. Patent 3,299,112 (1967).
48. Li, X., Ph.D. thesis, University of Minnesota, 1996.
49. Fowkes, F. M., *Chemistry and Physics of Interfaces II*, ACS Publications, Washington DC, 1971, p. 153.
50. Luzzati, V.; Mustacchi, H.; Skoulios, A.; Husson, F., *Acta Crystallogr. 13*:660 (1960).
51. Olsson, U., personal communication.

52. Mitchell, D. J.; Tiddy, G. J. T.; Waring, L.; Bostock, T.; MacDonald, M. P., *J. Chem. Soc., Faraday Trans. I 79*:975 (1983).
53. Hill, R. M.; He, M.; Davis, H. T.; Scriven, L. E., *Langmuir 10*:1724 (1994).
54. He, M.; Hill, R. M.; Doumaux, H. A.; Bates, F. S.; Davis, H. T.; Evans, D. F.; Scriven, L. E., in *Structure and Flow in Surfactant Solutions*, ACS Symposium Series 578 (Herb, C.; Prudhomme, R. K., eds.), American Chemical Society, Washington, DC, 1994, p. 192.
55. Strey, R.; Jahn, W.; Porte, G.; Bassereau, P., *Langmuir 6*:1635 (1990).
56. Laughlin, R. G., *The Aqueous Phase Behavior of Surfactants*, Academic Press: New York, 1994.
57. Hartshorne, N. H., *The Microscopy of Liquid Crystals*; Microscope Publications Ltd.: New York, 1974.
58. Marszall, L., in *Nonionic Surfactants* (Schick, M. J., ed.), Marcel Dekker, New York, 1987, p. 493.
59. Strey, R., *Colloid Polym. Sci. 272*:1005 (1994).
60. Owen, M. J.; Kendrick, T. C., *Macromolecules 3*:458 (1970).
61. Grigoras, S.; Lane, T. H., *J. Comput. Chem. 9*:25 (1988).
62. He, M.; Hill, R. M.; Lin, Z.; Scriven, L. E.; Davis, H. T., *J. Phys. Chem. 97*:8820 (1993).
63. Hill, R. M., U.S. Patent 5,623,017 (1997), to Dow Corning Corp.

Index